Inducible Gene Expression, Volume 2

Hormonal Signals

Progress in Gene Expression

Series Editor:

Michael Karin
Department of Pharmacology
School of Medicine
University of California, San Diego
La Jolla, CA 92093-0636

Books in the Series:

Gene Expression: General and Cell-Type-Specific
M. Karin, editor
ISBN 0-8176-3605-6

Inducible Gene Expression, Volume I: Environmental Stresses
and Nutrients
P.A. Baeuerle, editor
ISBN 0-8176-3728-1

Inducible Gene Expression, Volume II: Hormonal Signals
P.A. Baeuerle, editor
ISBN 0-8176-3734-6

Inducible Gene Expression, Volume 2

Hormonal Signals

P.A. Baeuerle

Editor

Birkhäuser
Boston • Basel • Berlin

P.A. Baeuerle
Institute of Biochemistry
Albert-Ludwigs-University Freiburg
Hermann-Herder-Str. 7
D-79104 Freiburg i.Br.
Germany

Library of Congress Cataloging-in-Publication Data

Inducible gene expression/P.A. Baeuerle, editor.
     p.   cm. – – (Progress in gene expreesion)
   Includes bibliographical references and index.
   Contents: v. 1.   Enviromental stresses and nutrients – – v.
2. Hormonal signals.
   ISBN-13: 978-1-4684-6839-7   e-ISBN-13: 978-1-4684-6837-3
   DOI: 10.1007/978-1-4684-6837-3
   1. Genetic regulation.   2. Gene expression.   I. Baeuerle, P.A.
(Patrick Alexander), 1957- . II. Series.
QH450.I53 1994
574.87'322– –dc20   94-27957
                                        CIP

Printed on acid-free paper.

Typeset by Alden Multimedia, Northampton, England.

9 8 7 6 5 4 3 2 1

Dedicated to
Edmund and Therese, My Parents

Contents

Preface

Cells have evolved multiple strategies to adapt the composition and quality of their protein equipment to needs imposed by changing conditions within the organism. Extracellular stimuli that inform cells about such needs are hormones, cytokines and neurotransmitters, which bind to specific cell surface receptors. Inside the cell, secondary signals are then produced which, ultimately, initiate the expression of proteins giving novel functional properties to the stimulated cells. This process can be controlled at a transcriptional, posttranscriptional, translational or posttranslational level. Extensive research over the past fifteen years has shown that transcriptional regulation is probably the most important strategy used to control the production of new proteins in response to hormonal signals. At the level of gene transcription, the initiation of mRNA synthesis is most frequently used to govern gene expression. The key elements controlling transcription initiation in eukaryotes are activator proteins (transactivators) that bind in a sequence-specific manner to short DNA sequences in the proximity of genes. The activator binding sites are elements of larger control units, called promoters and enhancers, which bind many distinct proteins that may synergize or negatively cooperative with the activators. The de novo binding of an activator to DNA or, if already bound to DNA, its functional activation is what ultimately turns on a high-level expression of genes.

In this second volume of *Inducible Gene Expression*, leading scientists in the field review eight eukaryotic transactivators that allow cells to respond to hormonal stimuli by the expression of new proteins. While the first volume covers systems responding to adverse environmental stresses, this book deals with physiological signals used by the organism to control and coordinate metabolic changes, cell proliferation, differentiation and development. In the first chapter, Sassone-Corsi and colleagues describe the transcriptional activators and modulates CREB and CREM, which respond to an intracellular increase of cAMP, a second messenger used by many distinct hormones. Nordheim and colleagues review the transactivators SRF and TCF which directly combine on their target DNA to induce gene expression in response to various growth-promoting stimuli. Girling and La Thangue describe a transactivator system, called DRTF1/E2F, which allows gene expression to be controlled by the cell cycle. The review by Fu deals with the so-called STAT transactivators. These can be directly phosphorylated by cell surface receptor-associated kinases in response to interferon or growth factors, and then migrate into the nucleus. Three ligand-induced trans-activators belonging to the steroid hormone super family are reviewed here, inviting readers to recognize similarities and differences in their way of action. Renkawitz and colleagues report on the glucocorticoid receptor, Pedrafita and Pfahl review the thyroid hormone receptors and Keaveney and Stunnenberg the retinoic acid receptors. Norris and Manley finish the volume with a chapter describing the transactivator dorsal. This protein, which is related to NF-κB (see Volume 1), is crucially involved in the early development of the Drosophila embryo.

The central idea behind bringing together these particular systems in one book is to invite readers to compare the molecular mechanisms used for intracellular signalling and inducible gene expression in response to hormonal signals. One intriguing insight gained from such a comparison is that specific DNA sequences of only fifteen base pairs or less can determine by which extracellular stimulus a gene is activated. I wish the readers much pleasure in listening to the tunes transactivators play on the genomic piano after having been inspired by distinct muses from outside the cell.

Patrick A. Baeuerle

List of Contributors

Michael A. Cahill, Institute for Molecular Biology, Hannover Medical School, D–30623 Hannover, Germany

Martin Eggert, Genetisches Institut, Justus-Liebig-Universität, Heinrich-Buff-Ring 58-62, D–35392 Giessen, Germany

Nicholas S. Foulkes, Laboratoire de Génétique Moléculaire des Eucaryotes, CNRS – U184 INSERM – Faculté de Médécine, 11, rue Humann F–67085 Strasbourg, France

Xin-Yuan Fu, Department of Pathology, School of Medicine, Yale University, 108 Lauder Hall, PO Box 308023, New Haven, CT 06520-8023, USA

Rowena Girling, Laboratory of Eukaryotic Molecular Genetics, MRC National Institute for Medical Research, The Ridgeway, Mill Hill, London NW7 1AA, United Kingdom

Ralf Janknecht, Institute for Molecular Biology, Hannover Medical School, D–30623 Hannover, Germany

Marie Keaveney, Gene Expression Programme, European Molecular Biology Laboratory, 1 Meyerhofstrasse, D–69117 Heidelberg, Germany

Enzo Lalli, Laboratoire de Génétique Moléculaire des Eucaryotes, CNRS – U184 INSERM – Faculté de Médécine, 11, rue Humann, F–67085 Strasbourg, France

Nicholas B. La Thangue, Laboratory of Eukaryotic Molecular Genetics, MRC National Institute for Medical Research, The Ridgeway, Mill Hill, London NW7 1AA, United Kingdom

Janet S. Lee, Laboratoire de Génétique Moléculaire des Eucaryotes, CNRS – U184 INSERM – Faculté de Médécine, 11, rue Humann, F–67085 Strasbourg, France

James L. Manley, Department of Biological Sciences, Sherman Fairchild Center for the Life Sciences, Columbia University, New York, NY 10027, USA

Denis Masquilier, Laboratoire de Génétique Moléculaire des Eucaryotes, CNRS – U184 INSERM – Faculté de Médécine, 11, rue Humann, F–67085 Strasbourg, France

Carlos A. Molina, Laboratoire de Génétique Moléculaire des Eucaryotes, CNRS – U184 INSERM – Faculté de Médecine, 11, rue Humann, F–67085 Strasbourg, France

Marc Muller, Genetisches Institut, Justus-Liebig-Universität, Heinrich, Buff-Ring 58-62, D–35392 Giessen, Germany

Alfred Nordheim, Institute for Molecular Biology, Hannover Medical School, Konstanty-Gutschow-Str.8, D–30623 Hannover, Germany

Jacqueline L. Norris, Department of Biological Sciences, Columbia University, New York, NY 10027, USA

Magnus Pfahl, LaJolla Cancer Research Foundation, 10901 N. Torrey Pines Rd., LaJolla, CA 92037, USA

F. Javier Piedrafita, LaJolla Cancer Research Foundation, 10901 N. Torrey Pines Rd., LaJolla, CA 92037, USA

Rainer Renkawitz, Genetisches Institut, Justus-Liebig-Universität, Heinrich-Buff-Ring 58-62, D–35392 Giessen, Germany

Paolo Sassone-Corsi, Laboratoire de Génétique Moléculaire des Eucaryotes, CNRS – U184 INSERM – Faculté de Médécine, 11, rue Humann, F–67085 Strasbourg, France

Florence Schlotter, Laboratoire de Génétique Moléculaire des Eucaryotes, CNRS – U184 INSERM – Faculté de Médécine, 11, rue Humann, F–67085 Strasbourg, France

Hendrik G. Stunnenberg, Gene Expression Programme, European Molecular Biology Laboratory, 1 Meyerhofstrasse, D–69117 Heidelberg, Germany

1

CREM, a master-switch in the nuclear response to cAMP signaling

JANET S. LEE, ENZO LALLI, DENIS MASQUILIER, FLORENCE SCHLOTTER, CARLOS A. MOLINA, NICHOLAS S. FOULKES AND PAOLO SASSONE-CORSI

Introduction

The regulation of gene expression by specific signal transduction pathways is tightly connected to the cell phenotype. The response elicited by a given transduction pathway will vary according to the cell type. The finding that most of the known nuclear oncogenes encode proteins involved in the regulation of gene expression inspired the concept that the aberrant expression of some key genes could cause cellular transformation or altered proliferation (Lewin, 1991). The study, and ultimately the understanding, of these processes will help us, it is hoped, to unravel the profound changes that cause cancer and by the same token the physiology of normal growth.

An important step towards the comprehension of how the function of transcription factors can be modulated has been the discovery that many constitute final targets of specific signals transduction pathways activated by various signals at the cell surface. Two major signal transduction systems are those utilizing either cAMP or diacylglycerol (DAG) as secondary messengers (Nishizuka, 1986; Berridge, 1987). Each pathway is also characterized by a specific protein kinase (protein kinase A and protein kinase C, respectively) and has as its ultimate target a DNA control element, the CRE (cAMP-responsive element) or the TRE (TPA-responsive element). Although initially characterized as distinct systems, accumulating evidence points towards extensive cross-talk between these two pathways in the cytoplasm (Cambier et al., 1987; Yoshimasa et al, 1987) and in the nucleus (Sassone-Corsi et al, 1990; Benbrook and Jones, 1990; Auwerx and Sassone-Corsi, 1991; Masquilier and Sassone-Corsi, 1992).

INDUCIBLE GENE EXPRESSION, VOLUME 2
P.A. Baeuerle, Editor
© 1995 Birkhäuser Boston

This review will focus primarily on the targets of the cAMP-mediated transduction response and their function within the neuroendocrine system.

The cAMP-dependent transduction pathway

Intracellular levels of cAMP are regulated primarily by adenylyl cyclase. This enzyme is in turn modulated by various extracellular stimuli through specific receptors and their interaction with G proteins (Gilman, 1987; McKnight et al, 1988; Borrelli et al, 1992). The binding of a specific ligand to a receptor results in the activation or inhibition of the cAMP-dependent pathway, ultimately affecting the transcriptional regulation of various genes through distinct promoter responsive sites (Montmayeur and Borrelli, 1991). Increased cAMP levels directly affect the function of the tetrameric protein kinase A (PKA) complex (Krebs and Beavo, 1979). Binding of cAMP to two PKA regulatory subunits releases the catalytic subunits, enabling them to phosphorylate target proteins (see Figure 1.1). A significant fraction of the catalytic subunit molecules migrate into the nucleus. A number of isoforms of both the regulatory and catalytic subunits have been identified suggesting a further level of complexity in this response (McKnight et al, 1988). In the nucleus, the phosphorylation state of transcription factors appears to directly modulate their function, and thus the expression of cAMP-inducible genes (Figure 1.1).

The analysis of promoter sequences of several genes has allowed the identification of promoter elements which can mediate the transcriptional response to increased levels of intracellular cAMP (Roesler et al, 1988; Borrelli et al, 1992). A number of sequences have been identified, of which the best characterised is the cAMP-responsive elements, the CRE. The CRE is recognized by a multitude of nuclear factors; their characteristics are described in the next sections.

The cAMP-responsive element

The CRE promoter elements mediates the response to increased levels of intracellular cAMP (Comb et al, 1986; Andrisani et al, 1987; Delegeane et al, 1987; Sassone-Corsi, 1988). A consensus CRE site constitutes an 8 base pair (bp) palindromic sequence (TGACGTCA). Several genes which are regulated by a variety of endocrinological stimuli contain similar sequences in their promoter regions although at different positions. A comparison of the CRE sequences identified to date, shows that the 5'-half of the palindrome, TGACG is the best conserved, whereas the 3'

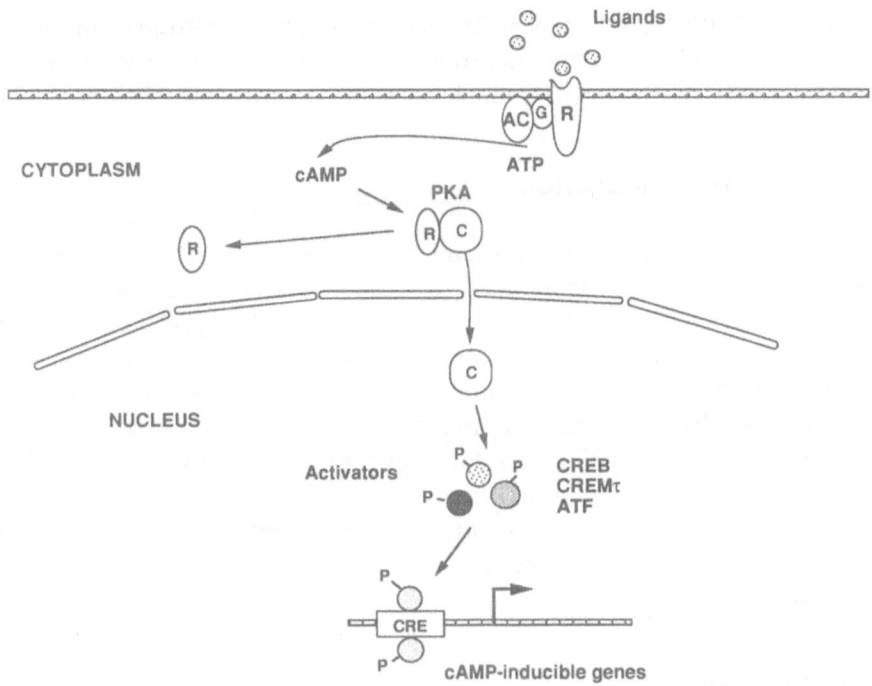

Figure 1.1 Schematic representation of the cAMP signal transduction pathway operating from the cell membrane, through the cytoplasm and into the nucleus. Ligands interacting with transmembrane receptors (R) stimulate the enzyme adenylyl cyclase (AC) via interactions with G-proteins (G). The subsequent rise in intracellular cAMP concentration results in the dissociation of the regulatory and catalytic subunits of PKA and the translocation of active catalytic subunits into the nucleus. PKA phosphorylates and thereby stimulates transcriptional activators binding to CREs (Activators, e.g. CREB, CREMτ and ATF-1) (Gonzalez et al, 1989; Rehfuss et al, 1991; deGroot et al, 1993) which induce transcription from the promoters of cAMP-responsive genes. These factors activate transcription as dimers (de Groot and Sassone-Corsi, 1993).

TCA motif is less constant (Borrelli et al, 1992). The binding site specificity appears to require 18–20 bp, since the five or so bases flanking the core consensus have been shown to dictate, in some cases, the permissivity of transcriptional activation (Deutsch et al, 1988). In many genes, the CRE sequence is located in the first 200 bp upstream from the cap site. In most cases there is only one CRE element per promoter, although there are notable exceptions. The promoter of the α-chorionic gonadotropin gene, for instance, contains two identical, canonical CREs in tandem, between positions −117/−142 (Delegeane et al, 1987). The promoter of the pituitary-specific transcription factor GHF-1/Pit-1, on the other hand, contains two different CREs between positions −200/ −150 which are separated by a 40 bp spacer (McCormick et al, 1990). The proto-oncogere c-*fos* contains a powerful CRE at position −60 (Sassone-

Corsi et al, 1988), but other CRE-like sequences are also present within the gene regulatory region (Berkowitz et al, 1989), however their precise function has yet to be determined.

Mechanism of activation

An important step toward the understanding of cAMP-regulated gene transcription has been made with the cloning of cDNAs encoding CRE-binding proteins, thereby allowing studies on the precise structure-function relationship of these factors. Several related genes have been found which encode CRE-binding proteins (Figure 1.2). cDNAs for one of these, CREB (CRE-binding protein), have been cloned from human placenta and rat brain libraries (Hoeffler et al, 1988; Gonzales et al, 1989). Other members of this family have been isolated from various sources (Maekawa et al, 1989; Hai et al, 1989; Ziff, 1990). These factors are ubiquitously expressed, suggesting that they have a housekeeping role (de Groot and Sassone-Corsi, 1993).

DNA binding

The CREB protein, as well as all other members of the family, belong to the leucine zipper group of transcriptional regulator (bZip, Figure 1.2) (Landschulz et al, 1988). CREB contains a heptad repeat of four leucines in the carboxy terminus which constitutes an α-helical coiled structure (Busch and Sassone-Corsi, 1990). It has been demonstrated that, as is the case of Fos, Jun and C/EBP, the leucine zipper is responsible for the dimerization of the protein and that dimerization is a prerequisite for DNA-binding. The model by Vinson et al (1989) suggests the presence of a bipartite DNA-binding domain, as dimerization ensures the correct orientation of the adjacent basic regions to allow their optimal contact with the recognition sequence. The basic region, 50% rich in lysine and arginine residues, is in fact divided into two sub-domains containing clusters of basic residues separated by a spacer of alanines, conserved among all leucine zipper transcription factors. In this model, the two basic regions of the dimer recognize the two halves of the palindromic recognition sequence. The positively charged amino acids in the basic region lie on one face of the two helices of a helix-bend-helix structure. The two positively charged α-helices lie in the major groove of the DNA helix positioned so that the positive charges are in contact with the negative charges of the phosphate backbone (Vinson et al, 1989). The structure of the DNA binding domain of CRE-binding proteins appears to be conserved.

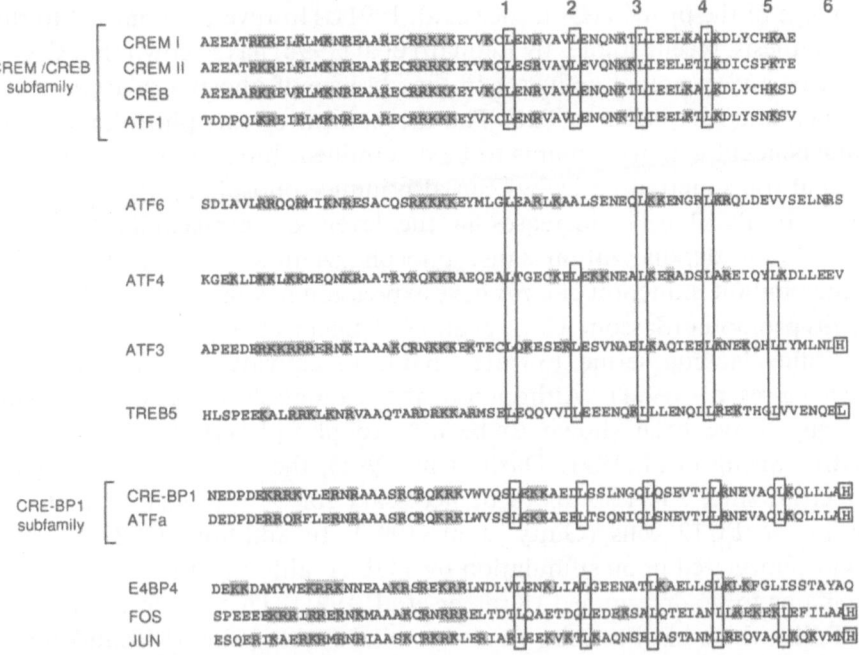

Figure 1.2 The bZip domain: a coiled-coil. Amino acid sequence alignment for the bZip domain of various regulatory proteins. The extent of the basic DNA binding region, the leucine zipper and the intervening spacer region are indicated. The component leucines of the leucine zipper are highlighted by boxing and are numbered. Basic amino acids are background shaded.

Activating the activators: the role of phosphorylation

An important insight into the molecular mechanisms by which the transcription of CRE-containing genes is induced, has come from experiments demonstrating that upon activation of the adenylyl cyclase pathway, a serine residue at position 133 of CREB is phosphorylated by PKA (Gonzalez and Montminy, 1989). Phosphorylation appears indispensable for activation, and the phosphoserine cannot be replaced by other negatively charged residues (Lee et al, 1990). Whether phosphorylation by PKA modulates DNA-binding by CREB is a somewhat controversial point. Indeed, Yamamoto et al indicate that PKA-mediated phosphorylation of CREB does not affect DNA-binding (Yamamoto et al, 1988). By contrast, it has been reported that phosphorylation of CREB by PKA causes a modest increase in binding to high affinity CRE sites and a stronger enhancement in binding to low affinity CREs (Nichols et al, 1992). However, the major effect of phosphorylation seems to occur at the level of the transactivation function of CREB. It has been proposed that this could happen by inducing a conformational change of the protein (Gonzalez et al, 1991). However, in contrast to this hypothesis, recent studies by Leonard et al (1992) have shown that CREB can be a very potent activator in the absence of phosphorylation in the pancreatic islet cell line Tu6. The mechanism of this phosphorylation-independent activity remains to be determined. Interestingly, alternative signal transduction pathways can also induce phosphorylation of serine 133. In PC12 cells, increases in the levels of intracellular Ca^{2+} by membrane depolarization cause phosphorylation of serine 133 and a concomitant induction of c-*fos* gene expression mediated by a CRE in the c-*fos* promoter (Sassone-Corsi et al, 1988; Sheng et al, 1990, 1991). CREB mutants lacking serine 133 are unable to activate c-*fos* transcription (Sheng et al, 1991). Although Ca^{2+} calmodulin-dependent (CAM) kinases have been shown to be able to phosphorylate serine 133 in vitro (Sheng et al, 1991; Dash et al, 1991), their role in vivo remains unclear, since PKA seems to be necessary for c-*fos* induction by Ca^{2+} influx in PC12 cells (Ginty et al, 1991). In addition, CREB is also phosphorylated upon stimulation by TGFβ1, although the target residue remains to be determined (Kramer et al, 1991). This is interesting since CREB, which has been shown also to bind to AP-1 sites (Masquilier and Sassone-Corsi, 1992), binds more efficiently to an AP-1 site after TGFβ1 stimulation.

Experiments by Hagiwara et al suggest a mechanism to explain the attenuation of CREB activity following induction by forskolin (Hagiwara et al, 1992). Their results indicate that after the initial burst of phosphorylation in response to cAMP, CREB is dephosphorylated in vivo by protein phosphatase-1 (PP-1) and transcription of the somatostatin gene is correspondingly reduced. However, it has been shown that both PP-1 and PP-2A can dephosphorylate CREB in vitro, resulting in a decreased binding to low affinity CRE sites in vitro (Nichols et al, 1992). Therefore, the precise role of PP-2A in the dephosphorylation of CREB remains to be determined.

The structure of the transcriptional activation domain of CRE-binding activators induces more than the phosphoacceptor region (see Figure 1.3). Serine 133 is located in a region of about 50 amino acids containing an abundance of phosphorylated serines and acidic residues, the phosphorylation box (P-Box) or kinase inducible domain (KID), which has been shown to be essential for transactivation by CREB (Lee et al, 1990) (see Figure 1.3). Although phosphorylation of serine 133 appears indispensable for activation by CREB, it is not sufficient for full activity. An acidic region just downstream of serine 133 (140-DLSSD) has been shown to be important for CREB function (Lee et al, 1990; Gonzalez et al, 1991). In addition, deletion of a region called α2, containing several

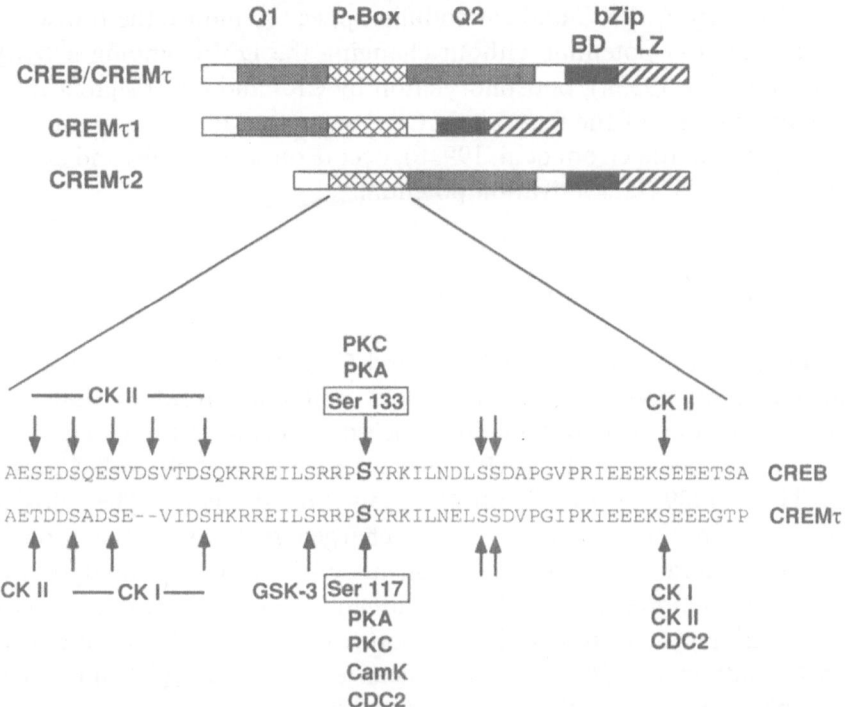

Figure 1.3 Comparison of the general structures of the activators CREB and CREMτ. Schematic representation of the structural features of CREB (Gonzales et al, 1991) and CREMτ (Foulkes et al, 1992). Q1 and Q2 indicate glutamine-rich activation domains. P-Box (phosphorylation box or kinase inducible domain) containing the PKA site (S133 in CREB, S117 in CREMτ), and putative phosphorylation sites for CK-II (S94, 97, 100, 103 and 107 in CREB, T94, S97, 100 and 105 in CREMτ), and the highly acidic downstream CKII site (S141 in CREB, S140 in CREMτ). Note that CREM contains two bZip domains, which are alternatively spliced in the different CREM isoforms (Foulkes et al, 1991).

sites that can be phosphorylated by CKII in vitro, caused a decrease in CREB activity, although differences in the magnitude of this decrease were reported (Lee et al, 1990; Gonzalez et al, 1991).

The recently described activator isoform of CREM (see below), CREMτ (Foulkes et al, 1992), can also mediate cAMP-induced transcription (Laoide et al, 1993) (Figure 1.3). CREMτ is phosphorylated by PKA in vitro as well as in vivo on serine 117, the counterpart of serine 133 in CREB (de Groot et al, 1993a). The results about CREMτ phosphorylation are important for several reasons. It has been shown that the CREM activator is phosphorylated by at least seven different kinases: PKA, PKC, casein kinase I and II, Ca^{2+}-dependent calmodulin, glycogen-synthase-3 and p34^{cdc2}. Multiple and cooperative phosphorylation events occur on the CREM protein, causing various effects at the functional level. While PKA, PKC and calmodulin appear to induce the transcriptional activation potential without changing the DNA-binding activity (de Groot et al, 1993a), phosphorylation by CKI and CKII significantly enhances binding of the factor to a CRE sequence. Phosphorylation by p34^{cdc2}, instead (de Groot et al, 1993b), occurs on several sites and causes a decrease in the transactivation potential.

The glutamine-rich activation domains

Flanking the P-box of both CREM and CREB, there are two regions which contain about three times more glutamine residues than the remainder of the protein. Glutamine-rich domains have been characterized in other factors, such as AP-2 and Sp1 (Williams et al, 1988; Courey and Tjian, 1989) as transcriptional activation domains. The current notion is that they constitute highly charged surfaces of the protein which can interact with other components of the transcriptional activation in conjunction with the P-box, although recent data strongly support the notion that their presence is an absolute requirement for transactivation (Laoide et al, 1993). The significance of the first glutamine-rich domain (Q1) is not completely clear since it has been reported to enhance CREB activity (Gonzalez et al, 1991), while Lee et al failed to find an effect when they deleted this region (Lee et al, 1990). However, this apparent contradiction might be caused by the different CREB isoforms studied by these two groups, since Gonzalez et al studied CREBα/341, while Lee et al have used CREBΔ/327 (see below) (Gonzalez et al, 1991; Lee et al, 1990). The current model to explain the activation of CREB suggests that upon phosphorylation of serine 133 by PKA, a conformational change is induced, which leads to exposure of the glutamine-rich activation domains (Gonzalez et al, 1991). The other regions which were identified as being important for CREB function might be involved in correctly spacing the phosphorylation site with respect to the glutamine-rich domains. Verification of this model awaits the determination of the crystal structure of unphosphorylated and phosphorylated CREB.

Interestingly, ATF-1 and CREM have also been reported to be activated by PKA. ATF-1 has been shown to activate transcription after co-transfection of the catalytic subunit of PKA (Rehfuss et al, 1991; Flint and Jones, 1991). Although ATF-1 can be phosphorylated by PKA in vitro, in vivo phosphorylation has yet to be demonstrated. Since ATF-1 lacks the Q1 domain and the CKII site C-terminal of the PKA site (Hai et al, 1989) (Figure 1.3), these results suggest that the KID region and the Q2 domain are sufficient to mediate cAMP-induced transcription.

Interestingly, two different CREM isoforms containing either Q1 (τ1) or Q2 (τ2) are both transcriptional activators (Figure 1.3). The Q2 domains appears to confer a slightly higher activation potential than the Q1 domain (Laoide et al, 1993). These results demonstrate that the Q1 and Q2 regions probably function in an additive manner to generate the full activation potential of CREMτ.

The allosteric conformational changes that are likely to be mediated by phosphorylation of the P-box allow the generation of an acidic face on the protein by a distant conformational change. Similar intramolecular mechanisms are implicated in the function of important enzymes such as protein kinase A.

The CREM gene

The discovery of the CREM gene opened a new dimension in the study of the transcriptional response to cAMP (Foulkes et al, 1981a). This is due to the remarkable dynamic and modular genomic structure of the gene, which offers clues to the understanding of the generation of functional diversity in transcription factors (Figure 1.4). CREM is the first gene known to encode multiple CRE-binding proteins with either antagonistic or activator function.

A remarkable, dynamic genomic structure

The CREM gene has been isolated from a mouse pituitary cDNA library screened at low stringency with oligonucleotides corresponding to the leucine-zipper and basic region of CREB. The logic behind this approach is that the adenylyl cyclase pathway plays an important role in the modulation of the hormonal regulation in the pituitary gland. The most striking feature about the CREM cDNA is the presence of two DNA-binding domains (Figure 1.4). The first is complete and contains a leucine zipper and basic region very similar to CREB; the second is located in the 3' untranslated region of the gene, out of phase with the main coding region, and contains a half basic region and a leucine zipper more divergent from CREB. Various mRNA isoforms have been identified which are obtained by differential cell-specific splicing. Alternative usage of the two DNA-binding domains has been demonstrated in various tissues and cell types, where quite different patterns of expression have been found (Foulkes et al, 1991a). This strongly contrasts with the expression of CREB and ATFs which is ubiquitous (Hai et al, 1989; Habener, 1990), suggesting they have a role as constitutive regulators.

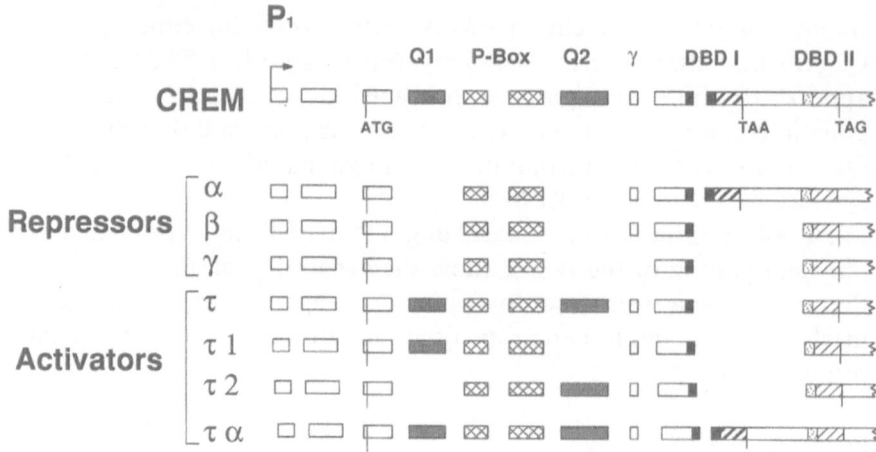

Figure 1.4 Schematic representation of CREM isoforms. A family of activators and repressors of cAMP-induced transcription are generated from the CREM gene by alternative splicing (CREMτ, τ1, τ2, α, β, γ) (Foulkes et al, 1991a, 1992; Laoide et al, 1993) as well as internal translation initiation (S-CREM) (Delmas et al, 1992). τ, τ1, τ2, activate transcription due to the presence of one or both glutamine-rich domains (Q1 and Q2). Isoforms α, β, γ lack the glutamine-rich domains and encode repressors of transcription (Laoide et al, 1993). S-CREM, generated from CREMτ mRNA by internal translation initiation contains one of the two glutamine-rich domain but acts as a repressor of transcription probably due to the lack of the kinase-inducible domain (KID). CREMα contains the first DNA-binding-domain (DBD) while the other isoforms encode the second DBD. γ is a specific domain in CREM, not present in the other CRE-binding proteins.

CREM expression appears to be finely regulated, both transcriptionally and posttranscriptionally. In fact, not only cell- and tissue-specific expression is observed, but also the production of isoforms with different functions. Three products with antagonistic activity have been the first to be described (Foulkes et al, 1991a). These isoforms reveal alternative usage of the two DNA binding domains (α and β isoforms, see Figure 1.4), as well as a small deletion of 12 aminoacids (γ isoform). The potential of even more complexity of CREM regulation is hinted at by its usage of alternative poly (A) addition sites which include or exclude the presence of ten AUUUA sequences in the 3′ untranslated region, elements thought to be involved in mRNA instability (Shaw and Kamen, 1986). The strict cell- and tissue-specific expression of CREM is indicative of a pivotal function in the regulation of cell-specific cAMP responses. This suggests that CREM occupies a central control point in the pituitary gland, since it is known that the physiology of this gland is finely regulated by a multiplicity of hormones whose signal transduction pathways involve adenylyl cyclase. Interestingly, other well described examples of cell-specific splicing include the genes encoding neuronal peptides and hormones in brain and pituitary cells (Leff et al, 1986). It thus appears clear, that cell-specific splicing is a crucial mechanism of CREM regulation, which modulates the DNA-binding specificity and the activity of the final CREM products.

Multiple products with diverse functions

The CREM products share a high homology with CREB, especially in the DNA-binding domains and in the P-box region (see Figure 1.3) (de Groot and Sassone-Corsi, 1993). In a hydrophobicity plot it appears that CREMα and CREMβ have a very similar profile, since the difference with CREMα resides only in the DNA-binding domain. Comparing these CREM isoforms to CREB it appears that the resulting proteins have the same basic structure although strikingly, the CREM antagonists are much smaller proteins (Borrelli et al, 1992). Indeed, sequence comparison indicates that the two glutamine-rich regions are absent in CREM antagonists, despite the perfect conservation of the P-box.

The CREM proteins specifically recognize CREs and show the same binding properties as CREB. This is not surprising, considering the high homology in the DNA-binding domains between these proteins. CREM proteins containing either DNA-binding domain I or II, heterodimerize with CREB (Foulkes et al, 1991a; Laoide et al, 1993), although it appears that CREMα-CREB heterodimer formation is more favoured than CREMβ-CREB. These notions suggest that CREM proteins might occupy CRE sites as CREM dimers or as CREM-CREB heterodimers, thus generating complexes with altered transcriptional functions. In fact CREM repressors act by impairing CRE-mediated transcription, and as such are considered antagonists of cAMP-induced expression. In transfection experiments, using CRE reporter plasmids, it has been demonstrated that CREMα, β and γ antagonist proteins block the transcriptional activation obtained by the joint action of CREB and the catalytic subunit of the cAMP-dependent protein kinase A (Mellon et al, 1989). These observations strongly support the notion that CREM antagonist proteins negatively modulate CRE promoter elements in vivo. An important question is how CREM proteins work. The two most likely hypotheses are as follows. According to the first scenario, CREM proteins dimerize and bind to CRE sites. Down-regulation is achieved by the occupation of these sites, which are unavailable for CREB. Similarly, if CREB is already bound, CREM proteins might squelch them because of their possible higher affinity for a specific site. According to the second model, CREM proteins are able to dimerize with CREB to generate nonfunctional heterodimers. Negative regulation is achieved by titrating active CREB molecules, and CREM dimers and CREB-CREM heterodimers bind to CRE sites, both hypotheses are justified, and both mechanisms may operate. However, results by Laoide et al (1993) indicate that the production of nonfunctional heterodimers is the most likely mechanism operating to obtain CREM-mediated antagonism of cAMP-induced transcription.

Antagonists and activators from the same gene

The first cDNA clones which have been characterized from the CREM gene encode antagonists of cAMP-induced transcription (Foulkes et al, 1991a). The central role of splicing in the regulation of this gene had been clearly hinted at by the presence of the two alternative DNA-binding domains which are used differentially in a cell-specific fashion. The CREM antagonists share extensive homology with CREB, but they lack two glutamine-rich domains, which have been shown to be necessary for transcriptional activation in CREB (see above and Figure 1.4). Interestingly the CREM gene also encodes an activator of transcriptional (Foulkes et al, 1992). In the adult testis (see below), an isoform (CREMτ) has been identified which resembles, in structure, one of the antagonist forms (CREMβ) but includes two exons that encode two glutamine-rich domains (see Figure 1.4). This form has been demonstrated to transactivate transcription from a CRE site. In adult testis, the CREMτ isoform is expressed alone, and it constitutes an abundant species in late spermatocytes and spermatids (see below). Importantly, the CREMτ transcript in testis translates only the full-length CREMτ protein (Delmas et al, 1993). This is of interest since it has been shown that the CREM activator transcript can also generate another protein with repressor function, S-CREM, by the alternative usage of an internal initiation AUG codon (Delmas et al, 1992).

CREM mRNA isoforms graphically illustrate how alternative splicing can modulate the function of a transcription factor in a tissue- and developmental-specific manner (Foulkes and Sassone-Corsi, 1992).

A gene with specialized neuroendocrine functions

Changes in intracellular levels of cAMP constitute a major regulatory mechanism of signal transduction in the central nervous system. To date, several nuclear effectors of this pathway have been characterized, although their functional relevance in brain has been unclear because of their widespread distribution and expression. We have reported the specific and anatomically distinct expression of the antagonist isoforms of the CREM gene in adult rat brain and the rapid induction of the α and β isoforms in supraoptic neurons upon physiological stimulation (Mellström et al, 1993). All known CREM isoforms are presented in total brain RNA after PCR amplification (Foulkes et al, 1991a). However, while more quantitative techniques such as RNase protection confirms the presence of both activator and repressor transcripts, in situ hybridization analysis shows that in neural tissues the antagonist isoforms have a well defined distribution pattern (Mellström et al, 1993) (Figure 1.5). In contrast, the activator CREM isoforms, which include the glutamine-rich domains, in common with CREB, have a more diffuse and general distribution.

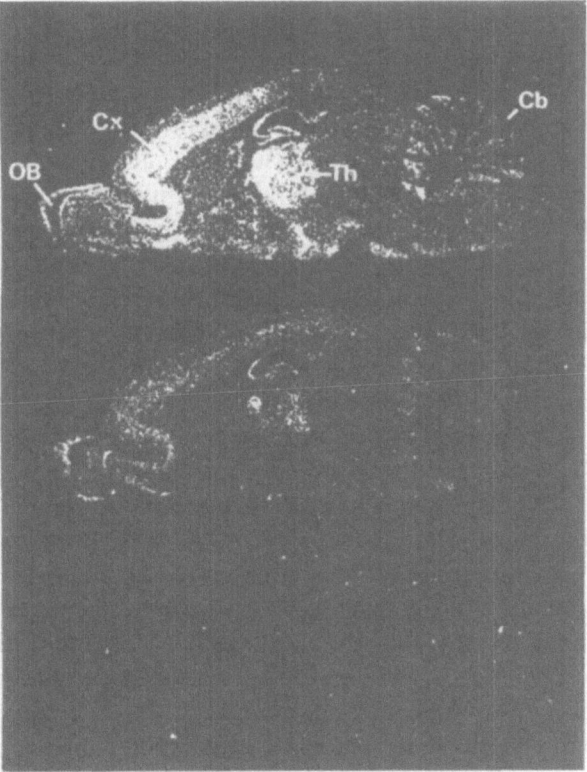

Figure 1.5 Distribution of CREM transcripts in rat brain. Parasagittal sections were hybridized with ^{35}S-labeled antisense probes (Mellström et al, 1993). The top section was hybridized with a probe able to recognize all CREM isoforms. The middle section was hybridized with a probe specific for the CREMα isoform and the lowest section was hybridized with a CREMτ-specific probe. Cerebral cortex (Cx), cerebellum (Cb), olfactory bulb (OB) and the thalamus (Th) are indicated.

A major point is that CREM differs from other members of the CRE/ATF family in that specific isoforms are induced upon physiological stimulation. To date, genes of the CRE/ATF class have been described as noninducible (Habener, 1990; Borrelli et al, 1992). Osmotic stimulation results in a differential accumulation of the two antagonist isoforms CREMα and CREMβ, but no change in CREMγ or the activator CREMτ. Consistent with previous reports (Habener, 1990; de Groot and Sassone-Corsi, 1993), no induction has been observed for CREB.

The induction of several genes in the supraoptic nucleus upon osmotic stimulation has been described previously, including the early response gene c-*fos* (Verma and Sassone-Corsi, 1987), which has been shown to undergo a rapid and transient induction (Sherman et al, 1986; Carter and Murphy, 1990; Sharp et al, 1991). Since CREM antagonists are able to negatively transregulate the activity of the c-*fos* promoter (Foulkes et al, 1991b), the temporal correlation between the onset of induction of CREMα and β and the decrease of c-*fos* transcript in supraoptic neurons would suggest a role for CREM antagonists as down-regulators of c-*fos* early induction in these neurons. Although the data available at present fall short of unambiguously demonstrating such a role for CREM (Foulkes et al, 1991b), they do provide a stimulating basis for future investigations.

A remarkable aspect of the distribution of CREM antagonists in brain is the high level of expression in the anterior thalamic nuclei. This region, forming a part of the forebrain limbic system, receives input from the hippocampus and the mammillary body of the hypothalamus and projects mainly to the cingulate cortex. This anatomical circuit has been associated with memory and integration of emotions. The significance of CREM expression in the anterior thalamus is unknown, but since this area has been reported to show no induction of early response genes after brain stimulation (Morgan et al, 1987; Sagar et al, 1988; Bullitt, 1989), it is tempting to speculate that the high basal expression of CREM antagonists could in part account for this phenomenon. In this respect, it is noteworthy that induction of c-*fos* in the thalamus after peripheral nociceptive or convulsive stimulation occurs in nuclei of the central, midline and ventral thalamic complexes (Sagar et al, 1988; Bullitt, 1989), which in general show a weak hybridization signal for CREM.

A second important observation is the presence of CREM antagonists in almost all the motor nuclei of the brain stem, while sensory nuclei are generally negative. Exceptions to this are the superior olive, which is associated with the auditory perceptions, and the mesencephalic trigeminal nucleus, equivalent to the dorsal root ganglia related to proprioceptive sensory information, in which CREM transcripts are present.

Conversely, CREM expression in motor nuclei includes the somatic motor nuclei: occulomotor, trochlear, abducens and hypoglossal; the special visceral, trigeminal and facial nuclei; and the general visceral motor nucleus of the vagus. Other positive motor nuclei are the red nucleus, the deep cerebellar nuclei and the pontine nucleus (Mellström et al, 1993).

Expression of CREM antagonist isoforms in several hypothalamic nuclei associated with homeostatic regulation is also noticeable. Such is the case for magnocellular neurons in the supraoptic hypothalamic nuclei which respond to osmotic stimulation by the differential temporal induction of two of the antagonist isoforms, CREMα and CREMβ. Other functionally related brain areas where the CREM antagonists are expressed are nuclei involved in visual processing: the suprachiasmatic nucleus, the dorsolateral geniculate nucleus, the lateroposterior thalamic nucleus and the medial terminal nucleus of the accessory optic tract. The last is supposedly involved in entrainment of endocrine rhythms by light, and fine adjustment of head-eye coordination. The presence of CREM in the pineal gland also points to a possible role for CREM in the processing of visual information and the establishment of circadian rhythms (see below).

These findings are a further demonstration of the physiological importance of the CREM gene in the CRE/ATF family. The discovery of the anatomically specific pattern of expression of distinct CREM isoforms, together with their potential for inducibility, sheds new light on the mechanisms whereby cAMP regulates gene expression in the brain.

Physiological roles of CRE-binding proteins

Although present knowledge clearly shows that CRE binding factors are important for cAMP-mediated transcriptional regulation, not much is known about the specific physiological roles of these proteins. This is of importance because of the crucial role played by the variations in cAMP levels in neuroendocrine regulation.

An interesting first clue for a physiological function of CREB comes from experiments using transgenic mice expressing a CREB mutant that cannot be phosphorylated by PKA (Struthers et al, 1991). Since cAMP serves as a mitogenic signal for the somatotroph cells of the anterior pituitary, the mutant cDNA is placed under the control of the somato-troph-specific promoter of the growth hormone gene. The pituitary glands of transgenic mice expressing this construct are atrophied and deficient in somatotroph cells. Moreover, the transgenic mice exhibit a dwarf phenotype. No other cell type in the pituitary is influenced by expression of the transgene. These effects might arise from repression of genes involved in proliferation and pituitary-specific gene expression, such as c-*fos* and GHF1/Pit-1, although the expression of these genes has not been analyzed in the transgenic animals (Struthers et al, 1991). It is noteworthy that the block of CREB function by the dominant repressor generates a transgenic phenotype equivalent to the one obtained by targeted cell death of the somatomammotrophs (Borrelli et al, 1989). This could be an indication that CRE-binding proteins are likely to have pivotal functions in normal pituitary development.

A number of reports indicate a role for CRE-binding proteins in spermatogenesis. This is not unexpected, since the metabolism of Sertoli and Leydig cells, the somatic cell-types directing the maturation of germinal cells, is regulated by the pituitary gonadotropins follicle-stimulating FSH hormone (FSH) and lutenizing hormone (LH), which in turn activate the adenylyl cyclase pathway. The most striking example of differential regulation of CRE binding proteins during spermatogenesis comes from recent studies on CREM (Foulkes et al, 1992). By studying the expression of the CREM gene during spermatogenesis it has been shown that it generates high levels of the CREMτ activator isoform by coordinate insertion of two glutamine-rich domains in the repressor isoform CREMβ (Figure 1.4). As a consequence of these insertions, CREM is switched into a powerful transcriptional activator. An abrupt developmental switch in CREM expression observed during spermatogenesis. Premeiotic germ cells express only the repressor forms in low amounts, while from the pachytene spermatocyte stage onwards CREMτ is expressed uniquely and in very high amounts (Foulkes et al, 1992).

CREM and spermatogenesis

Testis genes responsive to cAMP

The CREM activator isoform is very abundant in spermatids (Foulkes et al, 1992; Delmas et al, 1993). Thus, it is obvious to ask whether genes that are strongly activated at that stage of spermatogenesis differentiation contain CRE sites in their regulatory regions. In this respect it is noteworthy that CREM proteins are able to recognize a number of different CRE motifs (Laoide et al, 1993). Indeed, we have analyzed the sequences of the promoters of genes induced in haploid round spermatids and have found that several contain CREs (listed in the Table). Some promoters contain perfect consensus CRE motifs, as in the case of the transition protein 1 (MTP1), or quite divergent CREs, as in the thymosin promoter. DNA-

Table. Target CRE Sites for CREMτ in Testis*

CRE Sequence	Corresponding gene
TTGGC**TGACGTCA**GAGAG	Somatostatin
AATTGGG**TGAGGTCA**CTTAA	RT7
AGCTTCCTCTTT**GACTTCA**TAATTCCT	Protamine 1
TGGGCCGA**CAGGTCA**CAGTGGGG	Protamine 2
TATGTAG**TGACGTCA**CAAGAGAG	Transition protein 1
AGGACGT**GCTGGTCA**CCCCCAAA	Thymosin
TTGATCT**AGACGTCA**AAATTCCT	Tpx-1a
CTTCCTAA**TACGTCA**CAGCATCT	TPx-1b

*CREMτ binds to various CRE-like sequences present in promoters activated in round spermatids. Sequences of one strand of the oligonucleotides synthesized and used in DNA binding assays (Delmas et al., 1993), which correspond to the naturally occurring CRE-like sites. The eight base CRE sequence is shown in bold. The name and the reference of the corresponding gene is given. For the Tpx-1 promoter, two CRE motifs were found, Tpx-1a and Tpx-1b.

binding analyses reveal that all the oligonucleotides corresponding to CREs present in testis-specific genes are efficiently recognized by purified CREM protein (Delmas et al, 1993). Affinity for binding varies among the different CREs; for example, promoters of transition protein 1 (MTP1) and a testis-specific protein highly homologous to the acidic epididymal glycoprotein (tpx-1a), require a large excess of cold CRE oligonucleotide to compete binding, indicating a particularly strong affinity for CREM protein (Delmas et al, 1993). These results suggest that CREM activators may recognize a large variety of cellular targets in spermatids.

Spermatogenesis is a process of cellular differentiation in which diploid germ cell progenitors differentiate into haploid spermatozoa. This process is finely regulated by the hypothalamic-pituitary axis and requires the coordinated action of several hormones (Veldhuis, 1991). In response to hormone stimulation, testicular cells initiate a cascade of events inducing changes in cellular metabolism and gene expression. The transmission of pituitary hormonal stimuli from the cell surface to the cytoplasm and ultimately to the nucleus is mediated by the cAMP signaling pathway. Thus, nuclear factors involved in the regulation of gene expression by cAMP are likely to be of crucial importance during germ cell differentiation. Findings on CREM indicate that this gene plays a pivotal role in governing cAMP-dependent gene expression during spermatogenesis.

Hormonal control in testis: induction of the cAMP signal transduction pathway

Proliferation and differentiation of germ cells is ultimately dependent on two hormones produced by the gonadotrophs of the anterior pituitary, luteinizing hormone (LH) and follicle-stimulating hormone (FSH) (Steinberger, 1971; Lostroh, 1976). In the absence of these two hormones, spermatogenesis does not proceed beyond meiotic prophase. For instance, *hpg* mutant mice are deficient in gonadotropin releasing hormone, and spermatogenesis is interrupted at the diplotene stage (Cattanach, 1977). The developing male germ cells appear to lack receptors for LH and FSH, and consequently, it is believed that they receive hormonal signals via somatic cells present in the testis (Moore, 1978; Santen, 1987). LH and FSH receptors are located on the Leydig and Sertoli somatic cells, respectively (Steinberger, 1971; Lostroh, 1976). Stimulation of the Leydig cells by LH results in the secretion of testosterone in the interstitial compartment which then diffuses into the seminiferous tubules where Sertoli and germ cells are located. Upon FSH and testosterone stimulation, Sertoli cells secrete peptides and other components required for cell differentiation into the seminiferous tubules (Russell et al, 1980; Grootegoed et al, 1986; Jégou et al, 1992).

FSH and LH hormones are ligands of distinct transmembrane receptors which are coupled to stimulatory G proteins (Borrelli et al, 1992). Binding of these hormones to their receptors induces the activity of adenylyl cyclase which converts ATP to cAMP. The elevated intracellular level of cAMP stimulates the activity of protein kinase A (PKA) which subsequently phosphorylates various cellular proteins. Among these there are the CRE-binding proteins (Borrelli et al, 1992) (Figure 1.1).

The expression of testis-specific isoforms of PKA suggests that the cAMP pathway in testis has specialized functions which may involve the phosphorylation of testis-specific targets (Pariset et al, 1989; Oyen et al, 1988; 1990; Lonnerberg et al, 1992).

Nuclear effectors of the PKA pathway in testis

The expression of two members of the CRE-binding protein family has been described in testis, namely the CREM and CREB genes (Waeber et al, 1991; Foulkes et al, 1992; Ruppert et al, 1992). These genes share extensive regions of homology within the coding sequence and have a similar genomic structure (de Groot and Sassone-Corsi, 1993). As discussed above, however, the CREM gene is remarkable for several reasons.

The expression of the CREM gene during spermatogenesis shows some unique features (Figure 1.6 and Figure 1.7). CREM is expressed at low levels before meiosis, and only repressor isoforms are observed. In post-meiotic germ cells, an alternative splicing event causes a switch

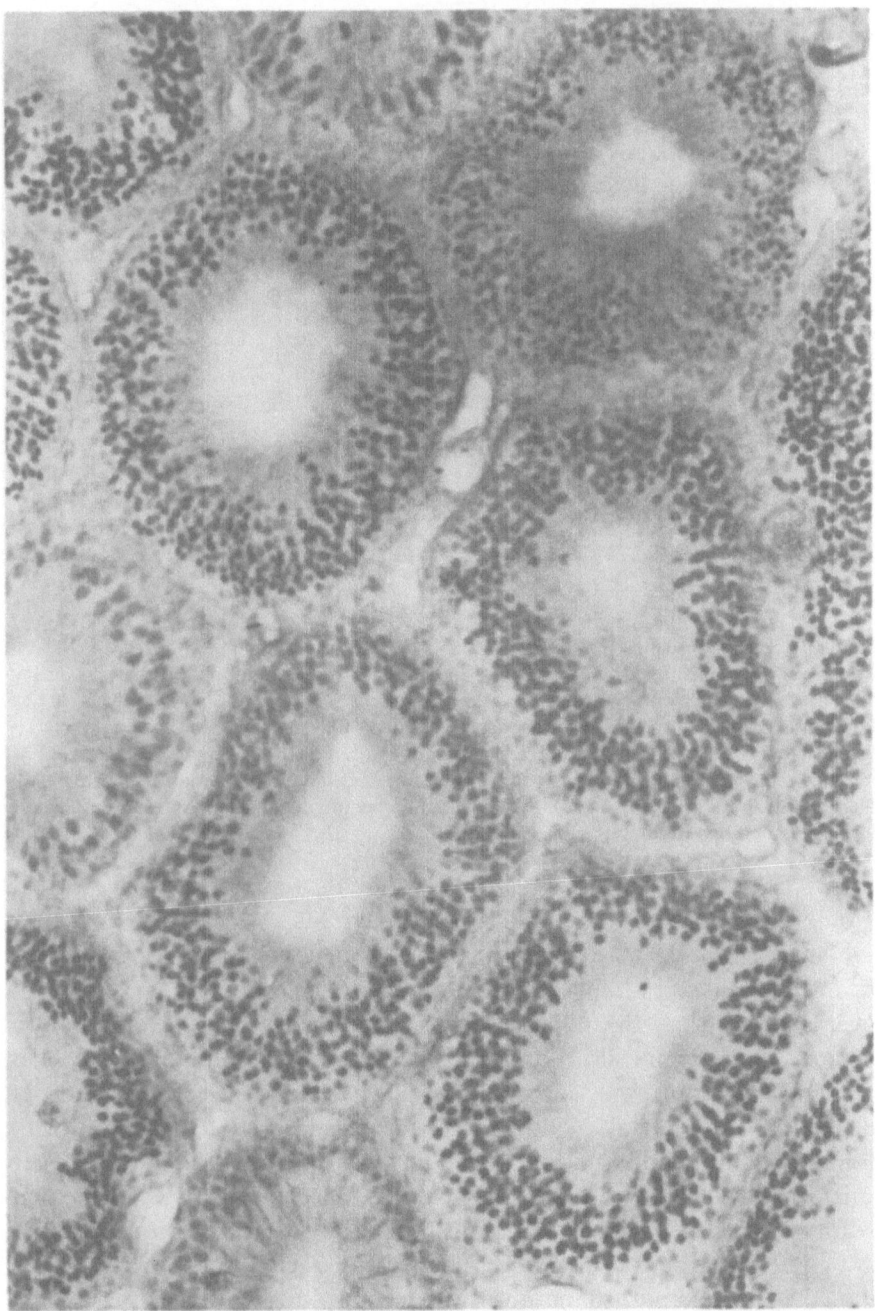

Figure 1.6 Peroxidase staining of rat seminiferous tubules showing expression of the CREM protein in spermatids. The CREM antibody used for this experiment was prepared against a bacterially produced CREMτ protein. Note the different intensity of staining in the various tubules, indicating that CREM expression is developmentally regulated.

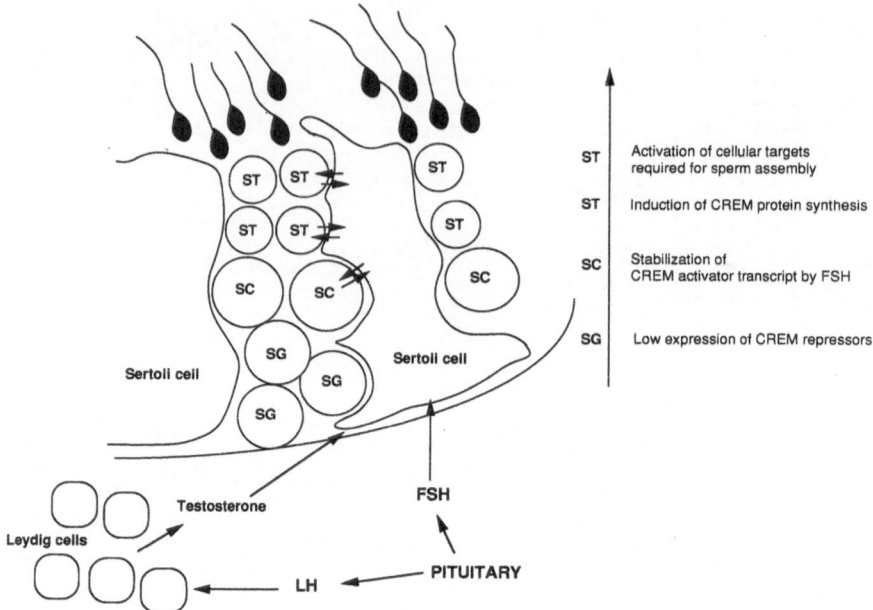

Figure 1.7 Schematic representation of a section of a seminiferous tubule where the CREM expression pattern is indicated. CREM expression is regulated at multiple levels during spermatogenesis. Premeiotic germ cells (Spermatogonia, SG) express a low level of CREM repressor isoforms. During meiotic prophase, the pituitary follicle-stimulating hormone (FSH) is responsible for the stabilization of CREM activator transcripts in spermatocytes (SC); CREM protein, on the other hand, is detected only after meiosis in haploid spermatids (ST) (arrows). In the haploid spermatids, CREM proteins activate a number of cellular genes expressed specifically during spermatid maturation (Delmas et al, 1993).

in CREM function, from transcriptional repressor to activator (Foulkes et al, 1992) (Figure 1.4). We have also observed a quantitative switch in CREM expression, since from the pachytene spermatocyte stage onwards the CREM transcript accumulates at very high levels. In contrast, the CREB gene is expressed at low levels in both somatic and germ cells. Somatic cells express the full-length CREB protein while truncated isoforms are detected in germ cells (Waeber et al, 1991; Ruppert et al, 1992). These isoforms lack the C-terminal bZip domain required for nuclear transport and DNA binding; thus, their functional relevance remains a mystery since the CREB truncated isoforms lack the ability to directly regulate gene expression (Waeber et al, 1991; Ruppert et al, 1992). The role of CREB in spermatogenesis is further questioned by the observation that, in mice where the CREB gene has been inactivated by homologous recombination, spermatogenesis proceeds normally (Hummler et al, 1994).

In the mouse, spermatogenesis is a cyclic process occurring every 12 days. The seminiferous epithelial cycle has been classified into XII stages representing different types of cellular associations found during the cycle (Oakberg, 1956). The expression of CRE-binding proteins appears to be dependent on the developmental stage of the seminiferous epithelium; for example, the CREM activator protein is detected mainly in the spermatids, at stage VII–VIII of the cycle while the CREB transcript is predominantly expressed in the Sertoli cells and at stage III–IV (Waeber et al, 1991; Delmas et al, 1993). These features reflect a strict regulation of gene expression and suggest that these proteins might function at restricted developmental stages of spermatogenesis.

These observations underline the importance of alternative splicing and of its regulation as part of the physiological mechanisms responsible for differentiation of germ cells.

CREM regulation and function in germ cells

The developmental switch in CREM expression occurring during meiosis raises two major questions: (1) What is the physiological mechanism controlling this switch? Is it only developmental, or is it hormonally regulated? (2) What is the function of this abundant CREM activator in germ cells?

FSH regulates CREM expression in testis

In seasonal breeders, such as the golden hamster, spermatogenesis ceases during winter. The analysis of CREM expression in the hamster reveals a seasonally-regulated appearance of the transcript corresponding to the activator isoform (Foulkes et al, 1993). Seasonal fluctuations are known to require an intact hypothalamic-pituitary axis, suggesting that CREM expression in testis may involve hormonal modulation. Since spermatogenesis is controlled by pituitary hormones, the involvement of the pituitary gland in the control of CREM expression has been investigated. The CREM switch is no longer observed in hypophysectomized animals, but can be restored by injection of FSH in a few hours (Foulkes et al, 1993). Since there are no FSH receptors on germ cells, while they are present on Sertoli cells, it is predicted that FSH stimulation is rapidly transmitted from Sertoli to germ cells (Figure 1.7). The action of FSH is specific since the switch is not induced by injection of other pituitary hormones, like LH and prolactin. A molecular analysis shows that induction of the CREM activator transcript is due to the use of an alternative polyadenylation site within the CREM transcript which truncates the 3' untranslated region. Omission of destabilizer elements thereby enhances transcript stability (Foulkes et al, 1993). This observation is of interest since it clearly demonstrates that FSH regulates the function of a cAMP-responsive nuclear factor in germ cells. Importantly, these experiments constitute the first report indicating the effect of hormonal stimulation on transcriptional factor message stability.

CREM, a regulator of gene expression in haploid germ cells

A first hint as to the role of CREM during spermatogenesis is indicated by its protein expression pattern. In seminiferous epithelium, CREM transcripts accumulate in spermatocytes and spermatids, but CREM protein is detected only in spermatids (Figure 1.6) (Delmas et al, 1993). Thus, CREM function is restricted to a specific type of germ cell, the haploid spermatid. The absence of CREM protein in spermatocytes reflects a strict translational control and indicates multiple levels of regulation of gene expression in testis. It will be extremely important to analyze further the mechanism of this translation delay and to define whether it is also hormone-dependent.

Phosphorylation by PKA activates CREM function allowing the relay of the hormonal signal from the cytoplasm to the nucleus (de Groot et al, 1993a). The CREM activator is efficiently phosphorylated by cAMP-dependent PKA activity endogenous to the spermatids, indicating that the CREM protein is a nuclear target for the cAMP pathway in haploid spermatogenic cells (Delmas et al, 1993).

The detection of CREM activator protein in spermatids coincides with the transcriptional activation of several genes containing a CRE motif in their promoter region (see Table). These genes encode mainly structural proteins required for spermatozoon assembly (transition protein, protamine, RT7, etc.), suggesting a role for CREM in the activation of genes required for the late phase of spermatid differentiation. This observation implies that the transcription of some key structural genes is directly linked to hormonal control and consequently to the level of cAMP present in seminiferous epithelium. A demonstration of the role of CREM in the expression of one of these genes, RT7, has been shown using in vitro transcription experiments. A CREM-specific antibody blocks RT7 in vitro transcription with nuclear extracts from seminiferous tubules but not with extracts from liver (Delmas et al, 1993). In conclusion, CREM might participate in testis- and developmental-specific regulation of genes containing a CRE in their promoter region, by expressing the repressor isoforms before meiosis, and high levels of the activator after meiosis.

Icer, a key player in the nuclear response to cAMP

During studies of CREM expression within the neuroendocrine system, an unexpected new facet emerged: transcription of the CREM gene is inducible by cAMP (Stehle et al, 1993; Molina et al, 1993; Masquilier et al, 1993). Furthermore, the kinetic of this induction is that of an early response gene (Bravo et al, 1988; Nathans et al, 1988). This important finding further reinforces the notion that CREM products play a pivotal role in the nuclear response to cAMP, since the expression of no other CRE-binding factor has been shown to be inducible, to date. For example, the recently characterised CREB promoter is GC-rich and reminiscent of the promoters of constitutively expressed, housekeeping genes (Cole et al, 1992; Meyer et al, 1993). Upon detailed analysis of the induced CREM products, there have been more surprises. The promoter that directs expression of the previously characterised CREM isoforms (P1) is not cAMP inducible. Instead, an alternative promoter lying within an intron near the 3′ end of the gene, directs cAMP induced transcription of a novel truncated CREM product, termed ICER (Induced cAMP Early Repressor) (Stehle et al, 1993; Molina et al, 1993) (Figure 1.8). ICER is the smallest bZip factor yet described, functions as a powerful repressor of cAMP-induced transcription and furthermore negatively autoregulates the ICER promoter (Molina et al, 1993). The expression of ICER was first described in the pineal gland where it is the subject of a dramatic circadian pattern of expression (Stehle et al, 1993). Dynamic ICER expression is a general feature of neuroendocrine systems (Stehle et al, 1993).

Inducibility of the CREM gene: use of an alternative intronic promoter.

Clues that the CREM gene is cAMP inducible first came from the demonstration that adrenergic signals direct CREM transcription in the pineal gland (Stehle et al, 1993). The inducibility phenomenon has been characterised in detail in the pituitary corticotroph cell line AtT20 (Molina et al, 1993). In unstimulated cells the level of CREM transcript is below the threshold of detectability. However, upon treatment with forskolin (or other cAMP analogs), within 30 minutes there is a rapid increase in CREM transcript levels which peak after two hours and then progressively decline to basal levels by five hours. This characteristic kinetic classifies CREM as an early response gene and thus, for the first time, directly implicates the cAMP pathway in the cell's early response. CREM inducibility is specific for the cAMP pathway since it is not inducible by TPA or dexamethasone treatment (Molina et al, 1993).

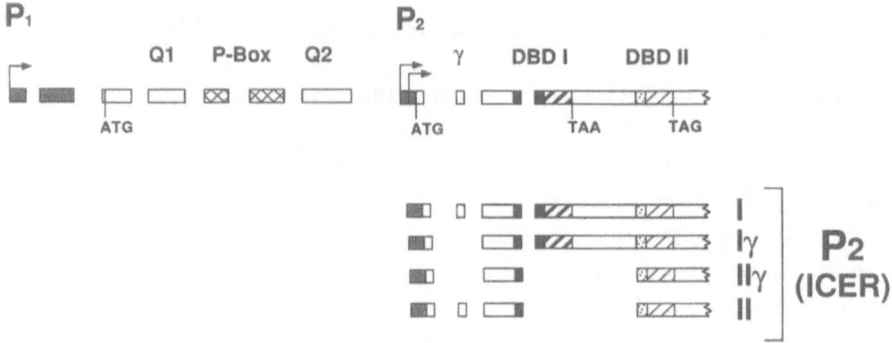

Figure 1.8 The ICER family. Schematic representation of the ICER transcript in relation to the CREM gene structure. Exons encoding the glutamine-rich domains (Q1 and Q2), the P-Box, the γ domain and the two alternative DNA binding domains (DBDI and DBDII) are shown. All the ICER transcripts are derived from an internal start-site of transcription (P2) located between the Q2 and γ-exon. A family of four types of ICER transcript is generated by alternative splicing of the DBD and γ-domain exons; ICER-I, ICER-Iγ, ICER-II, ICER-IIγ.

In order to further characterize the cAMP-induced CREM transcript, a battery of exon-specific probes have been used in a systematic northern, RNase protection and RT-PCR analysis. In this way it has been demonstrated that all previously characterized exons located 5' to the γ-exon are absent from the induced CREM transcript. The 5' boundary of the γ-exon is defined by a consensus splicing acceptor site. Thus, in order to identify the remaining exons constituting the 5' end, a RACE-PCR cloning strategy is used. By this approach a series of short, overlapping cDNA clones define a novel cDNA which is termed ICER (Stehle et al, 1993; Molina et al, 1993)). An 82 bp sequence lying 5' of the γ-exon boundary extends the CREM open reading frame upstream by eight amino acids to a consensus Kozak ATG codon and includes a short 5' untranslated region. Full-length cDNA clones, isolated from a pineal cDNA library, together with RNase protection assays, confirmed the results of RACE PCR while primer extension analysis demonstrated that the 5' end of the ICER clones correspond to a transcription start site (Stehle et al, 1993; Molina et al, 1993). The ICER cDNA clones also reveal different splicing of the two alternative DNA binding domains and of the γ-exon (Figure 1.8), as previously described for the CREMα and β isoforms (Figure 1.4). In addition, by northern blot and RNAse protection analyses it is apparent that the ICER transcripts employ the various polyadenylation sites in a cell-specific fashion, thus cAMP induction of the CREM gene generates a family of transcripts (Molina et al, 1993).

In order to locate the start point of transcription of the ICER transcripts relative to the promoter of the previously described isoforms, an overlapping series of phage clones encompassing the entire CREM gene, have been screened with a probe for the 5' ICER-specific exon. Hybridization locates the start of transcription (P2) within the 10 kb intron which lies between the Q2 glutamine-rich domain exon and the γ-exon (Figure 1.9) (Stehle et al, 1993; Molina et al, 1993).

In contrast to the promoter (P1) which generates all the previously characterised CREM isoforms, is GC-rich and not inducible by cAMP (N. Foulkes, in preparation), the P2 promoter has a normal A-T and G-C content and is strongly inducible by cAMP. It contains two pairs of closely-spaced CRE elements (cAMP-autoregulatory responsive elements, CAREs) (Figure 1.9) organized in tandem. Furthermore, the separation between the CAREs in each pair is only three nucleotides. These features make P2 unique among cAMP-regulated promoters and are suggestive of cooperative interactions among the factors binding to these sites (Molina et al, 1993). Previously, tandemly repeated pairs of CRE elements have been described in the promoters of the Pit-1 and α-CG genes (Delegeane et al, 1987; McCormick et al, 1990), but the

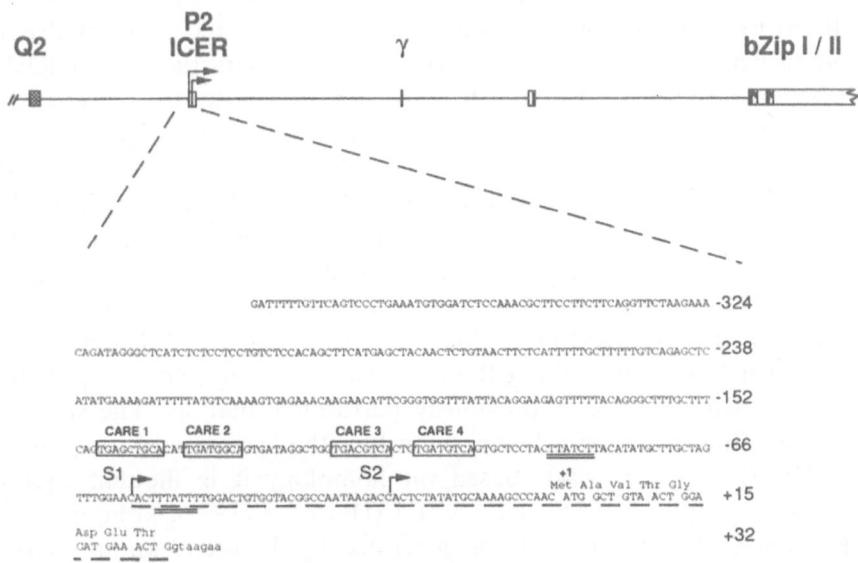

Figure 1.9 The ICER P2 promoter. Top section: schematic representation of the ICER exon and P2 promoter in the context of the genomic structure of the 3′ end of the CREM gene. Flanking exons encoding the Q2 glutamine-rich domain, the γ-domain and the 3′ terminal exon incorporating the alternative bZip I/II domains are indicated. Bottom section: 400 bp genomic sequence including the ICER 5′ exon. Dashed underlining delineates the ICER 5′ exon. Lower case sequence represents the beginning of the first intron in the ICER transcript. The two alternative start sites of transcription, S1 and S2, are shown. Putative TATA elements upstream of S1 and S2 are indicated by double underlining while the four CRE-like elements (CARE) are boxed and labeled. Position + 1 corresponds to the A of the Kozak ATG initiation codon.

individual elements are more widely spaced. The promoter directs transcription from two alternative start points (S1 and S2) and 23 bp upstream from each start site lie A-T rich elements which presumably function as TATA elements.

ICER, the smallest bZip factor

The ICER open reading frame is constituted by the C-terminal segment of CREM. The predicted open reading frame encodes a small protein of 120 amino acids with an expected molecular weight of 13.4 kD. This protein, compared with the previously described CREM isoforms, essentially consists of only the DNA binding domain, which is constituted by the leucine zipper and basic region. The provocative structure of ICER is suggestive of its function and makes it one of the smallest transcription factors described (Molina et al, 1993).

The intact DNA binding domain directs specific ICER binding to a consensus CRE element. Consistent with previous analysis of the other CREM isoforms, ICER fails to bind to Sp1 sites. Importantly, ICER is able to heterodimerize with CREMτ, as well as with the other CREM proteins and with CREB. ICER functions as a powerful repressor of cAMP-induced transcription in transfection assays using an extensive range of reporter plasmids carrying individual CRE elements of cAMP-inducible promoter fragments (Stehle et al, 1993). Interestingly, ICER-mediated repression is obtained at substoichiometric concentrations, similarly to the previously described CREM antagonists (Foulkes et al, 1991a). However, it should be noted that ICER is significantly more potent when compared to CREMα or CREMβ (Stehle et al, 1993).

The small size of the ICER products is striking and has possibly masked their presence in previously performed analysis. The smallest CRE-binding factor described previously is the liver regenerating factor (LRF) (Hsu et al, 1991). Based on homology it is thought that it represents a splicing variant of ATF3 (Hai et al, 1989), although it is also conceivable that it could be generated by the use of an alternative promoter as in the case of ICER. This 21kD bZip protein functions as an activator of transcription in a heterodimeric complex with c-Jun, and its expression is also stimulated by mitogens. The major feature of ICER proteins is that they lack the N-terminal domain shared by the other CREM isoforms (Figure 1.8). Specifically, the P-box domain which is retained in all the P1 promoter-generated CREM isoforms, even those acting as repressors. CREMα, β, and γ, phosphorylation by PKA has been shown to modulate the degree of repression activity (Laoide et al,

1993). On the contrary, ICER escapes from PKA-dependent phosphorylation and thus constitutes a new category of CRE binding factor, for which the principle determinant of their activity is their intracellular concentration and not their degree of phosphorylation.

Negative Autoregulation

Upon cotreatment with cycloheximide, the kinetics of CREM gene induction by forskolin are altered showing a significant delay in the post-induction decrease in the transcript; elevated levels persist for as long as 12 hours. This implicates a de novo synthesised factor which might downregulate CREM transcription (Molina et al, 1993). This observation combined with the presence of CRE elements in the P2 promoter (see above), suggests that the transient nature of the inducibility could be due to ICER. Consistently, the CARE elements in the P2 promoter have been shown to bind to the ICER proteins. Detailed studies have demonstrated that the ICER promoter is indeed a target for ICER negative regulation (Molina et al, 1993). Thus, there exists a negative autoregulatory mechanism controlling ICER expression (see Figure 1.10).

Rhythmic expression

Day-night switch in ICER expression in the pineal gland

A crucial step in the understanding of the physiological significance of ICER came with the observation that it is expressed in a rhythmic fashion in the pineal gland (Stehle et al, 1993). Crucial elements for the synchronization of biological rhythms in mammals are the pineal gland (Tamarkin et al, 1985; Reiter, 1991) and the suprachiasmatic nucleus (SCN) (Klein et al, 1991; Moore, 1983). Environmental lighting conditions are transduced by the pineal gland from a neuronal to an endocrine message, the rhythmic secretion of melatonin (Figure 1.11) (Tamarkin et al, 1985; Reiter, 1991; Klein, 1985). This hormone synthesis is controlled by the SCN, and is elevated at night and low during the day (Reiter, 1991; Klein, 1985). The cAMP-dependent signal transduction pathway serves as a relay to stimulate melatonin synthesis (Klein, 1985; Sudgen et al, 1985; Vanecek et al, 1985). Thus, from neuronal pathways including the retina and the SCN, the pineal gland acts as a temporal regulator for the function of the hypothalamic-pituitary-gonadal axis (Reiter, 1991). During the night, ICER constitutes an abundant transcript while

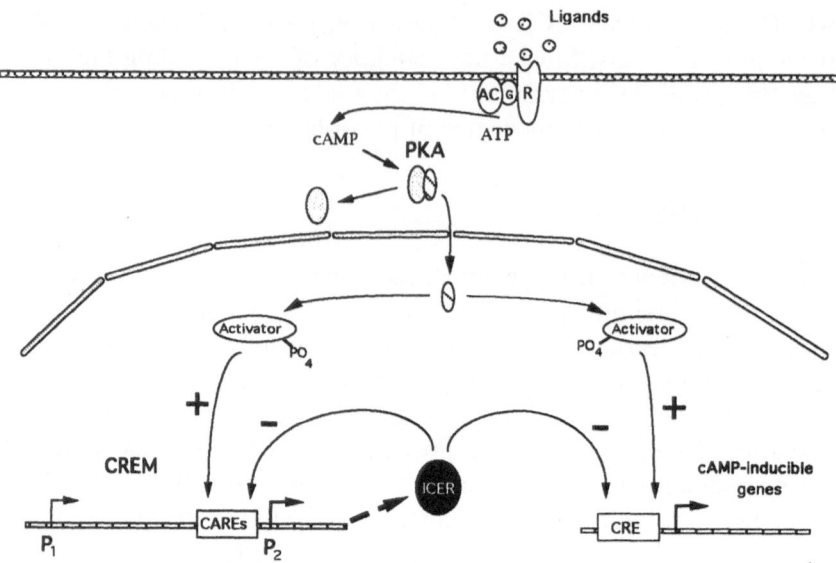

Figure 1.10 The role of ICER in the regulation of gene expression by cAMP. Schematic representation of the cAMP signal transduction pathway operating from the cell membrane, through the cytoplasm and into the nucleus. Ligands interacting with transmembrane receptors (R) stimulate the enzyme adenylyl cyclase (AC) via interactions with G-proteins (G). The subsequent rise in intracellular cAMP concentration results in the dissociation of the regulatory and catalytic subunits of PKA and the translocation of active catalytic subunits into the nucleus. PKA phosphorylates and thereby stimulates transcriptional activators binding to CREs (e.g. CREB, CREMτ and ATF1) which induce transcription from the promoters of cAMP-responsive genes. These factors activate activate transcription from the CREM P_2 promoter via the CARE elements and ultimately lead to a rapid increase in ICER protein levels. ICER represses cAMP-induced transcription, including that from its own promoter. The consequent fall in ICER protein levels eventually leads to a release of repression and permits a new cycle of transcriptional activation.

during the day it is present at low levels (Figure 1.12) (Stehle et al, 1993). An observation initially made by in situ hybridization subsequently has been elaborated by a RNAse protection assay. For this analysis rats were sacrificed at hourly time points during a 24 hour cycle of light (12 hours) and dark (12 hours). The transcript shows a very characteristic and reproducible kinetic of expression. By a series of physiological experiments, the mechanism controlling this pattern of ICER expression has been determined (Stehle et al, 1993). In animals maintained in constant darkness (DD) the temporal switch in ICER expression is conserved, which demonstrates that the ICER rhythm is driven by an endogenous clock mechanism (Figure 1.11). In contrast, under constant light treatment (LL) no ICER switch is observed. LL conditions are known to dissociate the coordination of the clock output (Figure 1.11) (Meijer, 1991) and to suppress melatonin synthesis (Tamarkin et al, 1979; Perlow

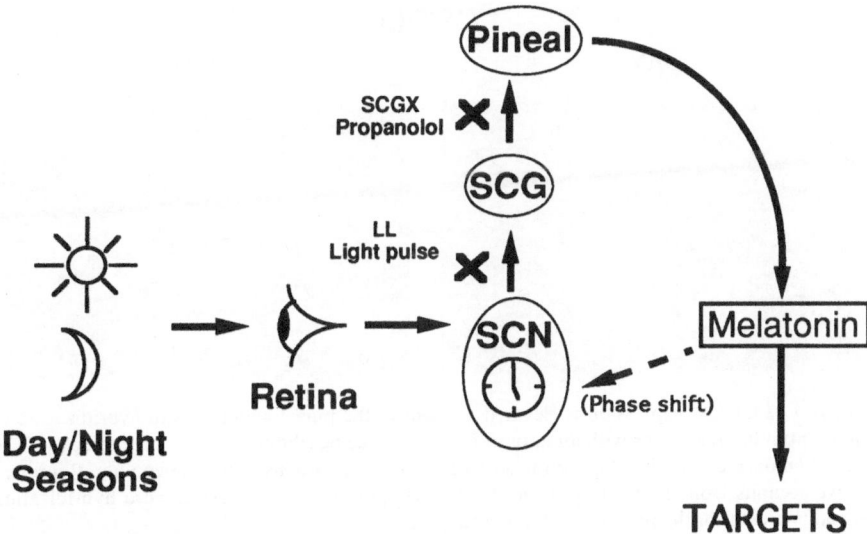

Figure 1.11 Regulation of pineal gland function. Schematic representation of the pathway whereby day/night and seasonal changes in lighting conditions are detected by the retina and interpreted by the suprachiasmatic nucleus (SCN) (the location of the biological clock). Rhythmic adrenergic signals originating from the SCN stimulate the pineal to produce the hormone melatonin. The nerve fibres linking the SCN and the pineal, pass through the superior cervical ganglion (SCG). While the specific targets for melatonin are still unclear, it is known to regulate the function of the SCN by phase shifting the biological clock. In this way melatonin production synchronises the biological clock with external time cues. Constant light (LL) or a light pulse at night blocks production of melatonin by inhibiting rhythmic stimulation by the SCN. Both surgical lesioning of the SCG (SCGX) and the adrenergic antagonist propanolol block the normal adrenergic signals from stimulating the pineal at the level of the SCG and the pineal gland respectively.

et al, 1980). In addition, a lightpulse at night, known to decrease melatonin production (Illnerova et al, 1979), also dramatically blocks ICER induction (Stehle et al, 1993).

The next step has been to define the signal whereby ICER circadian expression is driven by clock-distal elements. At night, postganglionic fibers originating from the superior cervical ganglia (SCG) release norepinephrine (Biownstein and Axelrod, 1974; Craft et al, 1984), which in turn regulates melatonin synthesis via β-adrenergic receptors (Klein, 1985). Injection of the β-adrenergic antagonist propanolol prior to the onset of darkness blocks accumulation of ICER mRNA and melatonin synthesis (Axelrod, 1988; Stehle et al, 1993). The same effect has been obtained by chronic denervation of pineal glands with bilateral removal of the SCG (SCGX). Furthermore the β-adrenergic agonist isoproterenol (Iso) induces ICER expression in ganglionectomized animals, concomitantly with melatonin synthesis (Axelrod, 1988; Stehle et al, 1993).

CREM

DAY   NIGHT

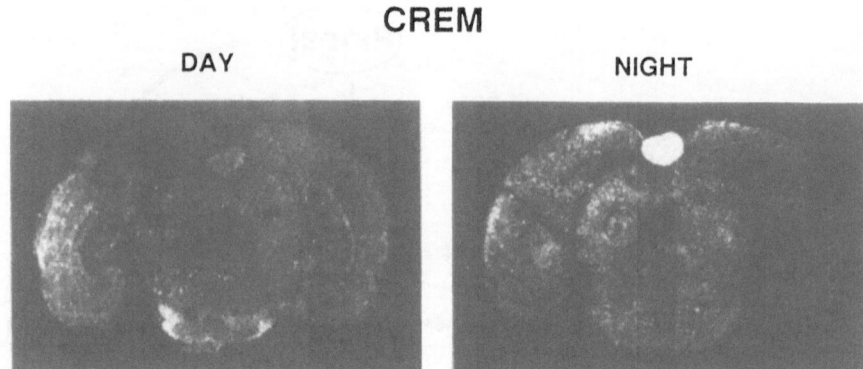

Figure 1.12 CREM expression is elevated at night in the pineal gland. In situ hybridization of midsagittal brain sections with an antisense CREM-specific riboprobe. Rats maintained in 12 h light/12 h dark conditions (light on 07:00) were sacrificed at consecutive timepoints. Representative sections from 12:00 (DAY) and 02:00 (NIGHT) are shown. An intense hybridization signal is present in the pineal gland at night.

Isoproterenol injections in intact rats are known to stimulate melatonin synthesis independently of the time of the day (Axelrod, 1988). Unexpectedly, while induction during the night is observed, ICER is not inducible upon isoproterenol administration during daytime (Stehle et al, 1993). This is in contrast with SCGX animals, where isoproterenol induces ICER expression both at day and night. ICER represents the first case of day/night disparity in cAMP-inducibility. In conclusion, signals from the SCN dictate the competence of CREM inducibility (Stehle et al, 1993).

The question of possible targets for downregulation by ICER in the gland can at the moment only be a matter for speculation. However it has been proposed that a reasonable target could be the enzyme that catalyses the rate-limiting step of melatonin synthesis (Takahashi, 1993); namely N-acetyl transferase (for which the gene has yet to be cloned) or factors that regulate its activity.

ICER expression also has been analyzed in a range of rat tissues by RNase protection assay (Stehle et al, 1993). Dramatically, although the overall amount of CREM message varies considerably between different tissues, in each sample the major protected fragment corresponds to ICER, with the exception of adult testis (Stehle et al, 1993) (where the vast majority of the CREM mRNA corresponds to the P1-derived CREMτ transcript, see above). An additional striking observation from this RNase protection analysis is a tissue-specific pattern of ICER expression. ICER is expressed at high levels predominantly in tissues of neuroendocrine origin, namely the pineal, pituitary and adrenal glands. In all other neuronal and non-endocrine tissues tested, ICER is expressed in uniformly low amounts (Stehle et al, 1993).

CREM: Transcriptional Clock?

In the light of the known properties of ICER, it is tempting to speculate on the molecular basis of the noninducibility in the pineal gland during the daytime. One possible explanation is that during the night the ICER protein accumulates and blocks cAMP-inducible transcription from the ICER promoter. With no de novo ICER production, the levels of ICER protein progressively decline until the beginning of the next night. By this time, the levels are sufficiently low to permit cAMP-directed activation. Activation is mediated by PKA which phosphorylates preexisting CRE-binding activators, and in turn restimulates the expression of ICER. In this scenario, the cyclic repression of ICER can be extended beyond the ICER promoter to include all other cAMP controlled genes (Figure 1.10). A particular gene's efficiency of repression by ICER would dictate at which time, relative to ICER protein induction, the gene can be activated once again. In this way, it is possible to envisage a cycle of transcriptional inducibility associated with rhythmic activation of the cAMP pathway. Thus, ICER could be thought of as a transcriptional clock.

Given the specific expression of ICER in neuroendocrine systems, the phenomenon first observed in the pineal gland may well apply to other tissues. In other cases, the refractory period of noninducibility following stimulation by cAMP may play a central role in interpreting and responding to repetitive signals impinging on neuroendocrine cells.

Conclusions and Perspectives

To date much of the research in transcription factor biology has been devoted to understanding the structural and functional relationship of these factors. Progress has been extremely rapid, and now the basic principles of transcription factor function are close to being elucidated. As a result of this work, new questions have been raised, and so it is clear we still have a long way to go before we understand completely how the promoters and enhancers of genes execute transcriptional control. However, a much greater challenge lies ahead and that is to relate transcriptional control mechanisms to the physiology and biology of the organism. The use of homologous recombination to inactivate specific gene products offers a powerful tool to address such questions. Paradoxically, however, in some cases it has complicated our understanding since it is clear that many important factors operate in the context of networks where there is considerable overlap of function, and thus it is likely that it will be necessary to knock out several genes in order to obtain a phenotype. Furthermore, it is possible that the phenotype obtained by loss of a single factor could reflect more than compensatory adjustments made by other factors in its network rather than the function of the target factor itself. In this context, it is important to stress that the inactivation by homologous recombination of a noninducible constitutive CRE-activator, the CREB gene, generates mice with no apparent mutant phenotype (Hummler et al, 1994). In contrast, CREM appears to play a central role in the physiology of the neuroendocrine system (Borrelli et al, 1992; Mellström et al, 1993). The generation of a mouse with an inactivated CREM gene should therefore represent another major step forward in our understanding since it is likely to provide an informative phenotype.

References

Andrisani OM, Hayes TE, Roos B, Dixon JE (1987): Identification of the promoter sequences involved in the cell-specific expression of the rat somatostatin gene. *Nucleic Acids Res* 15: 5715–5728

Auwerx J, Sassone-Corsi P (1991): IP-1: A dominant inhibitor of fos/jun whose activity is modulated by phosphorylation. *Cell* 64: 983–993

Axelrod, J (1988): The pineal gland: a neurochemical transducer *Science* 184: 1 091

Benbrook DM, Jones NC (1990): Heterodimer formation between CREB and Jun proteins. *Oncogene* 5: 295–302

Berkowitz LA, Riabowol KT, Gilman MZ (1989): Multiple sequence elements of a single functional class are required for cyclic AMP responsiveness of the mouse c-*fos* promoter. *Mol Cell Biol* 9: 4272–4281

Berridge MJ (1987): Inositol trisphosphate and diacylglycerol: two interacting second messengers. *Ann Rev Biochem* 56: 159–193

Biownstein M, Axelrod J (1974): Pineal gland: 24-hour rhythm in norepinephrine turnover. *Science* 184: 165

Borrelli E, Heyman R, Arias C, Sawchenko P, Evans RM (1989): Transgenic mice with inducible dwarfism. *Nature* 339: 538–541

Borrelli E, Montmayeur JP, Foulkes NS, Sassone-Corsi P (1992): Signal transduction and gene control: the cAMP pathway. *Critical Rev Oncogenesis* 3: 321–338

Bravo R, Zerial M, Toschi L, Schurmann M, Muller R, Hirai SI, Yaniv M, Almendral JM, Ryseck RP (1988): Identification of growth factor-inducible genes in mouse fibroblasts. *Cold Spring Harbor Symp Quant Biol* 53: 901–905

Bullitt E (1989): Induction of c-*fos*-like protein within the lumbar spinal cord and thalamus of the rat following peripheral stimulation. *Brain Res* 391–397

Busch SJ, Sassone-Corsi P (1990): Dimers, leucine zippers and DNA binding domains. *Trends Genet* 6: 36–40

Cambier JC, Newell NK, Justement LB, McGuire JC, Leach KL, Chen ZZ (1987): Ia binding ligand and cAMP stimulates nuclear translocation of PKC in β lymphocytes. *Nature* 327: 629–632

Carter DA, Murphy D (1990): Regulation of c-*fos* and c-*jun* expression in the rat supraoptic nucleus. *Mol Cell Neurobiol* 10: 435–446

Cattanach BM, Iddon CA, Charlton HM, Chiappa SA, Fink G (1977): Gonadotrophin releasing hormone deficiency in a mutant mouse with hypogonadism. *Nature* 269: 338–340

Cole TJ, Copeland NC, Gilbert DJ, Jenkins NA, Schütz G, Ruppert S (1992): The mouse CREB (cAMP responsive element binding protein) gene: structure, promoter analysis and chromosomal localization. *Genomics* 13: 974.

Comb M, Birnberg NC, Seasholtz A, Herbert E, Goodman HM (1986): A cyclic-AMP and Phorbol Ester-inducible DNA Element. *Nature* 323: 353–356

Courey AJ, Tjian R (1989): Analysis of Sp1 in vivo reveals multiple transcriptional domains, including a novel glutamine activation motif. *Cell* 55: 887–898

Craft, CM, Morgan, WW, Reiter RJ (1984): 24 hour changes in catecholamine synthesis in rat and hamster pineal glands. *Neuroendocrinology* 38: 193

Dash PK, Karl KA, Colicos MA, Prywes R, Kandel ER (1991): cAMP response element-binding protein is activated by Ca^{2+}/calmodulin-as well as cAMP-dependent protein kinase. *Proc Natl Acad Sci USA* 88: 5061–5065

deGroot RP, Sassone-Corsi P (1993): Hormonal control of gene expression: multiplicity and versatility of cyclic adenosine 3′, 5′-monophosphate-responsive nuclear regulators. *Mol Endocrinol* 7: 145–153

de Groot RP, den Hertog J, Vandenheede JR, Goris J, Sassone-Corsi P (1993a): Multiple and cooperative phosphorylation events regulate the CREM activator function. *EMBO J* 12: 3903–3911

de Groot RP, Derua R, Goris J, Sassone-Corsi P (1993b): Phosphorylation and negative regulation of the transcriptional activator CREM by $p34^{cdc2}$. *Mol Endocrinol* 7: 1495–1501

Delegeane A, Ferland L, Mellon PL (1987): Tissue specific enhancer of the human glycoprotein hormone a-subunit gene: dependence on cyclic AMP-inducible elements. *Mol Cell Biol* 7: 3994–4002

Delmas V, Laoide BM, Masquilier D, de Groot RP, Foulkes NS, Sassone-Corsi P (1992): Alternative usage of initiation codons in mRNA encoding the cAMP-responsive-element modulator (CREM) generates regulators with opposite functions. *Proc Natl Acad Sci USA* 89: 4226–4230

Delmas V, van der Hoorn F, Mellström B, Jégou B, Sassone-Corsi P (1993): Induction of CREM activator proteins in spermatids: down-stream targets and implications for haploid germ cell differentiation. *Mol Endocrinol* 7: 1502–1514

Deutsch PJ, Hoeffler JP, Jameson JL, Habener JF (1988): Cyclic AMP and phorbol ester-stimulated transcription mediated by similar DNA elements that bind distinct proteins. *Proc Natl Acad Sci USA* 85: 7922–7926

Flint KJ, Jones NC (1991): Differential regulation of three members of the ATF/CREB family of DNA-binding proteins. *Oncogene* 6: 2019–2026

Foulkes NS, Sassone-Corsi P (1992): More is better: activators and repressors from the same gene. *Cell* 68: 411–414

Foulkes NS, Borrelli E, Sassone-Corsi P (1991a): CREM gene: use of alternative DNA binding domains generates multiple antagonists of cAMP-induced transcription. *Cell* 64: 739–749

Foulkes NS, Laoide BM, Schlotter F, Sassone-Corsi P (1991b): Transcriptional antagonist CREM down-regulates c-*fos* cAMP-induced expression. *Proc Natl Acad Sci USA* 88: 5448–5452

Foulkes NS, Mellström B, Benusiglio E, Sassone-Corsi P (1992): Developmental switch of CREM function during spermatogenesis: from antagonist to transcriptional activator. *Nature* 355: 80–84

Foulkes NS, Schlotter F, Pévet P, Sassone-Corsi P (1993): Pituitary hormone FSH directs the CREM functional switch during spermatogenesis. *Nature* 362: 264–267

Gilman AG (1987): G Proteins: transducers of receptor-generated signals. *Ann Rev Biochem* 86: 615–649

Ginty DD, Glowacka D, Bader DS, Hidaka H, Wagner JA (1991): Induction of immediate early genes by Ca^{2+} influx requires cAMP-dependent protein kinase in PC12 cells. *J Biol Chem* 266: 17454–17458

Gonzalez GA, Montminy MR (1989): Cyclic AMP stimulates somatostain gene transcription by phosphorylation of CREB at ser 133. *Cell* 59: 675–680

Gonzalez GA, Menzel P, Leonard J, Fischer WH, Montminy MR (1991): Characterization of motifs which are critical for activity of the cyclic AMP-responsive transcription factor CREB. *Mol Cell Biol* 11: 1306–1312

Gonzalez GA, Yamamoto KK, Fischer WH, Karr K, Menzel P, Briggs III W, Vale WW, Montminy MR (1989): A cluster of phosphorylation sites on the cAMP-regulated nuclear factor CREB predicted by its sequence. *Nature* 337: 749–752

Grootegoed JA, Oonk RB, Toebosch AMW, Jansen R (1986): Extracellular factors that contribute to the development of spermatogenic cells. In: *Molecular and Cellular Endocrinology of the Testis*, Stefanini M, Conti M, Geremia R, Ziparo E, eds. Amsterdam: Excerpta Medica

Habener J (1990): Cyclic AMP response element binding proteins: a cornucopia of transcription factors. *Mol Endocrinol* 4: 1087–1094

Hagiwara M, Alberts A, Brindle P, Meinkoth J, Feramisco J, Deng T, Karin M, Shenolikar S, Montminy M (1992): Transcriptional attenuation following cAMP induction requires PP-1-mediated dephosphorylation of CREB. *Cell* 70: 105–113

Hai T-Y, Liu F, Coukos WJ, Green MR (1989): Transcription factor ATF cDNA clones: an extensive family of leucine zipper proteins able to selectively form DNA binding heterodimers. *Genes Dev* 3: 2083–2090

Heidaran MA, Kozak CA, Kistler WS (1989): Nucleotide sequence of the Stp-1 gene coding for rat spermatid nuclear transition protein 1 (TP1): homology with protamine P1 and assignment of the mouse Stp-1 gene to chromosome 1. *Gene* 75: 39–46

Hoeffler JP, Meyer TE, Yun Y, Jameson JL, Habener JF (1988): Cyclic AMP-responsive DNA-binding protein: structure based on a cloned placental cDNA. *Science* 242: 1430–1433

Hummler E, Cole TJ, Blendy JA, Ganss R, Aguzzi A, Schmid W, Beerman F, Schütz G (1994): Targeted mutation of the CREB gene: Compensation within the CREB/ATF Family of transcription factors. *Proc Natl Acad Sci* 91: 5647–5651

Illnervoa H, Vanecek J, Wetterberg L, Sääf JJ (1979): Effect of one minute exposure to light at night on rat pineal serotonin N-acetyltransferase and melatonin. *Neurochem* 32: 673

Jégou B, Syed V, Sourdaine P, Byers S, Gérard N, Velez de la Calle J, Pineau C, Garnier DH, Bauché F (1992): The dialogue between late spermatids and Sertoli cells in vertebrates: a century of research. In: *Spermatogenesis. Fertilization. Contraception. Molecular, Cellular and Endocrine Events in Male Reproduction*, Nieschlag E, Habenicht U-F, eds. New York: Springer Verlag.

Johnson P, Peschon JJ, Yelick PC, Palmiter RD, Hecht NB (1988): Sequence homologies in the mouse protamine 1 and 2 genes. *Biochim Biophys Acta* 950: 45–53

Klein DC (1985): Photoneural regulation of the mammalial pineal gland. In: *Photoperiodism, Melatonin and the Pineal Gland*, London: Pitman

Kramer IJM, Koornneef I, de Laat SW, van den Eijnden-van Raaij AJM (1991): TFG-β1 induces phosphorylation of the cyclic AMP responsive element binding protein in ML-CCL64 cells. *EMBO J* 10: 1083–1089

Krebs EG, Beavo JA (1979): Phosphorylation-dephosphorylation of enzymes. *Ann Rev Biochem* 48: 923–959

Landschulz WH, Johnson PF, McKnight SL (1988): The leucine-zipper: a hypothetical structure common to a new class of DNA binding proteins. *Science* 240: 1759–1764

Laoide BM, Foulkes NF, Schlotter F, Sassone-Corsi P (1993): The functional versatility of CREM is determined by its modular structure. *EMBO J* 12: 1179–1191

Lee CQ, Yun Y, Hoeffler JP, Habener JF (1990): Cyclic-AMP-responsive transcriptional activation involves interdependent phosphorylated subdomains. *EMBO J* 9: 4455–4465

Leff SE, Rosenfeld MG, Evans RM (1986): Complex transcriptional units: diversity in gene expression by alternative RNA processing. *Ann Rev Biochem* 55: 1091–1117

Leonard J, Serup P, Gonzalez G, Edlund T, Montminy M (1992): The LIM family transcription factor Isl-1 requires cAMP response element binding protein to promote somatostatin expression in pancreatic islet cells. *Proc Natl Acad Sci USA* 89: 6247–6251

Lewin B (1991): Oncogenic conversion by regulatory changes in transcription factors. *Cell* 64: 303–312

Lin SC, Morrison-Bogorad M (1991): Cloning and characterization of a testis-specific thymosin β_{10} cDNA. *J Biol Chem* 266: 23347–23353

Lonnerberg P, Parvinen M, Jahnsen T, Hansson V, Persson H (1992): Stage- and cell-specific expression of cyclic adenosine 3′, 5′-monophosphate-dependent protein kinases in rat seminiferous epithelium. *Biol Reprod* 46: 1057–1068

Lostroh AJ (1976): Hormonal control of spermatogenesis. In: *Regulation Mechanisms of Male Reproductive Physiology*, Spilman CH, Lobl TJ, Kirton KT, eds. Amsterdam: Excerpta Medica

Maekawa T, Sakura H, Kanei-Ishii C, Sudo T, Yoshimura T, Fujisawa J, Yoshida M, Ishii S (1989): Leucine zipper structure of the protein CRE-BP1 binding to the cyclic AMP response element in brain. *EMBO J* 8: 2023–2028

Masquilier D, Sassone-Corsi P (1992): Transcriptional cross-talk: nuclear factors CREM and CREB bind to AP-1 sites and inhibit activation by Jun. *J Biol Chem* 267: 22460–22466

Masquilier D, Foulkes NS, Mattei MG, Sassone-Corsi P (1993): Human CREM gene: Evolutionary conservation, chromosomal localization, and inducibility of the transcript. *Cell Growth & Differentiation* 4: 931–937

McCormick A, Brady H, Theill L, Karin M (1990): Regulation of the pituitary-specific homeobox gene GHF1 by cell-autonomous and environmental cues. *Nature* 345: 829–832

McKnight SG, Clegg CH, Uhler MD, Chrivia JC, Cadd GG, Correll LA, Otten AD

(1988): Analysis of the cAMP-dependent protein kinase system using molecular genetic approaches. *Rec Progr Horm Res* 44: 307–335

Meijer GH (1991): In: *Suprachiasmatic Nucleus/The Mind's Clock*, Klein DC, Moore RY, Reppert SM, eds. Oxford: Oxford University Press

Mellon PL, Clegg CH, Correll LA, McKnight SG (1989): Regulation of transcription by cyclic AMP-dependent protein kinase. *Proc Natl Acad Sci USA* 86: 4887–4891

Mellström B, Naranjo JR, Foulkes NS, Lafarga M, Sassone-Corsi P (1993): Transcriptional response to cAMP in brain: specific distribution and induction of CREM antagonists. *Neuron* 10: 655–665

Meyer TE, Weaber G, Lin J, Beckman W, Habener JF (1993): The promoter of the gene encoding 3′, 5′-cyclic adenosine monophosphate (cAMP) response element binding protein contains cAMP response elements: evidence for positive auto-regulation of gene transcription. *Endocrinology* 132: 770

Mizuki N, Sarapata DE, Garcia-Sanz JA, Kasahara M (1992): The mouse male germ cell-specific gene Tpx-1: molecular structure, mode of expression in spermato-genesis, and sequence similarity to two non-mammalian genes. *Mammalian Genome* 3: 274–280

Molina CA, Foulkes NS, Lalli E, Sassone-Corsi, P (1993): Inducibility and negative autoregulation of CREM: an alternative promoter directs the expression of ICER, and early response repressor. *Cell* 75: 875–886

Montmayeur JP, Borrelli E (1991): Transcription mediated by a cAMP-responsive promoter element is reduced upon activation of dopamine D2 receptors. *Proc Natl Acad Sci USA* 88: 3135–3139

Moore RY (1978): Neuroendocrine regulation of reproduction. In: *Reproductive Endocrinology*, Yen SSC, Jaffe RB, eds. Philadelphia: Sanders

Morgan JI, Cohen DR, Hempstead JL, Curran T (1987): Mapping patterns of c-*fos* expression in the central nervous system after seizure. *Science* 237: 192–197

Nathans D, Lau LF, Christy B, Hartzell S, Nakabeppu Y, Ryder K (1988): Genomic response to growth factors. *Cold Spring Harbor Symp Quant Biol* 53: 893

Nichols M, Weih F, Schmid W, DeVack C, Kowenz-Leutz E, Luckow B, Boshart M, Schütz G (1992): Phosphorylation of CREB affects its binding to high and low affinity sites: implications for cAMP induced gene transcription. *EMBO J* 11: 3337–3346

Nishizuka Y (1986): Studies and perspectives of protein kinase C. *Science* 233: 305–312

Oakberg J (1956): Duration of spermatogenesis in the mouse and timing of stages of the cycle of the seminiferous epithelium. *Am J Anat* 99: 504–516

Oyen O, Scott JD, Cadd GG, McKnight GS, Krebs EB, Hansson V, Jahnsen T (1988): A unique mRNA species for a regulatory subunit of cAMP-dependent protein kinase is specifically induced in haploid germ cells. *FEBS Lett* 229: 391–394

Oyen O, Myklebust F, Scott JD, Cadd GG, McKnight SG, Hansson V, Jahnsen T (1990): Subunits of cyclic adenosine 3′, 5′-monophosphate-dependent protein kinase show differential and distinct expression patterns during germ cell differentiation: alternative polyadenylation in germ cells gives rise to unique smaller-sized mRNA species. *Biol Reprod* 43: 46–54

Pariset C, Feinberg J, Dacheux JL, Oyen O, Jahnsen T, Weinman S (1989): Differential expression and subcellular localization for subunits of cAMP-dependent protein kinase during ram spermatogenesis. *J Cell Biol* 109: 1195–1205

Perlow MJ, Reppert SM, Tamarkin L, Wyatt RJ, Klein DC (1980): Photic regulation of the melatonin rhythm: monkey and man are not the same. *Brain Res* 182: 211

Rehfuss RP, Walton KM, Loriaux MM, Goodman RH (1991): The cAMP-regulated enhancer-binding protein ATF-1 activates transcription in response to cAMP-dependent protein kinase A. *J Biol Chem* 266: 18431–18434

Reiter RJ (1991): Pineal gland. Interface between the photoperiodic environment and the endocrine system. *Trends Endocr Met* 1: 13

Roesler WJ, Vanderbark GR, Hanson RW (1988): Cyclic AMP and the induction of eukaryotic gene expression. *J Biol Chem* 263: 9063–9066

Ruppert S, Cole TJ, Boshart M, Schmid E, Schu\dtz G (1992): Multiple mRNA isoforms of the transcription activator protein CREB: generation by alternative splicing and specific expression in primary spermatocytes. *EMBO J* 11: 1503–1512

Russell LD (1980): Sertoli-germ cell interrelations: a review. *Gamete Res* 3: 179–202

Sagar SM, Sharp FR, Curran T (1988): Expression of c-*fos* protein in brain: metabolic mapping at the cellular level. *Science* 240: 1328–1331

Santen RJ (1987): The testis. In: *Endocrinology and Metabolism*, Felig P, Baxter JD, Broadus AE, Frohman LA, eds. New York: McGraw-Hill

Sassone-Corsi P (1988): Cyclic AMP induction of early adenovirus promoters involves sequences required for E1A-transactivation. *Proc Natl Acad Sci USA* 85: 7192–7196

Sassone-Corsi P, Ransone LJ, Verma IM (1990): Cross-talk in signal transduction: TPA-inducible factor Jun/AP-1 activates cAMP responsive enhancer elements. *Oncogene* 5: 427–431

Sassone-Corsi P, Visvader J, Ferland L, Mellon PL, Verma IM (1988): Induction of proto-oncogene *fos* transcription through the adenylate cyclase pathway: characterization of a cAMP-responsive element. *Genes Dev* 2: 1529–1538

Sharp FR, Sagar SM, Hicks K, Lowenstein D, Hisanaga K (1991): c-*fos* mRNA, Fos and Fos-related antigen induction by hypertonic saline and stress. *J Neurosci* 11: 2321–2331

Shaw G, Kamen R (1986): A conserved AU sequence from the 3' untranslated region of GM-CSF mRNA mediates selective mRNA degradation. *Cell* 46: 659–667

Sheng M, McFadden G, Greenberg ME (1990): Membrane depolarization and calcium induce c-*fos* transcription via phosphorylation of transcription factor CREB. *Neuron* 4: 571–582

Sheng M, Thompson MA, Greenberg ME (1991): CREB: a Ca^{2+}-regulated transcription factor phosphorylated by calmodulin-dependent kinases. *Science* 252: 1427–1430

Sherman TG, McKelvy JF, Watson SJ (1986): Vasopressin mRNA regulation in individual hypothalamic nuclei: a northern and in situ hybridization analysis. *J Neurosci* 6: 1685–1694

Stehle JH, Foules NS, Molina CA, Simonneaux V, Pévet P, Sassone-Corsi P (1993): Adrenergic signals direct rhythmic expression of transcriptional repressor CREM in the pineal gland. *Nature* 365: 314–320

Steinberger E (1971): Hormonal control of mammalian spermatogenesis. *Physiol Rev* 51: 1–22

Struthers RS, Vale WW, Arias C, Sawchenko PE, Montminy MR (1991): Somatotroph hypoplasia and dwarfism in transgenic mice expressing a non-phosphorylatable CREB mutant. *Nature* 350: 622–624

Sudgen D, Vanecek J, Klein DC, Thomas TD (1985): Activation of protein kinase C potentiates isoprenaline-induced cyclic AMP accumulation in rat pinealocytes. *Nature* 314: 359

Takahashi JS (1993): Circadian clocks à la CREM *Nature* 365: 299

Tamarkin L, Reppert SM, Klein DC (1979): Regulation of pineal melatonin in the syrian hamster. *Endocrinology* 104: 385

Tamarkin L, Baird CJ, Almeida OFX (1985): Melatonin: a coordinating signal for mammalian reproduction? *Science* 227: 774

van der Hoorn FA, Tarnasky HA (1992): Factors involved in regulation of the RT7 promoter in a male germ cell-derived in vitro transcription system. *Proc Natl Acad Sci USA* 89: 703–707

Vanecek J, Sudgen D, Weller J, Klein DC (1985): Atypical synergistic $\alpha 1$ and β-adrenergic regulation of adenosine 3′,5′-*Endocrinology* 116: 2167

Veldhuis JD (1991): The hypothalamic-pituitary-testicular axis. In: *Reproductive Endocrinology* Yen SSC, Jaffe RB, eds. Philadelphia: W.B. Saunders

Verma IM, Sassone-Corsi P (1987): Proto-oncogene *fos*: complex but versatile regulation. *Cell* 51: 513–514

Vinson CR, Sigler P, McKnight SL (1989): Scissor-grip model for DNA recognition by a family of leucine zipper proteins. *Science* 246: 911–922

Waeber G, Meyer TE, LeSieur M, Hermann HL, Gérard N, Habener JF (1991): Developmental stage-specific expression of cyclic adenosine 3′,5′-monophosphate response element-binding protein CREB during spermatogenesis involves alternative exon splicing. *Mol Endocrinol* 5: 1418–1430

Williams T, Admon A, Luscher B, Tjian R (1988): Cloning and expression of AP-2, a cell-type-specific transcription factor that activates inducible enhancer elements. *Genes Dev* 2: 1557–1569

Yamamoto KK, Gonzales GA, Briggs III WH, Montminy MR (1988): Phosphorylation-induced binding and transcriptional efficiency of nuclear factor CREB. *Nature* 334: 494–498

Yoshimasa T, Sibley DR, Bouvier M, Lefkowitz RJ, Caron MG (1987): Cross-talk between cellular signalling pathways suggested by phorbol ester adenylate cyclase phosphorylation. *Nature* 327: 67–70

Ziff EB (1990): Transcription factors: a new family gathers at the cAMP response site. *Trends Genet* 6: 69–72

2

Signal uptake by the c-*fos* serum response element

MICHAEL A. CAHILL, RALF JANKNECHT, AND ALFRED NORDHEIM

Introduction

Detailed functional characterization of the c-*fos* serum response element (SRE) has identified this promoter segment as an important nuclear target structure for the uptake of mitogenic signals by the genome. The study of the SRE and its interacting proteins has permitted the retrograde connection with signaling steps activated upon stimulation of nonproliferating cells. The SRE is also targeted by nonmitogenic signaling events which similarly lead to the activation of c-*fos*. These include, e.g. heat treatment, neuronal stimulation, UV light, or reactive oxygen intermediates. Since SREs and SRE-like sequences are found in the promoters of many different genes, they may be crucial for the transmission of extra- and intracellular signals into responses at the level of gene activity.

The c-*fos* promoter: a paradigm of immediate early gene regulation

The transcription of a distinct set of genes is activated when quiescent cells are stimulated by treatment with individual growth factors or serum (Lau and Nathans, 1985; Almendral et al, 1988; Bravo, 1990). A subset of these mitogen inducible genes, estimated to comprise some 100 independent coding units, are called immediate early genes (IEGs) because of their rapid and transient transcriptional induction profile. Their stimulation occurs independently of protein synthesis and, in fact, transcription is superinduced by protein synthesis inhibitors. The c-*fos* gene is the best studied member of the IEGs and is stimulated in response to signals which induce proliferation and/or differentiation, as well as to the various cellular stresses (Greenberg and Ziff, 1984; Kruijer et al, 1984; Müller et

INDUCIBLE GENE EXPRESSION, VOLUME 2
P.A. Baeuerle, Editor
© 1995 Birkhäuser Boston

al, 1984; Mitchell et al, 1985; Cohen and Curran, 1989). For some IEGs, including c-*fos*, *junB*, *egr*-1 and *erg*-2, signal-mediated promoter induction is governed to a major extent by the action of an important transcriptional regulatory element, the SRE (Figure 2.1; Table) (Treisman, 1986; Gilman et al, 1986; Treisman, 1992). Two other important c-*fos* regulatory elements, the SIE (v-*sis* inducible element) (Hayes et al, 1987) and the −60 CRE (cAMP response element) (Gilman et al, 1986), are also targeted by signal cascades and thereby mediate induction by EGF/PDGF and cAMP/Ca^{2+}, respectively (Figure 2.1) (Rivera and Greenberg, 1990; Montminy, 1993; Nordheim et al, 1994).

The serum response element (SRE) as a signaling target

The SRE was originally identified by R. Treisman as a 5′ regulatory element mediating the induction of the c-*fos* proto-oncogene by serum (Treisman, 1985; Treisman, 1986; Treisman, 1992). Subsequently, the SRE has been recognized as a *cis*-element which confers gene activation by a multitude of different extracellular signals (Figure 2.1) (Treisman, 1992). Most of these signals have been shown to activate members of the mitogen-activated protein kinases (MAPKs), also referred to as ERKs (extracellular signal-regulated kinases (Figure 2.2). Upon signal transduction into the nucleus the SRE-binding proteins serum response factor (SRF) and ternary complex factor (TCF) (see below) can be phosphorylated, thereby

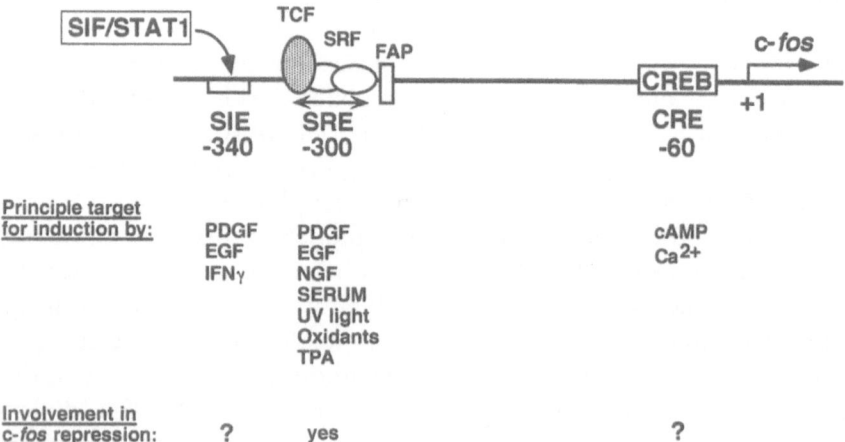

Figure 2.1. Regulatory cis-elements in the c-*fos* promoter, their cognate binding proteins and signals that cause their activation. CREB represents the cAMP response element binding protein and SIF denotes the v-*sis* inducible factor; for further explanation see text.

Table. Inducible genes containing SRE-like regulatory sequences

Gene	Species	Reference
c-*fos*	human, mouse	Gilman et al, 1986 Treisman, 1986
fos-B	murine	Lazo et al, 1992
fra-2	chicken	Yoshida et al, 1993
junB	murine	Perez-Albuerne et al, 1993
zif 268/egr-1 /*krox*-24	murine	Lemaire et al, 1988 Sukhatme et al, 1988 Christy and Nathans, 1989
krox-20/*egr*-2	murine	Chavrier et al, 1988 Joseph et al, 1988
pip92	murine	Charles et al, 1990
cyr61	murine	Latinkic et al, 1991
IL-2Rα	human	Phan-Dinh-Tuy et al, 1988
α *actin*	human, chicken	Miwa et al, 1987 Lee et al, 1991
β *actin*	human, murine	Orita et al, 1989 Stoflet et al, 1992
γ *actin*	*Xenopus*	Mohun et al, 1987
G0S19-1 /*G0S19-2*	murine	Blum et al, 1990
TSP1	human murine	Bornstein, 1992 Framson and Bornstein, 1993
PDGF α	human	Lin et al, 1992
Vinculin	human	Moiseyeva et al, 1993
RSV-LTR/ endogenous virus LTR	viral (avian)	Boulden and Sealy, 1990 Zachow and Conklin, 1992 Lang et al,1993 our unpublished data, 1993
CMV-IE promoter	viral (human, murine, primate)	Chang et al, 1993 our unpublished data, 1993

identifying the SRE ternary complex as direct nuclear target structure for signaling cascades (Schalasta and Doppler, 1990; Gille et al, 1992; Hill et al, 1993; Janknecht et al, 1993b; Marais et al, 1993; Rivera et al, 1993; Zinck et al, 1993). The different genes that contain SREs or SRE-related sequences in their regulatory regions are listed in the Table, and include developmentally regulated genes and a cytokine receptor gene, in addition to the immediate early genes and some viral promoters.

The primary nucleotide sequence of the c-*fos* SRE represents an imperfect dyad symmetry sequence of 20 base pairs (bp) which contains

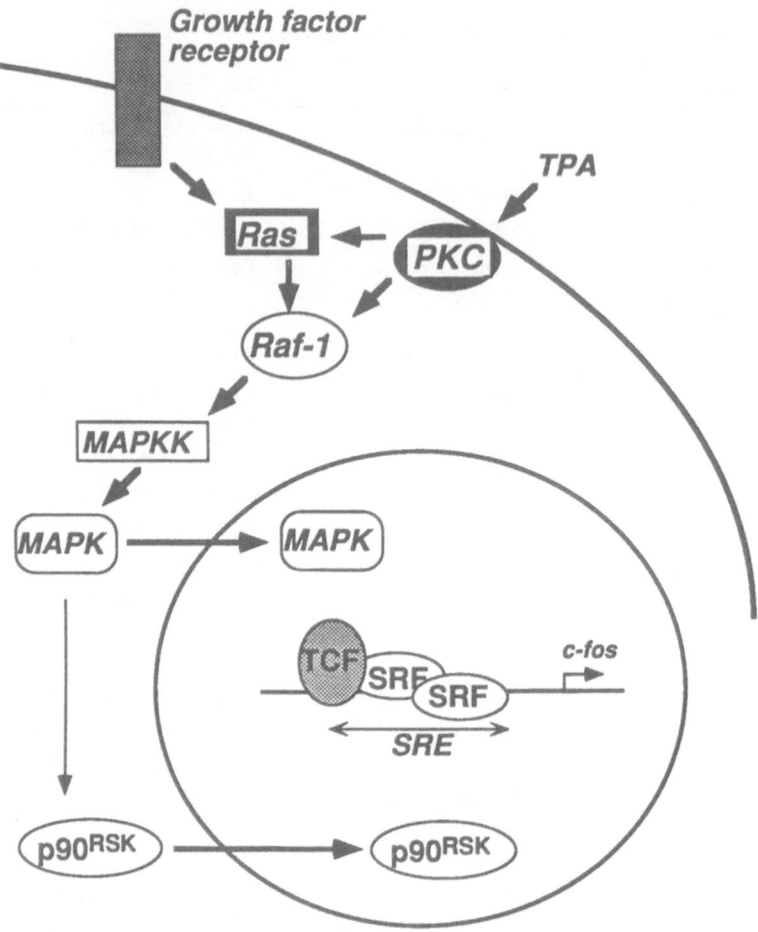

Figure 2.2. The MAPK signaling pathway directed towards ternary complex formed with the c-*fos* SRE.

a central $CC(A/T)_6GG$ element, often referred to as CArG box (Figure 2.3). The CArG box and its immediate flanking nucleotides provide the DNA helix geometry for specific recognition by the primary SRE-binding protein, the SRF (Treisman, 1986; Norman et al, 1988; Treisman and Ammerer, 1992). In addition, the 5′ boundary of the c-*fos* SRE is comprised of nucleotides, including a GGAT sequence, which represents a binding site for certain members of the Ets proto-oncoprotein family (Figure 2.3) (Janknecht and Nordheim, 1993; Wasylyk et al, 1993; Treisman, 1994). This binding site is occupied by members of the TCF subclass of Ets proteins that are recruited by the SRE:SRF binary complex to form a ternary complex but cannot bind the c-*fos* SRE in the absence of SRF (Shaw et al, 1989b; Schröter et al, 1990; Hipskind et

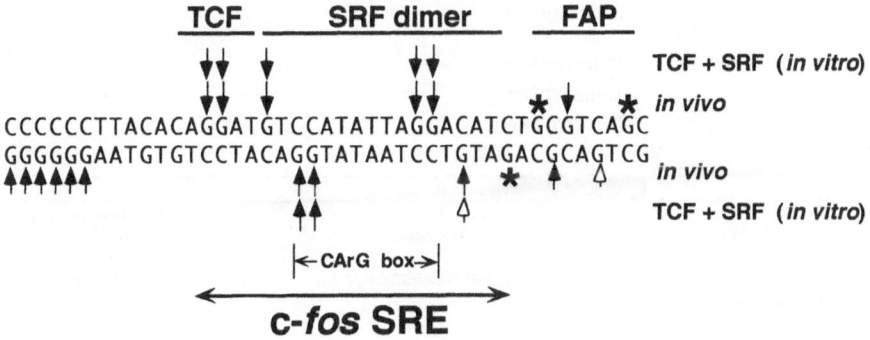

Figure 2.3. Protein occupancy at the human c-*fos* SRE. The primary nucleotide sequence of the SRE is shown, as well as the G residues found strongly protected (black arrows), weakly protected (white arrows), or hyperreactive (asterisks) upon exposure to DMS (dimethylsulfate) in footprint studies.

al, 1991; Dalton and Treisman, 1992; Janknecht and Nordheim, 1993; Treisman, 1994). Genomic footprinting studies (Figure 2.3) indicate the SRE to be occupied by a dimer of SRF and one molecule of TCF (Herrera et al, 1989; König, 1991). Interestingly, this in vivo occupancy has been found unaltered before, during and after c-*fos* induction, suggesting that posttranslational modification of preexisting SRE binding proteins occurs upon SRE activation. The genomic footprint also revealed continuous binding of an unknown protein immediately 3' to the SRE (Herrera et al, 1989) (Figure 2.3). This sequence resembles an AP-1/ATF or a CREB binding sequence and is termed FAP site (Velcich and Ziff, 1990). Its functional role in c-*fos* regulation is not understood.

Importantly, the SRE also has been demonstrated to be involved in basal level repression and in post-inductional down regulation of c-*fos* (Subramaniam et al, 1989; König et al, 1989; Shaw et al, 1989a; Nordheim et al, 1994), which will be discussed in more detail below.

The serum response factor (SRF)

The SRF protein was originally characterized by its ability to interact with the c-*fos* CArG box (Gilman et al, 1986; Prywes and Roeder, 1986; Treisman, 1986; Treisman, 1987). The *SRF* cDNA has been subsequently cloned from human (Norman et al, 1988) and *Xenopus* (Mohun et al, 1991). The human SRF molecule consists of 508 amino acids (Figure 2.4) and is related to a variety of other transcription factors, e.g. the yeast proteins MCM1 and ARG80, the plant homeotic gene products Agamous and Deficiens or the SRF-related proteins (RSRFs), by virtue of the MADS-box homology region extending from amino acid 143 to 197

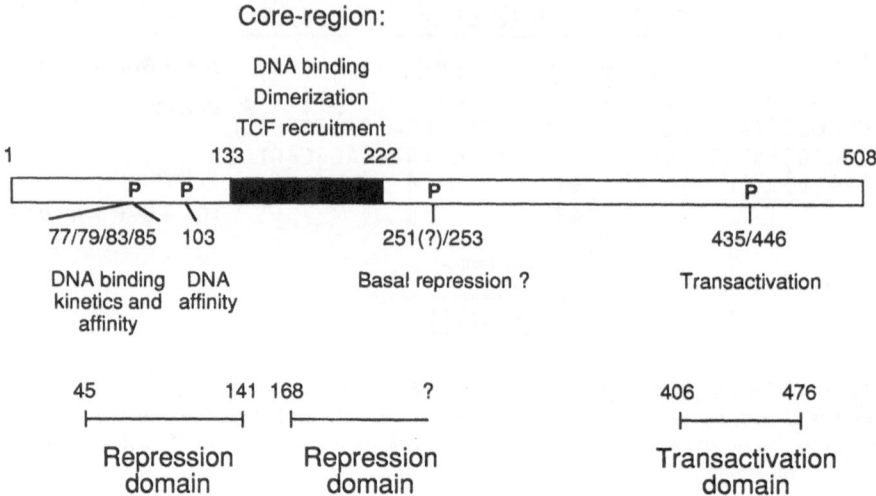

Figure 2.4. Structure of the human SRF protein. Identified phosphorylation sites (P) are indicated.

(Schwarz-Sommer et al, 1990; Treisman, 1992; Treisman and Ammerer, 1992). The MADS-box is part of the DNA binding core-region of SRF (amino acids 133–222) and is an important determinant of DNA specificity and affinity (Norman et al, 1988; Mueller and Nordheim, 1991; Sharrocks et al, 1993). Stable DNA binding of SRF requires dimerization which is dependent on the C-terminal half of the core-region (Norman et al, 1988; Sharrocks et al, 1993). Furthermore, the core-region is both necessary and sufficient for the recruitment of TCFs to the c-*fos* SRE (Mueller and Nordheim, 1991; Shaw, 1992). DNA binding of SRF might be augmented by homeodomain proteins such as Phox1 which itself can bind to the AT-rich core of the CArG box. However, even Phox1 mutants incapable of DNA binding enhanced the affinity of SRF to the c-*fos* SRE, suggesting that the Phox1 protein induces an altered conformation in SRF which leads to a higher DNA affinity (Grueneberg et al, 1992). Whether Phox1 has any biological relevance to DNA binding by SRF is unclear.

Transcriptional Regulatory Domains of SRF

SRF can function as a transcriptional activator in vitro (Norman et al, 1988; Hipskind and Nordheim, 1991; Zhu and Prywes, 1992) and in vivo (Gutman et al, 1991). A transactivation domain has been identified near the C-terminus (Figure 2.4) by use of GAL4-SRF fusion proteins (Johansen and Prywes, 1993; Liu et al, 1993). Deletion of this transactivation domain reduces the transcriptional potency of the SRF/TCF complex approximately twofold (Hill et al, 1993; Ernst et al, 1994). In addition, the C-terminal transactivation domain is part of a region involved in the interaction of SRF with the HTLV-1 Tax protein. Via this interaction, Tax1 may stimulate immediate early genes such as c-*fos*, *egr*-1 or *egr*-2 without signal mediated activation of the SRF/TCF complex (Fujii et al, 1992; Suzuki et al, 1993). In contrast to the C-terminus, the N-terminal amino acids 45–141 of SRF repress transcription in the context of GAL4 fusion proteins (Johansen and Prywes, 1993; Ernst et al, 1994), however this effect may be dampened in the full length SRF molecule due to shielding by the C-terminal region (Ernst et al, 1994). Another transcriptional repression domain may be localized downstream of amino acid 168 (Johansen and Prywes, 1993). Whether these domains repress transcription in the endogenous SRF molecule remains to be determined.

Phosphorylation of SRF

Several phosphorylation sites have been identified within SRF (Figure 2.4). In the N-terminal region, Ser77, Ser79, Ser83 and Ser85, which are classical casein kinase II (CKII) sites, are phosphorylated in vivo and can be phosphorylated by CKII in vitro (Manak and Prywes, 1991; Janknecht et al, 1992; Marais et al, 1992). Phosphorylation at these sites enhances the on- and off-rates of DNA binding and leads to a slight increase of DNA affinity, while it does not influence the in vitro transcriptional activity of SRF or its ability to interact with TCFs (Janknecht et al, 1992; Marais et al, 1992). Since the c-*fos* SRE is constitutively occupied by SRF in vivo (Herrera et al, 1989; König, 1991), it is unclear how altered characteristics of DNA binding affect c-*fos* induction by SRF/TCF. However, microinjection of CKII elicits c-*fos* induction which can be abrogated by the simultaneous injection of SRE oligonucleotides or antibodies against SRF (Gauthier-Rouvière et al, 1991).

Further phosphorylation within the N-terminal region occurs at Ser103 (Janknecht et al, 1992; Rivera et al, 1993). In contrast to phosphorylation in the region 77–85, phosphorylation at Ser103 does not drastically alter DNA binding kinetics, yet seems to enhance DNA affinity. Ser103 is localized within consensus sequences for several protein kinases, e.g. protein kinase A (PKA), CKII or calmodulin/Ca^{2+}-dependent protein kinase II (CaM KII), but only the S6 kinase $pp90^{rsk}$ has been shown to phosphorylate Ser103 in vitro (Rivera et al, 1993). Since this kinase is regulated by growth factors, it could be responsible for hyperphosphorylation at Ser103 detectable upon growth factor stimulation of cells (Rivera et al, 1993). Whether this hyperphosphorylation affects transcriptional activation of the SRF/TCF complex is unknown at present.

Ser435 and Ser446, both localized within the C-terminal transactivation domain of SRF, are targets for a DNA-activated protein kinase in vitro. Mutation of these two serines to alanine reduces the efficacy of the C-terminal transactivation domain in the context of GAL4 fusion proteins (Liu et al, 1993), but no studies have been performed to elucidate whether this also affects transcription mediated by the SRF/ TCF complex. Phosphorylation at Ser253, and perhaps Ser251, may be involved in repression of transcription, since mutation of these sites to alanine alleviates repression of transcription at the c-*fos* SRE (Janknecht et al, 1993a). Both Ser251 and Ser 253 are potential targets for CKII, but their mutation to alanine did not drastically affect DNA binding properties, in contrast to the CKII sites within the N-terminus.

In addition to phosphorylation, SRF is posttranslationally modified by glycosylation (Schröter et al, 1990). N-acetylglucosamine has been shown to be attached to four serine residues within the C-terminal region of SRF (Reason et al, 1992), but the function of this glycosylation remains elusive.

Ternary complex factors (TCFs)

TCFs have been identified as protein activities which interact with the c-*fos* SRE only upon recruitment by SRF (Shaw et al, 1989b; Schröter et al, 1990; Treisman, 1994). Three TCFs are characterized at the molecular level, Elk-1 (Hipskind et al, 1991), Net/Erp (Wasylyk et al, 1993) and SAP-1 (Dalton and Treisman, 1992), the latter existing in the two splice variants SAP-1a and SAP-1b. In addition, a splice variant of Elk-1 is known (ΔElk-1) which does not form a ternary complex due to a lack of a region required for interaction with SRF (Rao and Reddy, 1993). The originally described HeLa cell factor p62[TCF] (Shaw et al, 1989b) is encoded by Elk-1 (Hipskind et al, 1991; Pingoud et al, 1994), and these cells have since been shown to contain at least four different TCF proteins (Pingoud et al, 1994). All TCFs characterized at the cDNA sequence level constitute a subclass of the Ets oncoprotein family (Janknecht and Nordheim, 1993; Wasylyk et al, 1993) and share three homologous regions: an N-terminal ETS-domain, and the B- and C-regions (Figure 2.5).

TCF Functional Domains: the ETS-domain

The DNA binding ETS-domain is approximately 85 amino acids long and is localized at the N-terminus of TCFs, in contrast to all other known Ets proteins (Janknecht and Nordheim, 1993; Wasylyk et al, 1993;

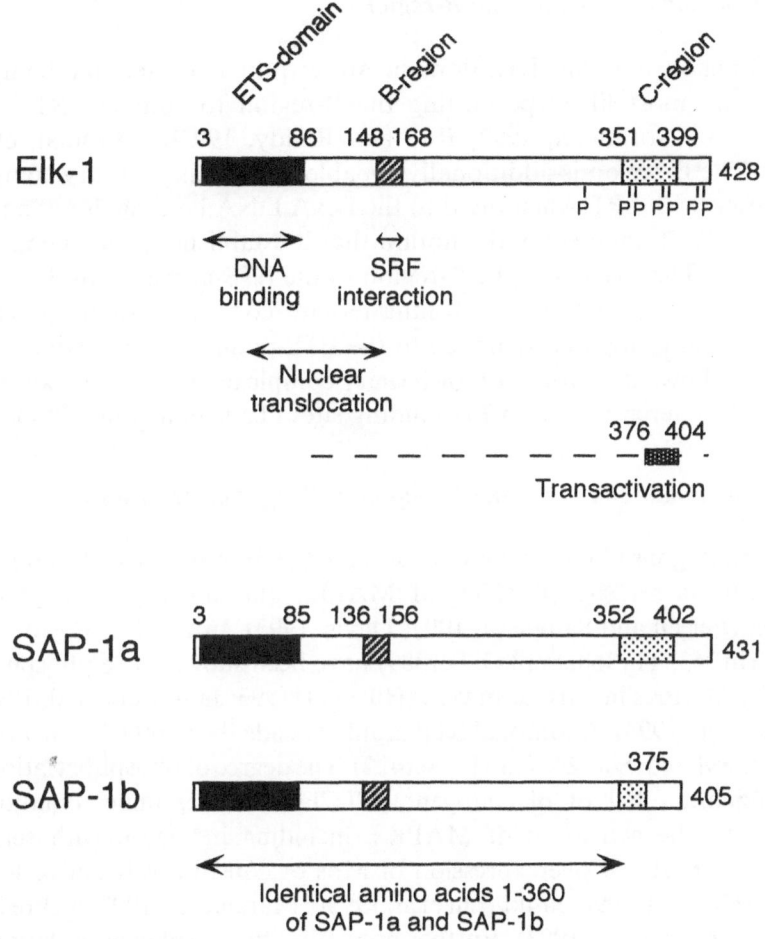

Figure 2.5. Comparison of human TCFs. Identified phosphorylated amino acids (P) within Elk-1 (Ser324, Thr363, Thr368, Ser383, Ser389, Thr417 and Ser422) are indicated.

Treisman, 1994). This domain is able to interact with some Ets target sites containing a GGAAT core (Janknecht and Nordheim, 1992; Rao and Reddy, 1992a; Treisman et al, 1992); however, TCFs fail to establish stable contacts with the GGAT Ets binding site of the c-*fos* SRE in the absence of SRF. Truncating Elk-1 or SAP-1 from the C-terminus beyond the B-region results in molecules capable of directly binding to the c-*fos* SRE, indicating that the B-region impedes DNA binding by the ETS-domain (Dalton and Treisman, 1992; Treisman et al, 1992; Rao and Reddy, 1992b; Janknecht et al, 1994). Furthermore, both nuclear localization of TCFs and ternary complex formation are dependent on the ETS-domain (Janknecht et al, 1994; Treisman, 1994).

TCF Functional Domains: the B-region

The B-region and the ETS-domain are required for ternary complex formation, most likely permitting the B-region to contact SRF (Janknecht and Nordheim, 1992; Rao and Reddy, 1992b; Treisman et al, 1992). The B-region is additionally capable of directing ternary complex formation with SRF when fused to the LexA DNA binding domain (Hill et al, 1993), strengthening the notion that it establishes protein contacts with SRF. The spacing of the B-region to the ETS-domain can be varied to a great degree without abolishing ternary complex formation. Thus, the B-region is apparently linked to the ETS-domain by a flexible hinge and this allows the formation of ternary complexes at CArG boxes with adjacent, differently spaced Ets binding sites (Treisman et al, 1992).

TCF Functional Domains: the C-region as Target of MAPKs

The homologous C-region contains several $^S/_TP$ motifs which resemble the consensus target site (P)XSTP of MAPKs and other proline-directed kinases (Pelech and Sanghera, 1992; Davis, 1993). In Elk-1, some of these sites (Thr363, Thr368, Ser383, Ser389) have been shown to be phosphorylated by MAPKs in vitro or in vivo (Hill et al, 1993; Janknecht et al, 1993b; Marais et al, 1993). Additional STP motifs outside the C-region can also be phosphorylated (Ser324, Thr417, Ser422). The degree of phosphorylation of recombinant Elk-1 or of endogenous TCFs increases under conditions leading to the activation of MAPKs, including induction with serum, TPA, and EGF, or overexpression of Mos or constitutively active Raf-1 kinase (Hill et al, 1993; Janknecht et al, 1993b; Marais et al, 1993; Nebreda et al, 1993; Zinck et al, 1993). Furthermore, abolition of phosphorylation at STP motifs leads to a reduction of transcriptional induction by the Elk-1 molecule at the c-*fos* SRE, or at an Ets binding site (the E74 site) independently of SRF. The same has been observed for fusions of Elk-1 to the DNA binding domains of GAL4 or LexA when assayed with respective reporter constructs (Hill et al, 1993; Janknecht et al, 1993b; Marais et al, 1993). The most severe reduction of transcriptional activity has been observed upon mutation of Ser383 to alanine, which is localized within Elk-1 amino acids 376–404 which are indispensable for transactivation mediated upon MAPK induction. Amino acids 83–375 and 405–428 are not absolutely required for transactivation, but improve the transcriptional activation by amino acids 376–404 (Janknecht et al, 1994). The SAP-1b protein, which lacks amino acids homologous to Elk-1 amino acids 376–404, may therefore represent a transcriptionally inactive molecule antagonizing the action of Elk-1 or other TCFs. Since SAP-1a contains potential phosphorylation sites homologous to those identified in Elk-1 and bears a strong resemblance to Elk-1 in the C-terminal region, it may also represent a transcription factor that can be stimulated by MAPKs. Interestingly, in the quail fibroblast cell line QT6, transactivation mediated by Elk-1 via an Ets binding site seems to be independent of the C-region and also of activation of MAPKs (Bhattacharya et al, 1993).

At the c-*fos* SRE, both SRF and TCF contribute to transcription induction. Deletion of the SRF transactivation domain reduces the efficiency of transactivation by the TCF/SRF complex (Hill et al, 1993; Ernst et al, 1994). However, studies with Elk-1 molecules devoid of the C-region reveal that TPA induction is completely abolished, indicating that the transactivation domain of SRF cannot exert its effect on c-*fos* induction alone. However, a residual inducibility towards serum is still noted which demonstrates that additional signaling pathway(s) may be utilized (Ernst et al, 1994). One target in SRF that may affect this residual inducibility is Ser103 which is hyperphosphorylated upon growth factor induction, possibly by pp90^{rsk} (Rivera et al, 1993).

Other proteins capable of binding the c-*fos* SRE

In addition to the best characterized SRE binding proteins SRF and TCF several other factors have been described to specifically interact with SRE sequences. These include p62^{DBF}/YY1 (Ryan et al, 1989; Gualberto et al, 1992; Natesan and Gilman, 1993), E12 (Metz and Ziff, 1991), NF-IL6 (Metz and Ziff, 1991), and Phox1 (Grueneberg et al, 1992). The functional importance of the binding to the SRE by these proteins is at present unclear. However, as discussed above, genomic footprinting data are consistent with the binding of SRF and TCF in vivo (Herrera et al, 1989; König, 1991), whereas so far no genomic footprinting data have been obtained revealing genomic occupancy by any of these alternative factors. Therefore we have concentrated on the SRF/TCF complex. However, it should be born in mind that if a strong transcriptional activator displaced SRF for only a small percentage of the time it could mediate the observed transcriptional induction, yet remain undetected in a genomic footprint.

Signal pathways upstream of TCF/SRF

MAPK and SRE Regulation

The last few years have witnessed significant progress in our understanding of signaling events upstream of the SRE. At this point we change

focus to provide an overview of the large body of research dealing with
these pathways, with emphasis on those aspects that potentially contribute to signal uptake by the SRE. We concentrate on the MAPK
pathway since most stimuli causing induction of SRE-regulated genes
function via activation of MAPK. The MAPK pathway is conserved in
evolution from yeast to mammals (Neimann, 1993; Ammerer, 1994;
Marshall, 1994). Generally, MAPKs are activated by MAPK kinases
(MAPKKs), which in turn are activated by MAPKK kinases
(MAPKKKs) (Figure 2.6). In yeast separate MAPK pathways exist in
parallel to regulate separate functions. It remains to be established which
of the molecules shown in Figure 2.6 are involved in signaling toward the
SRE.

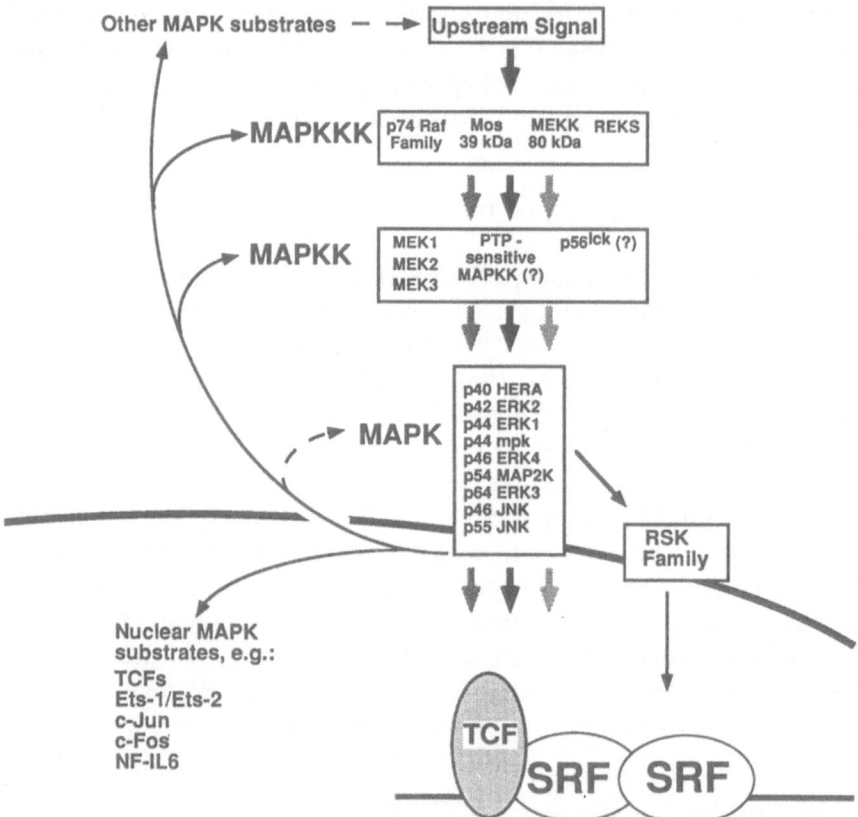

Figure 2.6. The mammalian MAPK pathway. MAPKKK and MAPKK represents all known
proteins displaying those activities. REKS is an uncloned MAPKKK activity from *Xenopus* and
PTP-sensitive MAPKK is a protein tyrosine phosphatase sensitive MAPKK is a protein tyrosine
phosphatase sensitive MAPKK activity (Jaiswal et al, 1993). MAPK depicts all known or
suspected mammalian MAPK-like proteins (Pelech and Sanghera, 1992), including the proline
directed Jun kinase (JNK) activities described (Hibi et al, 1993; Dérijard et al, 1994).

MAPKKK

The currently identified MAPKKK activities are Raf, Mos, and MEKK. There are three identified *raf* family genes in mammals: c-*raf*-1, A-*raf*-1, and B-*raf*. The A-*raf*-1 and B-*raf* genes exhibit relatively restricted expression patterns, whereas expression of c-*raf*-1 is more ubiquitous (Heidecker et al, 1992). Raf has been identified biochemically and genetically as being downstream of Ras in signal transduction cascades (Moodie and Wolfman, 1994; Marshall, 1994) and has been shown to target SREs (Kaibuchi et al, 1989; Jamal and Ziff, 1990). Raf physically contacts the Ras protein via a cysteine finger-like domain in the N-terminal regulatory domain of Raf (Zhang et al, 1993). Although phosphatidylcholine-specific phospholipase C (PC-PLC) (Cai et al, 1994) and PKCζ (Berra et al, 1993) have been reported to be involved, it remains unclear how GTP-bound activated Ras mediates activation of Raf (Moodie and Wolfman, 1994). Mos is related to the Raf family, and together this group constitutes the only members of the *src* superfamily which are serine/threonine rather than tyrosine kinases (Heidecker et al, 1992). The *mos* proto-oncogene is expressed primarily in germ tissue, and the *Xenopus* Mos protein has been shown to phosphorylate and activate MAPKK in both *Xenopus* oocytes and, when exogenously expressed, in mammalian fibroblasts (Nebreda et al, 1993). MEKK is thought to operate in the same cells as Raf. It may transduce signals from serpentine membrane receptors via heterotrimeric G proteins into the MAPK pathway by analogy to yeast homologs (Lange-Carter et al, 1993). Another MAPKKK might be represented by REKS (*ras* p21-dependent ERK kinase stimulator), a 150–200 kDa MEK stimulating activity identified in *Xenopus* oocytes (Itoh et al, 1993).

MAPKK

Three MAPKK cDNAs representing two genes have been cloned from humans: MEK1 (MKK1a), MEK2, and MEK3 (MKK1b) (Seger et al, 1992; Zheng and Guan 1993a; 1993b). MEK3 is identical to MEK1 except for a splicing variation which deletes part of kinase domain V. Mammalian MEK1 shares over 98% identity with MEK1 from rat (Otsu et al, 1993; Wu et al, 1993) and mouse (Crews et al, 1992). Likewise, MEK2 from rat and human are highly conserved, and approximately 80% similar to MEK1 (Otsu et al, 1993; Wu et al, 1993). MEK1 and MEK2 phosphorylate ERK1 and ERK2 on threonine and tyrosine regulatory residues in the amino acid motif TEY adjacent to the MAPK active site (Zheng and Guan, 1994). No substrate is known for

MEK3, which complexes but does not phosphorylate ERK1 and ERK2. MEK3 may represent a MEK antagonist or may target other MAPKs (Zheng and Guan, 1993b). Unlike MEK2, MEK1 is negatively regulated by threonine phosphorylation (Rossomando et al, 1994).

There are likely to be more MAPKK activities. Jaiswal et al observed a 50–60 kDa MAPK stimulating activity from PC12 cells which is sensitive to both serine/threonine and tyrosine phosphatase treatment, unlike the cloned MEKs (Jaiswal et al, 1993). The *src* family tyrosine kinase p56^{lck} directly tyrosine phosphorylates and activates MAPK in vitro (Ettehadieh et al, 1992); however, the in vivo significance or extent of this potential signal pathway is unclear.

MAPK

Seven MAPK-like molecules had been recognized in mammalian cells (Figure 2.6) (Pelech and Sanghera, 1992), although only p42 ERK2, p44 ERK1, and p64 ERK3 had been cloned. Recently described Jun kinase (JNK) activities of 46 kDa and 55 kDa phosphorylate Jun at MAPK consensus sites. The JNK activities are not reactive with anti-ERK antibodies and contain no detectable phosphotyrosine (Hibi et al, 1993). Molecular cloning of one of these factors reveals it to be a distant member of the MAPK family (Dérijard et al, 1994). Comparison of the cloned MAPK family members shows that some, but not all, contain the regulatory sequence TEY phosphorylated by MEK1 and MEK2, suggesting alternative modes of activation of the various MAPK family members.

The protein synthesis inhibitor anisomycin induces expression of c-*fos* mRNA at subinhibitory concentrations. Neither ERK1 nor ERK2 are activated by anisomycin; however, the involvement of a MAPK family member is suspected (Kardalinou et al, 1994). Precedent for induction of c-*fos* independently of ERK1 or ERK2 is found in macrophages constitutively expressing the v-Raf protein (Büscher et al, 1993).

MAPK targets

Various proteins in addition to Elk-1 are thought to be nuclear targets of MAPKs (Blenis, 1993; Davis, 1993; Coffer et al, 1994). In particular the ribosomal S6 kinase pp85–90^{rsk} (RSK) family (Erikson, 1991) has been proposed to activate c-*fos* transcription by phosphorylation of SRF (Rivera et al, 1993). ERK1 and ERK2 can partially activate RSK in vitro, and the first of two peaks of RSK activity after cell stimulation accompanies ERK1/ERK2 activation. The second peak is coincident with the induction of a 63 kDa kinase, which may be a MAPK family member (Blenis, 1993). The other S6 kinase family, pp70–85^{S6k}, may be stimulated by MAPK-related proteins (Kardalinou et al, 1994). MAPK and RSK also phosphorylate the c-Fos transrepression domain (Chen et al, 1993), and may thereby contribute to c-*fos* post induction repression (below).

Signal transduction upstream of MAPKKK

Members of several structural classes of membrane receptors transduce signals into the MAPK pathway. Detailed descriptions of events for each receptor-type can be found in recent reviews on lymphocyte antigen receptors (Weiss and Littman, 1994), cytokine receptors (Kishimoto et al, 1994), tyrosine kinase receptors (Fantl et al, 1993), and G-coupled serpentine receptors (Hepler and Gilman, 1992; Inglese et al, 1993; Parmentier et al, 1993).

Binding of the cognate ligand to antigen, cytokine, or tyrosine-kinase receptors induces formation of homo- or heteromeric multiprotein receptor complexes, which activate intrinsic or recruited tyrosine kinase activities on the cytoplasmic portion of the receptor complex. Several types of protein interaction motif direct the association of proteins involved in signal transduction. These motifs include the SH2 and SH3 (Src homology 2 and 3) (Birge and Hanafusa, 1993; Williamson, 1994), and possibly the PH (Pleckstrin homology) (Musacchio et al, 1993; Parker et al, 1994) domains. Each molecule recruited to the receptor-complex can potentially activate or repress separate signal pathways, and different subsets of these pathways are accessible to unique receptor types.

Ras activation

There are four mammalian Ras proteins (H-Ras, N-Ras, K-RasA, K-RasB) and many related small GTPases (Boguski and McCormick, 1993). Principally, Ras is activated indirectly by phosphatidylinositol-specific phospholipase Cγ (PI-PLCγ), and by phosphatidylinositol-3 kinase (PI-3 kinase), and directly by guanine nucleotide exchange factor (GEF) proteins like Sos. It is deactivated by GTPase activating proteins (GAPs). Activation of Ras stimulates the Raf/MAPK pathway (Lowy and Willumsen, 1993; Moodie and Wolfman, 1994).

There seem to be at least two requirements for Ras activity in the mitogenic response of a resting cell. PDGF-mediated activation of Ras occurs via PI-PLCγ and PI-3 kinase, both of which activate PKC

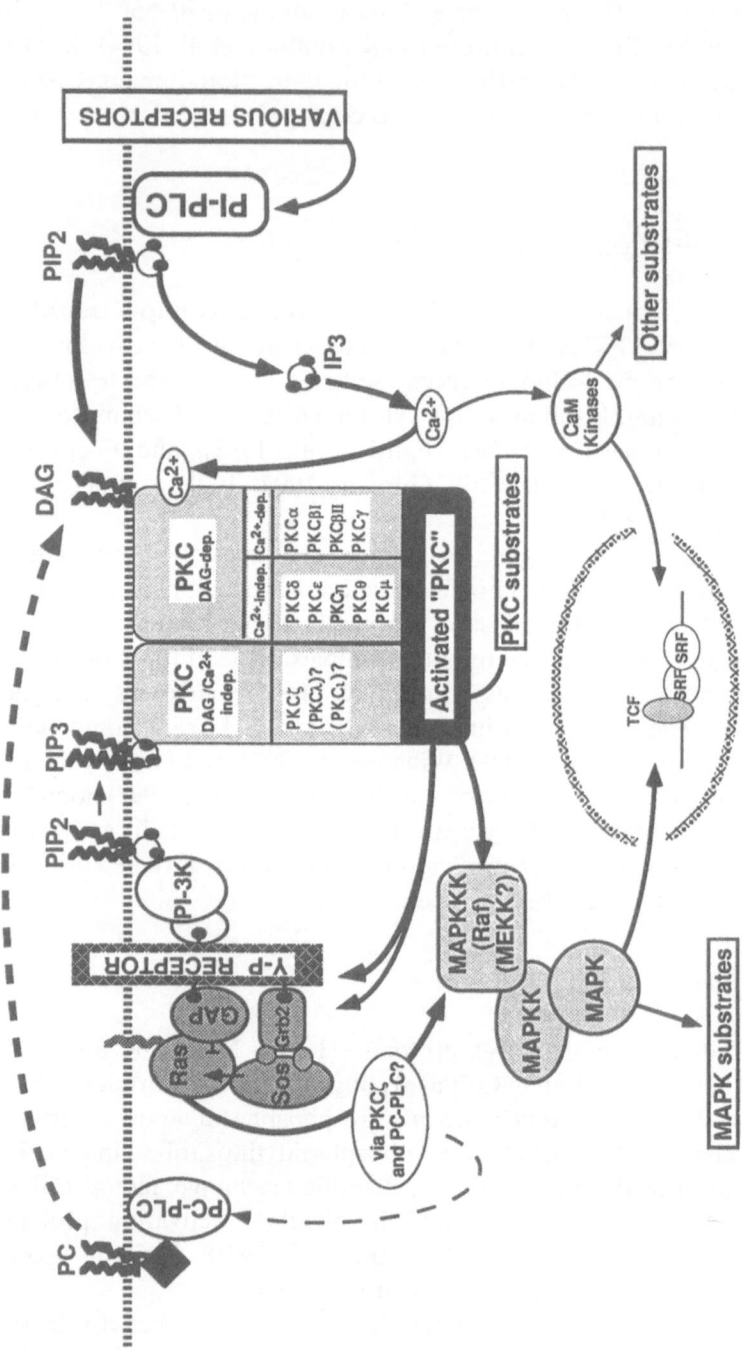

Figure 2.7. The activation of PKC isoforms and calmodulin dependent (CaM) kinases. PKC isoforms are either dependent (dep) or independent (indep) on the presence of Ca^{2+} and DAG. Abbreviations are: IP_3, inositol-(3,4,5)-trisphosphate; PIP_2, phosphatidylinositol-(4,5)-diphosphate; PIP_3, phosphatidylinositol-(3,4,5)-trisphosphate; PC-PLC, phosphatidylcholine-specific phospholipase C; PI-PLC, phosphatidylinositol-specific phospholipase C; PI-3K, phosphatidylinositol-3-kinase; Y-P Receptor, a receptor complex containing tyrosine phosphate. Stippled line indicates uncertainty concerning the pathway involved.

isoforms (Figure 2.7). It has been proposed that these differences are related to the progression or competence natures of these growth hormones (Moodie and Wolfman, 1994).

Sos is recruited to the membrane upon activation by several receptor classes and converts inactive GDP-Ras into activated GTP-Ras. Sos contains proline-rich domains which complex the SH3 domains of the adapter Grb2. Via an SH2-phosphotyrosine interaction, Grb2 is recruited to some activated receptors, sometimes by yet other adapter proteins. Additional GEFs and adapters also exist (Boguski and McCormick 1993; McCormick, 1993). Thus Ras is activated by the creation of a binding site for the Grb2/Sos complex on the inner surface of the membrane, increasing the concentration of GEF activity for membrane-bound Ras (Figure 2.7).

PKC isoforms in signal transduction

PKC has been found to operate both upstream and downstream of Ras in Raf activation under different circumstances possibly reflecting the activities of different isoforms (Satoh et al, 1992; Hug and Sarre, 1993; Lowy and Willumsen, 1993). Rat brain PKC phosphorylates and activates Raf in vitro, while Raf activation correlates with the activated or inhibited status of PKC in vivo (Carroll and May, 1994); however, as outlined above, the relationship between PKC and Raf activation is unclear. There are currently 12 PKC isoforms recognized in mammals (Figure 2.7), some of which are universally expressed whereas others exhibit restricted expression patterns (Hug and Sarre, 1993; Dekker and Parker, 1994). Although some functional redundancy among PKC isoforms may be expected, this is not comprehensive. For instance whereas PKCs $-\alpha$, $-\beta I$, and $-\varepsilon$ can activate both the SRE and a TPA response element upon phorbol ester treatment, PKCγ activates only the SRE (Hata et al, 1993).

The ubiquitous PKCζ may be necessary for serum stimulation of mouse fibroblasts and for signal transduction by activated Ras in *Xenopus* oocytes (Berra et al, 1993). Injection of either peptides corresponding to the PKCζ pseudosubstrate site, or antibodies specific for PKCζ, interfered with these processes. Catalytically inactive dominant negative PKCζ behaved similarly. Models of Raf activation by Ras should consider the possible involvement of the Ca^{2+}/DAG-independent PKCζ downstream of Ras (Figure 2.7). PKCζ is activated by phosphatidylinositol (3,4,5)-triphosphate (PIP$_3$), a product of PI-3 kinase (Nakanishi et al, 1993). PI-3 kinase is recruited to phosphotyrosine-containing receptors via the SH2-containing p85 subunit

(Figure 2.7); however its role in signal transduction is poorly understood (Hiles et al, 1992; Hu et al, 1993; Joly et al, 1994; Kapeller et al, 1994).

In the classical pathway of PKC activation (Figure 2.7), PI-PLCs convert phosphatidylinositol (4,5)-diphosphates (PIP_2) into diacyl glycerols (DAG) and inositol trisphosphate (IP_3). IP_3 causes cytoplasmic Ca^{2+} levels to rise, and together with DAG this activates PKC. The N-terminal regulatory domain of certain PKC isoforms bind both Ca^{2+} and/or DAG, which activates the C-terminal activation domain (Zidovetski and Lester, 1992; Hug and Sarre, 1993). Additionally, phosphatidyl-cholines (PC) can be hydrolyzed to generate DAG by PC-PLC (Cai et al, 1993), and by the heterogeneous phospholipase class PLD, which is associated with sustained PKC activation and events after immediate early gene induction (Asaoka et al, 1992).

Some PKC isoforms are Ca^{2+} independent, while PKCζ and possibly PKCλ are also DAG-independent (Figure 2.7). Treating cells with phorbol esters, which mimic DAG, depletes the cell of not only DAG-dependent PKC activity, but also PKCζ in some but not all cases (Hug and Sarre, 1993; Dekker and Parker, 1994). Phorbol ester exhaustion of PKC activity has often been used to supposedly eliminate the involvement of PKC in subsequent stimulation processes, a practice which must be reassessed in terms of the behavior of DAG-independent PKC isoforms (Hug and Sarre, 1993; Dekker and Parker, 1994). The cause of PKCζ exhaustion by phorbol esters in some cells is unclear, but may involve availability of adapter/target proteins mediating the effects of PKC (Hug and Sarre, 1993). It cannot currently be excluded that isoforms of PKC are involved in each pathway leading to MAPKKK activation in Figure 2.7 and Figure 2.8.

Ca^{2+}/Calmodulin

As outlined above, PI-PLCs release DAG and IP_3. Via cognate receptors, IP_3 induces sequential Ca^{2+} influxes from both intra and extracellular reservoirs which can persist for several hours (Berridge, 1993). Cytoplasmic Ca^{2+} binds the protein calmodulin (CaM), and the Ca^{2+}/CaM complex activates a variety of signal transducing molecules (Figure 2.7) including CaM KII and the phosphatase calcineurin (Schulman and Lou, 1989). Calcineurin may be an antagonist of the activity of PKA since their substrate specificities overlap (Cohen, 1992). Thus Ca^{2+}, cAMP, and the MAPK pathway have ample ability to interact, depending on specific factors present. CaM KII can also phosphorylate transcription factors binding the c-*fos* CRE (Figure 2.1), and there are consensus CaM KII sites

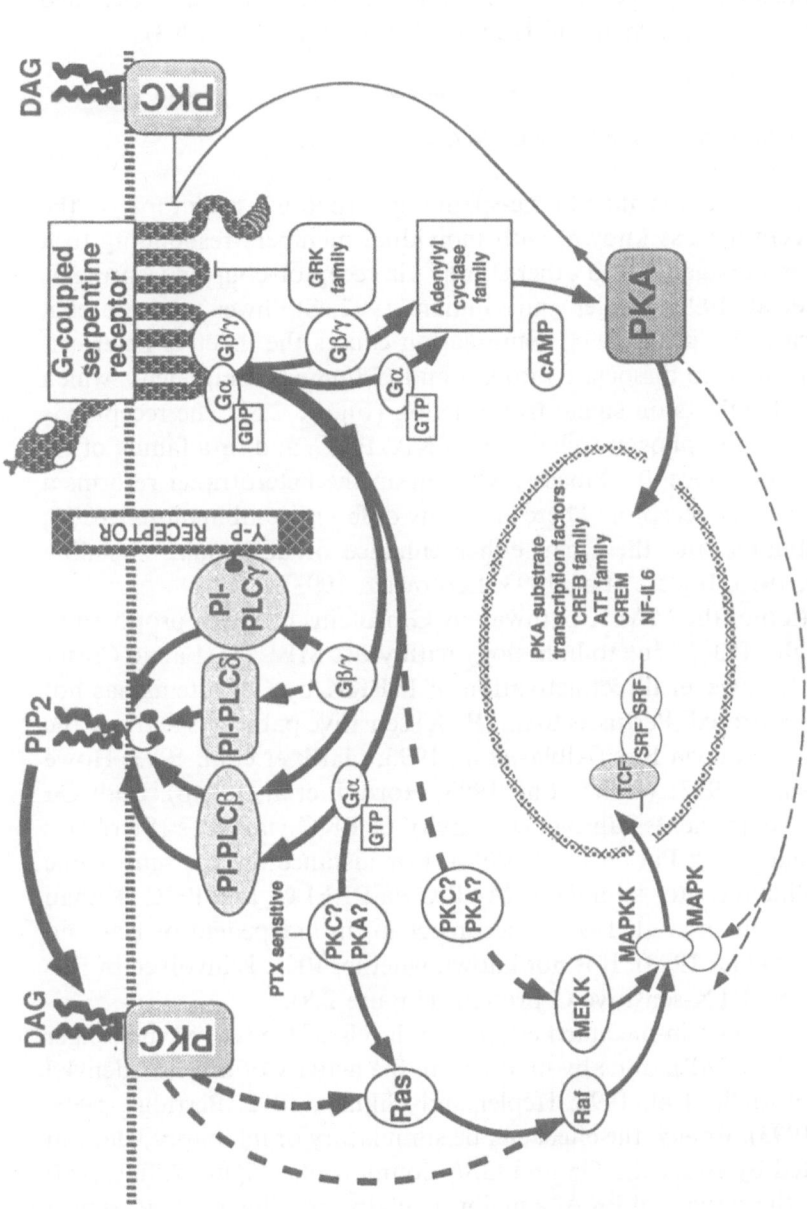

Figure 2.8. G-proteins target the SRE via divergent pathways. PH domain Gβ/γ interactions with some GRKs, PI-PLCβ, PI-PLCδ and PI-PLCγ are depicted as proposed by Parker et al (1994). PKC? denotes uncertainly as to whether PKC is involved in the pathway concerned. Abbreviations: GRK, G protein-coupled receptor kinase (Inglese et al, 1993; Lefkowitz, 1993). PTX, pertussis toxin. Other abbreviations follow Figure 2.7.

in SRF, including one conserved within the MADS-box (Schwarz-Sommer et al, 1990). No evidence for phosphorylation of SRF by CaM KII has been published; however the SRE can contribute to Ca^{2+}-mediated activation of c-*fos* (Aoyagi and Izumo, 1993; Bading et al, 1993).

G Proteins Influence cAMP, PKC, MEKK and Ras

The family of "seven membrane-spanning serpentine receptors" is the largest receptor class known, with individual members responding to a host of cognate ligands and other stimuli via receptor-coupled G proteins (Borrelli et al, 1992; Hepler and Gilman, 1992; Sternweis and Smrcka, 1992; Parmentier et al, 1993). Stimulation causes the inactive heterotri-meric G-protein to dissociate into Gα and Gβ/*amma* components which then mediate effects on signal transduction (Figure 2.8). The receptor is desensitized upon phosphorylation by PKA, PKC, and/or a family of G-protein coupled receptor kinases, whereupon the heterotrimer reforms a complex with the receptor. There are many different combinations of α, β, and γ subunits, and these can either enhance or antagonize signaling (Berridge, 1993; Inglese et al, 1993; Lefkowitz, 1993).

Activation of the MAPK pathway by G proteins has been proposed to act through MEKK due to homology with yeast MEKKs (Lange-Carter et al, 1993) however direct activation of MEKK by G Proteins has not been demonstrated. Pertussis toxin (PTX) sensitive pathways activate the MAPK pathway via Ras (Alblas et al, 1993; Gardner et al, 1993; Howe and Marshall, 1993; Winitz et al, 1993; Hordijk et al, 1994b). Both Gα and Gβ/γ components can induce activity of PKC via PLCs (Sternweis and Smrcka, 1992; Parker et al, 1994). For instance the M_1 muscarinic acetylcholine receptor stimulates MAPK via PI-PLCs and PKC (Kahan et al, 1992; Quain et al, 1993), via a mechanism indepedent of Ras and Raf (Winitz et al, 1993). It is not known whether PKC is involved in Ras activation by PTX-sensitive G proteins (Figure 2.8).

Another G protein-mediated effect is on levels of the second messenger cyclic AMP (cAMP), done by modulating the activity of diverse adenylyl cyclases (Borrelli et al, 1992; Hepler and Gilman, 1992; Berridge, 1993; Iyengar, 1993). Briefly, the effect can be stimulatory or inhibitory, and can be mediated by either the Gα and Gβ/γ components (Figure 2.8). cAMP stimulates the activity of PKA, a major regulator of cellular activity which activates the CREB/ATF group of transcription factors that bind the c-*fos* CRE (Borrelli et al, 1992). SRF may be directly activated by PKA, since serine 103 is the phosphoacceptor in a consensus PKA site (above). PKA has recently been shown to influence the MAPK pathway, being able to enhance or curtail the activation of MAPK in different circumstances (Marx, 1993; Marshall, 1994; Hordijk et al, 1994a, and Frödin et al, 1994). cAMP is also hydrolyzed by cAMP phosphodiesterases, which can be regulated by unknown kinases, allowing potential for regulatory interaction between kinase pathways (Cohen, 1992)

SRE-mediated down regulation of c-*fos* transcription

Having considered the SRE at the levels of anatomy, major binding proteins, and upstream signaling, we finally turn to another important aspect of SRE function, namely the postinductional repression (deinduction) of the activated SRE (Nordheim et al, 1994). Both induction and deinduction processes are likely to involve SRE binding proteins whose participation may be controlled by posttranslational modifications. While TCF phosphorylation can cause c-*fos* induction, TCF dephosphorylation might be responsible for c-*fos* deinduction. This notion is supported by two lines of evidence. (1) Down regulation of c-*fos* transcription and TCF dephosphorylation coincide, and (2) both deinduction of c-*fos* and TCF dephosphorylation are blocked by the protein phosphatase inhibitor okadaic acid (Zinck et al, 1983). Therefore, TCFs are probably dephosphorylated in vivo, which readily occurs in vitro by protein phosphatase 2A. The in vivo phosphorylation status of TCFs, which correlates with c-*fos* activity, may therefore be regarded as a dynamic equilibrium between phosphorylation by MAPKs and dephosphorylation of TCFs by protein phosphatases. It will be most interesting to see whether the autoregulatory role of Fos and Fos-related proteins (e.g. Fra-1) in c-*fos* down regulation (Nordheim et al, 1994) is exerted by influencing TCF dephosphorylation. Current evidence suggests that the phosphorylation status of TCFs could be dictated by the stimulation and deactivation of MAPKs. Dual specificity (serine/threonine and tyrosine) phosphatases, 3CH134/ CL100 (Charles et al, 1992; Keyse and Emslie, 1992) and PAC1 (Rohan et al, 1993), have been shown to dephosphorylate and down regulate MAPKs (Sun et al, 1993; Ward et al, 1994). Since the corresponding phosphatase genes are themselves inducible by stimuli which induce c-*fos* (Charles et al, 1992; Keyse and Emslie, 1992; Rohan et al, 1993), this could easily explain the transient nature of c-*fos* stimulation. Consistent with such a notion is that constitutively active PAC1 opposes the activating effect of MAPKs on the c-*fos* SRE (Ward et al, 1994). However, the existence of a regulated TCF phosphatase is not excluded.

Perspectives

The study of the SRE as a key regulatory element of the immediate early gene c-*fos* has established a connection between growth factor signaling and gene regulation. Identification and functional characterization of the major SRE binding proteins, SRF and TCFs, has uncovered a high degree of evolutionary conservation, both in structural and mechanistic terms, with regard to signal controlled gene regulation. The ternary complexes involving mammalian SRF and TCFs on the one hand, and yeast MCM1 and its associated factors α1, α2, STE12 and SFF on the other hand, highlight this fact (Treisman and Ammerer, 1992).

The mechanistic conservation at the level of gene control is paralleled by the striking degree of conservation found between signaling pathways, especially the receptor tyrosine kinase (RTK)-stimulated MAPK pathway, in yeast, insects, nematodes and vertebrates (Neiman, 1993; Marshall, 1994). It is interesting to see the involvement of these signaling pathways not only in mitogenic signaling but also in signal transduction governing growth arrest, cell-cell communication, cell type specific gene control, and cell differentiation. Accordingly, RTK-stimulated MAPK pathways contributing to animal development have been identified in the *torso* pathway of the *Drosophila melanogaster* terminal system (Perrimon, 1993), the sevenless pathway of *D. melanogaster* required for epidermal and photoreceptor R7 development (Brunner et al, 1994; Diaz-Benjumea and Hafen, 1994), the *Caenorhabditis elegans* Let-23 pathway required for vulval development (Sternberg, 1993), and mesoderm induction during *Xenopus laevis* gastrulation (Amaya et al, 1991; MacNicol et al, 1993). A direct involvement of SRE-like sequences in a developmental context has indeed been described with the muscle-specific transcription of the cardiac actin gene in *Xenopus* embryos (Mohun et al, 1989).

Although much has been learned concerning the regulation of transcription and the immediate early response by the study of c-*fos* and other SRE-containing genes, some recent results prompt an almost heretical question. How certain are we that SRE driven genes are directly involved in events leading to mitogenesis? Murine pro-B BAF-303 hematopoetic cells expressing mutated forms of the GM-CSF receptor β-subunit can proliferate in response to GM-CSF, albeit slowly, without apparent induction of the Ras/Raf pathway or c-*fos* expression (Sato et al, 1993). Derivatives of the same cells expressing a mutated interleukin 2 (IL-2) β-receptor enter S phase but do not progress further in response to IL-2, also in the apparent absence of c-*fos* and *junB* induction (Shibuya et al, 1993). Failure of these mutant receptors to activate the Ras/Raf pathway probably results in negligible activation of many SRE driven genes, raising the possibility that these genes are not required to progress through G1 into S phase, although these results could be explained in several ways. In the case of Fos, whose function has been studied in detail, overexpression can clearly be mitogenic, and c-*fos* is a proto-oncogene. Also, anti-sense inhibition of c-Fos expression in cultured cells prevents DNA replication (Holt et al, 1986). Yet it could be argued that Fos is only indirectly involved in the mitogenic response. The inhibitory effects of anti-sense *fos* RNA could be abrogated by increasing the concentration of serum to levels perhaps more representative of in vivo growth conditions (Holt, 1993). Knock-out mice lacking Fos protein mature relatively normally, with the exception of some bone defects and immune deficiencies. Therefore Fos is apparently not crucial for mitogenesis in the majority of tissues, although redundancy in the AP-1 family may largely account for this (Wang et al, 1992).

The SRE confers immediate early kinetics on several genes, but is also responsive to a host of environmental assaults. SREs in the *junB* and c-*fos* genes (see Table) probably contribute to the regulation by AP-1 of several genes (Li and Jaiswal, 1992; Bergelson et al, 1994) which are involved in the response to oxidative and chemical stress. Increased metabolism following mitogen stimulation is also accompanied by the generation of oxygen radicals and metabolic waste products, and is preceded by induction of SRE driven genes. Might SREs thus stimulate the production of a set of proteins whose principal function is to detoxify waste products or rectify environmental challenges to cellular homeostasis? Answers to such questions await further study. Insight gained into the process of signal uptake by SREs sets the stage for future research addressed toward understanding c-*fos* and immediate early gene induction with regard to the requirements of a cell as a component of a complex organism.

Acknowledgements

This review is written in partial fulfillment of the doctoral thesis by Michael A. Cahill. We appreciate the critical comments on this manuscript by Marie Henriksson, Lars-Gunnar Larsson, and the members of our group. B. Defize provided useful discussions on MAPK activation by G-proteins. Financial support by the DFG (grant No 120/7-2), the Mildred-Scheel-Stiftung (grant W37/92 No 1), and the Fonds der Chemischen Industrie is gratefully acknowledged.

References

Alblas J, van Corven EJ, Hordijk PL, Milligan G, Moolenaar WH (1993): G_i-mediated activation of the $p21^{ras}$-mitogen-activated protein kinase pathway by α_2-adrenergic receptors expressed in fibroblasts. *J Biol Chem* 268: 22235–22238

Almendral JM, Sommer D, MacDonald-Bravo H, Burckhardt J, Perera J, Bravo R (1988): Complexity of the early genetic response to growth factors in mouse fibroblasts. *Mol Cell Biol* 8: 2140–2148

Amaya E, Musci TJ, Kirschner MW (1991): Expression of a dominant negative mutant of the FGF receptor disrupts mesoderm formation in Xenopus embryos. *Cell* 66: 257–270

Ammerer G (1994): Sex, stress and integrity: importance of MAP kinases in yeast. *Curr Opin Genet Dev* 4: 90–95

Aoyagi T, Izumo S (1993): Mapping of the pressure response element of the c-*fos* gene by direct DNA injection into beating hearts. *J Biol Chem* 268: 27176–27179

Asaoka Y, Nakamura S-i, Yoshida K, Nishizuka Y (1992): Protein kinase C, calcium and phospholipid degradation. *Trends Biochem Sci* 17: 414–417

Bading H, Ginty DD, Greenberg ME (1993): Regulation of gene expression in hippocampal neurons by distinct calcium signaling pathways. *Science* 260: 181–186

Bergelson S, Pinkus R, Daniel V (1994): Induction of AP-1 (Fos/Jun) by chemical agents mediates activation of glutathione S-transferase and quinone reductase gene expression. *Oncogene* 9: 565–571

Berra E, Diaz-Meco MT, Dominguez I, Municio MM, Sanz L, Lozano J, Chapkin RS, Moscat J (1993): Protein kinase C ζ isoform is critical for mitogenic transduction. *Cell* 74: 555–563

Berridge MJ (1993): Inositol trisphosphate and calcium signalling. *Nature* 361: 351–325

Bhattacharya G, Lee L, Reddy ESP, Rao VN (1993): Transcriptional activation domains of *elk*-1, Δ*elk*-1 and SAP-1 proteins. *Oncogene* 8: 3459–3464

Birge RB, Hanafusa H (1983): Closing in on SH2 specificity. *Science* 262: 1522–1524

Blenis J (1993): Signal transduction via the MAP kinases: proceed at your own RSK. *Proc Natl Acad Sci USA* 90: 5889–5892

Blum S, Forsdyke RE, Forsdyke DR (1990): Three human homologs of a murine gene encoding an inhibitor of stem cell proliferation. *DNA Cell Biol* 9: 589–602

Boguski MS, McCormick F (1993): Proteins regulating Ras and its relatives. *Nature* 366: 643–654

Bornstein P (1992): Thrombospondins: structure and regulation of expression. *FASEB J* 6: 3290–3299

Borrelli E, Montamayeur J-P, Foulkes NS, Sassone-Corsi P (1992): Signal transduction and gene control: the cAMP pathway. *Crit Rev Oncogen* 3(4): 321–338

Boulden A, Sealy L (1990): Identification of a third protein factor which binds to the Rous Sarcoma Virus LTR enhancer: possible homology with the serum response factor. *Virol* 174: 204–216

Bravo R (1990): Genes induced during the G_0/G_1 transition in mouse fibroblasts. *Cancer Biology* 1: 37–46

Brunner D, Oellers N, Szarbad J, Biggs WH III, Zipursky SL, Hafen E (1994): A gain-of-function mutation in Drosophila MAP kinase activates multiple receptor tyrosine kinase signaling pathways. *Cell* 76: 875–888

Büscher D, Sbarba PD, Hipkind RA, Rapp UR, Stanley ER, Baccarini M (1993): v-

raf confers CSF-1 independent growth to a macrophage cell line and leads to immediate early gene expression without MAP-kinase activation. *Oncogene* 8: 3323–3332

Cai H, Erhardt P, Troppmair J, Diaz-Meco MT, Sithanandam G, Rapp UR, Moscat J, Cooper GM (1993): Hydrolysis of phosphatidylcholine couples Ras to activation of Raf protein kinase during mitogenic signal transduction. *Mol Cell Biol* 13: 7645–7651

Carroll MP, Stratford MW (1994): Protein kinase C-mediated serine phosphorylation directly activates Raf-1 in murine hematopoietic cells. *J Biol Chem* 269: 1249–1256

Chang Y-N, Jeang K-T, Chiou C-J, Chan Y-J, Pizzorno M, Hayward S (1993): Identification of a large bent DNA domain and binding sites for serum response factor adjacent to the NF1 repeat cluster and enhancer region in the major IE94 promoter from Simian Cytomegalovirus. *J Virol* 67: 516–529

Charles CH, Abler AS, Lau LF (1992): cDNA sequence of a growth factor-inducible immediate early gene and characterization of its encoded protein. *Oncogene* 7: 187–190

Charles CH, Simske, JS, O'Brien TP, Lau LF (1990): PiP92: a short-lived, growth factor-inducible protein in BALB/c 3T3 and PC12 cells. *Mol Cell Biol* 10: 6769–6774

Chavrier P, Zerial M, Lemaire P, Almendral J, Bravo R, Charnay P (1988): A gene encoding a protein with zinc fingers is activated during G_0/G_1 transition in cultured cells. *EMBO J* 7: 29–36

Chen R-H, Abate C, Blenis J (1993): Phosphorylation of the c-Fos transrepression domain by mitogen-activated protein kinase and 90-kDA ribosomal S6 kinase. *Proc Natl Acad Sci USA* 90: 10956–10956

Christy B, Nathans D (1989): Functional serum response elements upstream of the growth factor inducible *zif268*. *Mol Cell Biol* 9: 4889–4895

Coffer P, de Jonge M, Mettouchi A, Binetruy B, Ghysdael J, Kruijer W (1994): *junB* promoter regulation: Ras mediated transactivation by c-Ets-1 and c-Ets-2. *Oncogene* 9: 911–921

Cohen DR, Curran T (1989): The structure and function of the *fos* proto-oncogene. *Oncogenesis* 1: 65–88

Cohen P (1992): Signal integration at the level of protein kinases, protein phosphatases and their substrates. *Trends Biochem Sci* 17: 408–413

Crews CM, Alessandrini A, Erikson RL (1992): The primary structure of MEK, a protein kinase that phosphorylates the *ERK* gene product. *Science* 258: 478–480

Dalton S, Treisman R (1992): Characterization of SAP-1, a protein recruited by serum response factor to the c-*fos* serum response element. *Cell* 68: 597–612

Davis RJ (1993): The mitogen-activated protein kinase signal transduction pathway. *J Biol Chem* 268: 14553–14556

Dekker LV, Parker PJ (1994): Protein kinase C — a question of specificity. *Trends Biochem Sci* 19: 73–77

Dérijard B, Hibi M, Wu I-H, Barret T, Su B, Deng T, Karin M, Davis RJ (1994): JNK1: A Protein kinase stimulated by UV light and Ha-Ras that binds and phosphorylates the c-Jun activation domain. *Cell* 76: 1–20

Diaz-Benjumea FJ, Hafen E (1994): The sevenless signalling cassette mediates *Drosophila* receptor function during epidermal development. *Development* 120: 569–578

Erikson RL (1991): Structure, expression, and regulation of protein kinases involved in the phosphorylation of ribosomal protein S6. *J Biol Chem* 266: 6007–6010

Ernst WH, Janknecht R, Cahill MA, Nordheim A (1994): Transcriptional activation and repression mediated by the serum response factor. Manuscript submitted

Ettehadieh E, Sanghera JS, Pelech SL, Hess-Bienz D, Watts J, Shastri N, Aebersold R (1992): Tyrosyl phosphorylation and activation of MAP kinases by p56^{lck}. *Science* 255: 853–855

Fantl WJ, Johnson DE, Williams LT (1993): Signalling by receptor tyrosine kinases. *Annu Rev Biochem* 62: 453–481

Feig LA (1993): The many roads that lead to Ras. *Science* 260: 767–768

Framson P, Bornstein P (1993): A serum response element and a binding site for NF-Y mediate the serum response of the human thrombospondin Z gene. *J Biol Chem* 268: 4989–4996

Frödin M, Peraldi P, Van Obberghen E (1994): Cyclic AMP activates the mitogen-activated protein kinase cascade in PC12 cells. *J Biol Chem* 269: 6207–6214

Fujii M, Tsuchiya H, Chuhjo T, Akizawa T, Seiki M (1992): Interaction of HTLV-1 Tax1 with p67^{SRF} causes the aberrant induction of cellular immediate early genes through CArG boxes. *Genes Dev* 6: 2066–2076

Gardner AM, Vaillancourt RR, Johnson GL (1993): Activation of mitogen-activated protein kinase/extracellular signal-regulated kinase kinase by G protein and tyrosine kinase oncoproteins. *J Biol Chem* 268: 17896–17901

Gauthier-Rouvière C, Basset M, Blanchard J-M, Cavadore J-C, Fernandez A, Lamb NJC (1991): Casein kinase II induces c-*fos* expression via the serum response element pathway and p67^{SRF} phosphorylation in living fibroblasts. *EMBO J* 10: 2921–2930

Gille HG, Sharrocks AD, Shaw PE (1992): Phosphorylation of transcription factor p62^{TCF} by MAP kinase stimulates ternary complex formation at the c-*fos* promoter. *Nature* 358: 414–417

Gilman MZ, Wilson RN, Weinberg RA (1986): Multiple protein binding sites in the 5' flanking region regulate c-*fos* expression. *Mol Cell Biol* 6: 4305–4314

Greenberg ME, Ziff EB (1984): Stimulation of 3T3 cells induces transcription of the c-*fos* proto-oncogene. *Nature* 311: 433–438

Grueneberg DA, Natesan S, Alexandre C, Gilman MZ (1992): Human and *Drosophila* homeodomain proteins that enhance the DNA-binding activity of serum response factor. *Science* 257: 1089–1095

Gualberto A, LePage D, Pons G, Mader SL, Park K, Atchison ML, Walsh K (1992): Functional antagonism between YY1 and the serum response factor. *Mol Cell Biol* 12: 4209–4214

Gutman A, Wasylyk C, Wasylyk B (1991): Cell-specific regulation of oncogene-responsive sequences of the c-*fos* promoter. *Mol Cell Biol* 11: 5381–5387

Hata A, Akita Y, Suzuki K, Ohno S (1993): Functional divergence of protein kinase C (PKC) family members. *J Biol Chem* 268: 9122–9129

Hayes TE, Kitchen AM, Cochran BH (1987): Inducible binding of a factor to the c-*fos* regulatory region. *Proc Natl Acad Sci USA* 84: 1217–1225

Heidecker G, Kölch W, Morrison DK, Rapp UR (1992): The role of Raf-1 phosphorylation in signal transduction. *Adv Cancer Res* 58: 53–73

Hepler JR, Gilman AG (1992): G proteins. *Trends Biochem Sci* 17: 383–387

Herrera RE, Shaw PE, Nordheim A (1989): Occupation of the c-*fos* serum response element *in vivo* by a multi-protein complex is unaltered by growth factor induction. *Nature* 340: 68–70

Hibi M, Lin A, Smeal T, Minden A, Karin M (1993): Identification of an oncoprotein- and UV-responsive protein kinase that binds and potentiates the c-Jun activation domain. *Genes Dev* 7: 2135–2148

Hiles ID, Otsu M, Volinia S, Fry MJ, Gout I, Dhand R, Panayotou G, Ruiz-Larrea F, Thompson A, Totty NF, Hsuan JJ, Courtneidge SA, Parker PJ, Waterfield MD (1992): Phosphatidylinositol 3-kinase: structure and expression of 110 kd catalytic subunit. *Cell* 70: 419–429

Hill CS, Marais R, John S, Wynne J, Dalton S, Treisman R (1993): Functional analysis of a growth factor-responsive transcription factor complex. *Cell* 73: 395–406

Hipskind RA, Nordheim A (1991): Functional dissection *in vitro* of the human c-*fos* promoter. *J Biol Chem* 266: 19583–19592

Hipskind RA, Rao VN, Mueller CGF, Reddy ESP, Nordheim A (1991): Ets-related protein Elk-1 is homologous to the c-*fos* regulatory factor p62^{TCF}. *Nature* 354: 531–534

Holt JT (1993): Antisense rescue defines specialized and generalized functional domains for c-Fos protein. *Mol Cell Biol* 13: 3821–3830

Holt JT, Gopal TV, Moulton AD, Nienhuis AQ (1986): Inducible production of c-*fos* anti-sense RNA inhibits 3T3 cell proliferation. *Proc Natl Acad USA* 83: 4794–4798

Hordijk PL, Verlaan I, Jalink K, van Corven EJ, Moolenaar WH (1994a): cAMP abrogates the p21^{ras}-mitogen activated protein kinase pathway in fibroblasts. *J Biol Chem* 269: 3534–3538

Hordijk PL, Verlaan I, van Corven EJ, Moolenaar WH (1994b): Protein tyrosine phosphorylation induced by lysophosphatidic acid in Rat-1 fibroblasts. *J Biol Chem* 269: 645–651

Howe LR, Marshall CJ (1993): Lysophosphatidic acid stimulates mitogen-activated protein kinase activation via a G-protein-coupled pathway requiring p21^{ras} and p74^{raf-1}. *J Biol Chem* 268: 20717–20720

Hu P, Mondino A, Skolnik EY, Schlessinger J (1993): Cloning of a novel, ubiquitously expressed human phosphatidylinositol 3-kinase and identification of its binding site on p85. *Moll Cell Biol* 13: 7677–7688

Hug H, Sarre TF (1993): Protein kinase C isoenzymes: divergence in signal transduction? *Biochem J* 291: 329–343

Inglese J, Freedman NJ, Koch WJ, Lefkowitz RJ (1993): Structure and mechanism of the G protein-coupled receptor kinases. *J Biol Chem* 268: 23735–23738

Itoh T, Kaibuchi K, Masuda T, Yamamoto Y, Matsuura Y, Maeda A, Shimizu K, Takai Y (1993): A protein factor for ras p21-dependent activation of mitogen-activated protein (MAP) kinase through MAP kinase kinase. *Proc Natl Acad Sci USA* 90: 975–979

Iyengar R (1993): Multiple families of G$_S$-regulated adenylyl cyclases. *Advances in Second Messenger and Phosphoprotein Research* 28: 27–36

Jaiswal RK, Murphy MB, Landreth GE (1993): Identification and characterization of a nerve growth factor-stimulated mitogen-activated protein kinase activator in PC12 cells. *J Biol Chem* 268: 7055–7063

Jamal S, Ziff EB (1990): Transactivation of c-*fos* and β-actin genes by raf as a step in early response to transmembrane signals. *Nature* 344: 463–466

Janknecht R, Nordheim A (1992): Elk-1 protein domains required for direct and SRF-assisted DNA-binding. *Nucleic Acids Res* 20: 3317–3324

Janknecht R, Nordheim A (1993): Gene regulation by Ets proteins. *Biochim Biophys Acta* 1155: 346–356

Janknecht R, Ernst WH, Houthaeve T, Nordheim A (1993a): C-terminal phosphorylation of the serum-response factor. *Eur J Biochem* 216: 469–475

Janknecht R, Ernst WH, Pingoud V, Nordheim A (1993b): Activation of ternary complex factor Elk-1 by MAP kinases. *EMBO J* 12: 5097–5104

Janknecht R, Hipskind RA, Houthaeve T, Nordheim A, Stunnenberg HG (1992): Identification of multiple SRF N-terminal phosphorylation sites affecting DNA binding properties. *EMBO J* 11: 1045–1054

Janknecht R, Zinck R, Ernst WH, Nordheim A (1994): Functional dissection of the transcription factor Elk-1. *Oncogene* 9: 1273–1278

Johansen F-E, Prywes R (1993): Identification of transcriptional activation and inhibitory domains in serum response factor (SRF) by using GAL4-SRF constructs. *Mol Cell Biol* 13: 4640–4647

Joly M, Kazlauskas A, Fay FS, Corvera S (1994): Disruption of PDGF receptor trafficking by mutation of its PI-3 kinase binding sites. *Science* 263: 684–687

Joseph L, J Le Beau MM, Jamieson GA, Acharya S, Shows TB, Rowley JD, Sukhatme VP (1988): Molecular cloning, sequencing, and mapping of *EGR2*, a human early growth response gene encoding a protein with "zinc-binding finger" structure. *Proc Natl Acad Sci USA* 85: 7164–7168

Kahan C, Seuwen K, Meloche S, Pouysségur J (1992): Coodinate biphasic activation of p44 Mitogen-Activated protein kinases and S6 kinase by growth factors in Hamster fibroblasts. *J Biol Chem* 267: 13369–13375

Kaibuchi K, Fukumoto Y, Oku N, Hori Y, Yamamoto T, Toyoshima K, Takai Y (1989): Activation of the serum response element and 12-*O*-tetra-decanoyl-13-acetate response element by the activated c-*raf*-1 protein in a manner independent of protein kinase C. *J Biol Chem* 264: 20855–20858

Kapeller R, Prasad KVS, Janssen O, Hou W, Schaffhausen BS, Rudd CE, Cantley LC (1994): Identification of two SH3-binding motifs in the regulatory subunit of phosphatidylinositol 3-kinase. *J Biol Chem* 269: 1927–1933

Kardalinou E, Zhelev N, Hazzalin CA, Mahadevan LC (1994): Anisomycin and rapamycin define an area upstream of $p70/85^{S6k}$ containing a bifurcation to histone H3-HMG-like protein phosphorylation and c-*fos*-c-*jun* induction. *Mol Cell Biol* 14: 1066–1074

Keyse SM, Emslie EA (1992): Oxidative stress and heat shock induce a human gene encoding a protein-tyrosine phosphatase. *Nature* 359: 644–647

Kishimoto T, Taga T, Akira S (1994): Cytokine signal transduction. *Cell* 76: 253–262

König H (1991): Cell-type specific multiprotein complex formation over the c-fos serum response element *in vivo*: ternary complex formation is not required for the induction of c-fos. *Nucleic Acids Res* 19: 3607–3611

König H, Ponta H, Rahmsdorf U, Büscher M, Schönthal A, Rahmsdorf HJ, Herrlich P (1989): Autoregulation of *fos*: the dyad symmetry element as the major target of repression. *EMBO J* 8: 2559–2566

Kruijer W, Cooper JA, Hunter T, Verma IM (1984): Platelet-derived growth factor induces rapid but transient expression of the c-*fos* gene and protein. *Nature* 312: 711–716

Lang A, Fincham VJ, Wyke JA (1993): Factors influencing physiological variations in the actitvity of the Rous Sarcoma Virus long terminal repeat. *Virol* 196: 564–575

Lange-Carter CA, Pleiman CM, Gardner AM, Blumer KJ, Johnson GL (1993): A divergence in the MAP kinase regulatory network defined by MEK kinase and Raf. *Science* 260: 315–319

Latinkic BV, O'Brien TP, Lau LF (1991): Promoter function and structure of the growth factor-inducible immediate early gene *cyr61*. *Nucl Acids Res* 19: 3261–3267

Lau LF, Nathans D (1985): Identification of a set of genes expressed during the G_0/G_1 transition of cultured mouse cells. *EMBO J* 4: 3145–3151

Lazo PS, Dorfman K, Noguchi T, Mattei MG, Bravo R (1992): Structure and mapping of the *fosB* gene. FosB downregulates the activity of the *fosB* promoter. *Nucl Acids Res* 20: 343–350

Lee T-C, Chow K-L, Fang P, Schwartz RJ (1991): Activation of skeletal α-actin gene transcription: The cooperative formation of serum response factor-binding complexes over positive *cis*-acting promoter serum response elements displaces a negative-acting nuclear factor enriched in replicating myoblasts and nonmyogenic cells. *Mol Cell Biol* 11: 5090–5100

Lefkowitz RJ (1993): G protein-coupled receptor kinases. *Cell* 74: 409–412

Lemaire P, Revelant O, Bravo R, Charnay P (1988): Two mouse genes encoding potential transcription factors with identical DNA-binding domains are activated by growth factors in cultured cells. *Proc Natl Acad Sci USA* 85: 4691–4695

Li Y, Jaiswal K (1992): Identification of JunB as third member in human antioxidant response element-nuclear proteins complex. *Biochem Biophys Res Comm* 188: 992–996

Lin X, Wang Z, Gu L, Deuel TF (1992): Functional analysis of the human platelet-derived growth factor A-chain promoter region. *J Biol Chem* 267: 25614–25619

Liu S-H, Ma J-T, Yueh AY, Lees-Miller SP, Anderson CW, Ng S-Y (1993): The carboxyl-terminal transactivation domain of human serum response factor contains DNA-activated protein kinase phosphorylation sites. *J Biol Chem* 268: 21147–21154

Lowy DR, Willumsen BM (1993): Function and regulation of Ras. *Annu Rev Biochem* 62: 851–891

MacNicol AM, Muslin AJ, Williams LT (1993): Raf-1 kinase is essential for early Xenopus development and mediates the induction of mesoderm by FGF. *Cell* 73: 571–583

Manak JR, Prywes R (1991): Mutation of serum response factor phosphorylation sites and the mechanism by which its DNA-binding activity is increased by casein kinase II. *Mol Cell Biol* 11: 3652–3659

Marais RM, Hsuan JJ, McGuidan C, Wynne J, Treisman R (1992): Casein kinase II phosphorylation increases the rate of serum response factor-binding site exchange. *EMBO J* 11: 97–105

Marais R, Wynne J, Treisman R (1993): The SRF accessory protein Elk-1 contains a growth factor-regulated transcriptional activation domain. *Cell* 73: 381–393

Marshall CJ (1994): MAP kinase kinase kinase, MAP kinase kinase and MAP kinase. *Curr Opin Genet Dev* 4: 82–89

Marx J (1993): Two major signal pathways linked. *Science* 262: 988–990

McCormick F (1993): How receptors turn Ras on. *Nature* 363: 15–16

Metz R, Ziff E (1991): The helix-loop-helix protein rE12 and the C/EBP-related factor rNFIL-6 bind to neighboring sites within the c-*fos* serum response element. *Oncogene* 6: 2165–2178

Mitchell RL, Zokas L, Schreiber RD, Verma IM (1985): Rapid induction of the expression of proto-oncogene *fos* during human monocyte differentiation. *Cell* 40: 209–217

Miwa T, Boxer L, Kedes L (1987): CArG boxes in the human cardiac α-actin gene are core binding sites for positive *trans*-acting regulatory factors. *Proc Natl Acad Sci USA* 84: 6702–6706

Mohun TJ, Chambers AE, Towers N, Taylor MV (1991): Expression of genes encoding the transcription factor SRF during early development of *Xenopus laevis*: identification of a CArG box-binding activity as SRF. *EMBO J* 10: 933–940

Mohun TJ, Garrett N, Treisman RH (1987): Xenopus cytoskeletal actin and human c-fos gene promoters share a conserved protein binding site. *EMBO J* 6: 667–673

Mohun TJ, Taylor MV, Garrett N, Gurdon JB (1989): The CArG promoter sequence is necessary for muscle-specific transcription of the cardiac actin gene in *Xenopus* embryos. *EMBO J* 8: 1153–1161

Moisseyeva EP, Weller PA, Zhidkova NI, Corben EB, Patel B, Jasinka I, Koteliansky VE, Critchley DR (1993): Organization of the human gene encoding the cytoskeletal protein vinculin and the sequence of the vinculin promoter. *J Biol Chem* 268: 4318–4325

Montminy M, (1993): Trying on a new pair of SH2s. *Science* 261: 1694–1695

Moodie SA, Wolfman A (1994): The 3Rs of life: Ras, Raf and growth regulation. *Trends Gen* 10: 44–48

Mueller CGF, Nordheim A (1991): A protein domain conserved between yeast MCM1 and human SRF directs ternary complex formation. *EMBO J* 10: 4219–4229

Müller R, Bravo R, Burckhardt J, Curran T (1984): Induction of c-*fos* gene and protein by growth factors precedes activation of c-*myc*. *Nature* 312: 716–720

Musacchio A, Gibson T, Rice P, Thompson J, Saraste M (1993): The pH domain: a common piece in the structural patchwork of signalling proteins. *Trends Biochem Sci* 18: 343–348

Nakanishi H, Brewer KA, Exton JG (1993): Activation of the ζ-isozyme of protein kinase C by phosphatidylinositol 3,4,5-trisphosphate. *J Biol Chem* 268: 13–16

Natesan S, Gilman MZ (1993): DNA bending and orientation-dependent function of YY1 in the c-*fos* promoter. *Genes Dev* 7: 2497–2509

Nebreda AR, Hill C, Gomez N, Cohen P, Hunt T (1993): The protein kinase *mos* activates MAP kinase kinase in vitro and stimulates the MAP kinase pathway in mammalian somatic cells in vivo. *FEBS Lett* 333: 183–187

Neiman AM (1993): Conservation and reiteration of a kinase cascade. *Trends Gen* 9: 390–394

Nordheim A, Janknecht R, Hipskind R (1994): Transcriptional regulation of the human c-*fos* proto-oncogene. In: *Cellular and Viral fos Genes*, Angel P, Herrich P, eds. CRC Press, Boca Raton (in press)

Norman C, Runswick M, Pollock R, Treisman R (1988): Isolation and properties of cDNA clones encoding SRF, a transcription factor that binds to the c-*fos* serum response element. *Cell* 55: 989–1003

Orita S, Makino K, Kawamoto T, Niwa H, Sugiyama H, Kakanuga T (1989): Identification of a site that mediates transcriptional response of the human β actin gene to serum factors. *Gene* 75: 13–19

Otsu M, Terada Y, Okayama H (1993): Isolation of two members of the rat MAP kinase kinase gene family. *FEBS* 320: 246–250

Parker PJ, Hemmings BA, Gierschik P (1994): PH domains and phospholipases — a meaningful relationship? *Trends Biochem Sci* 19: 54–55

Parmentier M, Libert F, Perret J, Eggerickx D, Ledent C, Schurmans S, Raspe E, Dumont JE, Vassart G (1993): Cloning and characterization of G protein-coupled receptors. *Advances in Second Messenger and Phosphoprotein Research* 28: 11–18

Pelech SL, Sanghera JS (1992): Mitogen-activated protein kinases: versatile transducers for cell signaling. *Trends Biochem Sci* 17: 233–238

Perez-Albuerne ED, Schatteman G, Sanders LK, Nathans D (1993): Transcriptional regulatory elements downstream of the junB gene. *Proc Natl Acad Sci USA* 90: 11960–11964

Perrimon N (1993): The Torso receptor protein-tyrosine kinase signaling pathway: an endless story. *Cell* 74: 219–222

Phan-Dinh-Tuy F, Tuil D, Schweighoffer F, Pinset C, Kahn A, Minty A (1988): The CC.Ar.GG box: a protein-binding site common to transcription-regulatory proteins of the cardiac actin, c-*fos*, and interleukin-2 receptor genes. *Eur J Biochem* 173: 507–515

Pingoud V, Zinck R, Hipskind RA, Janknecht R, Nordheim A (1994): Heterogeneity of ternary complex factors in HeLa cell nuclear extracts. *J Biol Chem*: in press

Prywes R, Roeder RG (1986): Inducible binding of a factor to the c-*fos* enhancer. *Cell* 47: 777–784

Qian N-X, Winitz S, Johnson GL (1993): Epitope-tagged $G_q\alpha$ subunits: expression of GTPase-deficient α subunits persistently stimulates phosphatidylinositol-specific phospholipase C but not mitogen-activated protein kinase activity regulated by the m_1 muscarinic acetylchloline receptor. *Proc Natl Acad Sci USA* 90: 4077–4081

Rao VN, Reddy ESP (1992a): A divergent *ets*-related protein, Elk-1, recognizes similar c-*ets*-1 proto-oncogene target sequences and acts as a transcriptional activator. *Oncogene* 7: 65–70

Rao VN, Reddy ESP (1992b): *elk*-1 domains responsible for autonomous DNA binding, SRE:SRF interaction and negative regulation of DNA binding. *Oncogene* 7: 2335–2340

Rao VN, Reddy ESP (1993): Δelk-1, a variant of *Elk*-1, fails to interact with the serum response factor and binds to DNA with modulated specificity. *Cancer Res* 53: 215–220

Reason AJ, Morris HR, Panico M, Marais R, Treisman RH, Haltiwanger RS, Hart GW, Kelly WG, Dell A (1992): Localization of *O*-GlcNAc modification on the serum response transcription factor. *J Biol Chem* 267: 16911–16921

Rivera VM, Greenberg ME (1990): Growth factor-induced gene expression: the ups and downs of c-*fos* regulation. *New Biol* 2: 751–758

Rivera VM, Miranti CK, Misra RP, Ginty DD, Chen R-H, Blenis J, Greenberg ME (1993): A growth factor-induced kinase phosphorylates the serum response factor at a site that regulates its DNA-binding activity. *Mol Cell Biol* 13: 6260–6273

Rohan PJ, Davis P, Moskaluk CA, Kearns M, Krutzsch H, Siebenlist U, Kelly K (1993): PAC-1: a mitogen-induced nuclear protein tyrosine phosphatase. *Science* 259: 1763–1766

Rossomando AJ, Dent P, Sturgill TW, Marshak DR (1994): Mitogen-activated protein kinase kinase 1 (MKK1) is negatively regulated by threonine phosphorylation. *Mol Cell Biol* 14: 1594–1602

Ryan WA Jr, Franza BR Jr, Gilman MZ (1989): Two distinct cellular phosphoproteins bind to the c-*fos* serum response element. *EMBO J* 8: 1785–1792

Sato N, Sakamaki K, Terada N, Arai K-i, Miyajima A (1993): Signal transduction by the high affinity GM-CSF receptor: two distinct cytoplasmic regions of the common β subunit responsible for different signaling. *EMBO J* 12: 4181–4189

Satoh T, Nakafuku M, Kaziro Y (1992): Function of Ras as a Molecular Switch in Signal Transduction. *J Biol Chem* 267: 24149–24152

Schalasta G, Doppler C (1990): Inhibition of c-*fos* transcription and phosphorylation of the serum response factor by an inhibitor of phospholipase C-type reactions. *Mol Cell Biol* 10: 5558–5561

Schröter H, Mueller CGF, Meese K, Nordheim A (1990): Synergism in ternary complex formation between the dimeric glycoprotein $p67^{SRF}$, polypeptide $p62^{TCF}$ and the c-*fos* serum response element. *EMBO J* 9: 1123–1130

Schulman H, Lou LL (1989): Multifunctional Ca^{2+}-calmodulin-dependent protein kinase: domain structure and regulation. *Trends Biochem Sci* 14: 62–66

Schwarz-Sommer Z, Huijser P, Nacken W, Saedler H, Sommer H (1990): Genetic control of flower development by homeotic genes in *Antirrhinum majus*. *Science* 250: 931–936

Seger R, Seger D, Lozeman FJ, Ahn NG, Graves LM, Campbell JS, Ericsson L, Harrylock M, Jensen AM, Krebs EG (1992): Human T-cell mitogen-activated protein kinase kinases are related to yeast signal transduction kinases. *J Biol Chem* 267: 25628–25631

Sharrocks AD, Gille H, Shaw PE (1993): Identification of amino acids essential for DNA binding and dimerization in $p67^{SRF}$: implications for a novel DNA-binding motif. *Mol Cell Biol* 13: 123–132

Shaw PE (1992): Ternary complex formation over the c-*fos* serum response element: $p62^{TCF}$ exhibits dual component specificity with contacts to DNA and an extended structure in the DNA-binding domain of $p67^{SRF}$. *EMBO J* 11: 3011–3019

Shaw PE, Frasch S, Nordheim A (1989a): Repression of c-*fos* transcription is mediated through $p67^{SRF}$ bound to the SRE. *EMBO J* 8: 2567–2574

Shaw PE, Schröter H, Nordheim A (1989b): The ability of a ternary complex to form over the serum response element correlates with serum inducibility of the human c-*fos* promoter. *Cell* 56: 563–572

Shibuya H, Yoneyama M, Ninomiya-Tsuji J, Matsumoto K, Taniguchi T (1993): IL-2 and EGF receptors stimulate the hematopoietic cell cycle via different signaling pathways: demonstration of a novel role for c-*myc*. *Cell* 70: 57–67

Sternberg PW, Golden A, Han M (1993): Role of the *raf* proto-oncogene during *Caenorhabditis elegans* vulval development. *Phil Trans Roy Soc (Lond) B* 340: 259–265

Sternweis PC, Smrcka AV (1992): Regulation of phospholipase C by G proteins. *Trends Biochem Sci* 17: 502–506

Stoflet ES, Schmidt LJ, Elder PK, Korf GM, Foster DN, Strauch AR, Getz MJ (1992): Activation of a muscle-specific actin gene promoter in serum-stimulated fibroblasts. *Mol Biol Cell* 3: 1073–1083

Subramaniam M, Schmidt LJ, Crutchfield CE III, Getz MJ (1989): Negative regulation of serum-responsive enhancer elements. *Nature* 340: 64–66

Sukhatme VP, Cao X, Chang LC, Tsai-Morris C-H, Stamenkovich D, Ferreira PCP,

Cohen DR, Edwards SA, Shows TB, Curran T, LeBeau MM, Adamson ED (1988): A zinc finger-encoding gene coregulated with c-fos during growth and differentiation, and after cellular depolarization. *Cell* 53: 37–43

Sun H, Charles CH, Lau LF, Tonks NK (1993): MKP-1 (3CH134), an immediate early gene product, is a dual specificity phosphatase that dephosphorylates MAP kinase in vivo. *Cell* 75: 487–493

Suzuki T, Hirai H, Fujisawa J-I, Fujita T, Yoshida M (1993): A trans-activator Tax of human T-cell leukemia virus type 1 binds to NF-κB p50 and serum response factor (SRF) and associates with enhancer DNAs of the NF-κB site and CArG box. *Oncogene* 8: 2391–2397

Treisman R (1985): Transient accumulation of c-*fos* RNA following serum stimulation requires a conserved 5' element and c-*fos* 3' sequences. *Cell* 42: 889–902

Treisman R (1986): Identification of a protein-binding site that mediates transcription response of the c-*fos* gene to serum factors. *Cell* 46: 567–574

Treisman R (1987): Identification and purification of a polypeptide that binds the c-*fos* serum response element. *EMBO J* 6: 2711–2717

Treisman R (1992): Structure and function of serum response factor. In: *Transcriptional Regulation*, McKnight SL, Yamamoto KR, eds. Cold Spring Harbor: Cold Spring Harbor Laboratory Press

Treisman R (1994): Ternary complex factors: growth factor regulated transcriptional activators. *Curr Opin Genet Dev* 4: 96–101

Treisman R, Ammerer G (1992): The SRF and MCM1 transcription factors. *Curr Opin Genet Dev* 2: 221–226

Treisman R, Marais R, Wynne J (1992): Spatial flexibility in ternary complexes between SRF and its accessory proteins. *EMBO J* 11: 4631–4640

Velcich A, Ziff EB (1990): Functional analysis of an isolated *fos* promoter element with AP-1 site homology reveals cell type-specific transcriptional properties. *Mol Cell Biol* 10: 6273–6282

Wang Z-Q, Ovitt C, Grigoriadis E, Möhle-Steinlein U, Rüther U, Wagner EF (1992): Bone and haematopoietic defects in mice lacking c-*fos*. *Nature* 360: 741–745

Ward Y, Gupta S, Jensen P, Wartmann M, Davis RJ, Kelly K (1994): Control of MAP kinase activation by the mitogen-induced threonine/tyrosine phosphatase PAC1. *Nature* 367: 651–654

Wasylyk B, Hahn SL, Giovane A (1993): The Ets family of transcription factors. *Eur J Biochem* 211: 7–18

Weiss A, Littman DR (1994): Signal transduction by lymphocyte antigen receptors. *Cell* 76: 263–274

Williamson MP, (1994): The structure and function of proline-rich regions in proteins. *Biochem J* 297: 249–260

Winitz S, Russell M, Qian N-X, Gardner A, Dwyer L, Johnson GL (1993): Involvement of Ras and Raf in the G_i-coupled acetylcholine muscarinic m2 receptor activation of mitogen-activated protein (MAP) kinase kinase and MAP kinase. *J Biol Chem* 268: 19196–19199

Wu J, Harrison JK, Vincent LA, Haystead C, Haystead TAJ, Michel H, Hunt DF, Lynch KR, Sturgill TW (1993): Molecular structure of a protein-tyrosine/threonine kinase activating p42 mitogen-activated protein (MAP) kinase: MAP kinase kinase. *Proc Natl Acad Sci USA* 90: 173–177

Yoshida T, Suzuki T, Sato H, Nishina H, Iba H (1993): Analysis of *fra2* gene expression. *Nucl Acids Res* 21: 2715–2721

Zachow K, Conklin KF (1992): CArG, CCAAT, and CCAAT-like protein binding sites in avian retrovirus long terminal repeat enhancers. *J Virol* 66: 1959–1970

Zhang X-f, Settleman J, Kyriakis JM, Takeuchi-Suzuki E, Elledge SJ, Marshall MS, Bruder JT, Rapp UR, Avruch J (1993): Normal and oncogenic p21^{ras} proteins bind to the amino-terminal regulatory domain of c-Raf-1. *Nature* 364: 308–313

Zheng C-F, Guan K-L (1993a): Cloning and characterization of two distinct human extracellular signal-regulated kinase activator kinases, MEK1 and MEK2. *J Biol Chem* 268: 11435–11439

Zheng C-F, Guan K-L (1993b): Properties of MEKs, the kinases that phosphorylate and activate the extracellular signal-regulated kinases. *J Biol Chem* 268: 23933–23939

Zheng, C-F, Guan K-L (1994): Activation of MEK family kinases requires phosphorylation of two conserved Ser/Thr residues. *EMBO J* 13: 1123–1131

Zhu H, Prywes R (1992): Identification of a coactivator that increases activation of transcription by serum response factor and GAL4-VP16 *in vitro*. *Proc Natl Acad Sci USA* 89: 5191–5295

Zidovetzki R, Lester DS (1992): The mechanism of activation of protein kinase C: a biophysical perspective. *Biochim Biophys Acta* 1134: 261–272

Zinck R, Hipskind RA, Pingoud V, Nordheim A (1993): c-*fos* transcriptional activation and repression correlate temporally with the phosphorylation status of TCF. *EMBO J* 12: 2377–2387

3

DRTF1/E2F: A Molecular Switch in Cell Cycle Control

ROWENA GIRLING AND NICHOLAS B. LA THANGUE

Introduction

The cell cycle is a series of phases through which a cell must pass in order to divide. It is usually considered to be composed of four phases: the gap before DNA replication (G1), the DNA synthesis phase (S), the gap after DNA replication (G2) and the mitotic phase (M). In fact these phases represent a series of complex events which are regulated at different levels. One important level of control is the regulation of the transcriptional activity of particular genes during the cell cycle.

RNA polymerase II mediates the transcription of protein-encoding genes and requires three groups of protein factors: general transcription factors, sequence-specific transcription factors, and coactivators. The general transcription factors (TFIIA, TFIIB, TFIID, TFIIE, TFIIF, TFIIH and TFIIJ), when reconstituted with RNA polymerase II in vitro, can support a basal level of transcription (Drapkin et al, 1993). This basal level can be stimulated by sequence-specific transcription factors which typically have a DNA-binding domain that recognises a DNA sequence and an activation domain that interacts with the transcription apparatus (Drapkin et al, 1993). Some sequence-specific transcription factors can interact with general transcription factors, but others require the activity of coactivators which connect the regulatory domains of sequence-specific factors to the general transcription apparatus (Drapkin et al, 1993). To date, the best characterised coactivators are those that are tightly associated and co-purify with the TATA binding protein (TBP), called TBP-associated factors (TAFs) (Goodrich and Tjian, 1994).

During cell cycle progression RNA polymerase II-mediated transcription of certain genes is coordinated with cell cycle events. For example,

INDUCIBLE GENE EXPRESSION, VOLUME 2
P.A. Baeuerle, Editor
© 1995 Birkhäuser Boston

the levels of *dihydrofolate reductase* (DHFR), DNA polymerase α and p34^{cdc2} mRNA peak as cells progress through the G1/S transition reflecting an increase in transcriptional activity (Farnham and Schimke, 1985; Wahl et al, 1988; Dalton, 1992). Recent developments have begun to elucidate one of the potential mechanisms involved in this coordination process. Here we will discuss the evidence that the sequence-specific transcription factor, DRTF1/E2F, plays an important role in this process.

The term DRTF1/E2F describes two related, if not identical, sequence-specific transcription factors which bind to the E2F DNA sequence (-TTTCGCGC-): differentiation-regulated transcription factor 1 (DRTF1), and E2F. E2F was originally defined in HeLa cells as an activity stimulated upon adenovirus type 5 lytic infection, which bound cooperatively to the two E2F sites in the E2A promoter (Kovesdi et al, 1986), the interaction requiring the sites to be appropriately spaced and oriented (Hardy and Shenk, 1989). Independently, DRTF1 was identified in F9 embryonal carcinoma (EC) stem cells as a single E2F-site-binding activity down-regulated as these cells differentiate to parietal endoderm (PE) cells (La Thangue and Rigby, 1987). Given the information now available for both of these activities, it seems likely that DRTF1 is the murine equivalent of human E2F. Thus DRTF1/E2F will be used here in a generic sense to describe DNA-binding activity that interacts with the E2F site. However, the reader should bear in mind that the activity DRTF1/E2F is in fact an expanding group of related transcription factors (Helin et al, 1992; Kaelin et al, 1992; Shan et al, 1992; Girling et al, 1993; Ivey-Hoyle et al, 1993; Lees et al, 1993; Girling et al, 1994), and it is possible that distinct forms of this activity exist in cells.

DRTF1/E2F: a Cell Cycle-Regulated DNA-Binding Activity

Several lines of evidence originally suggested the involvement of DRTF1/E2F in coordinating transcription with cell cycle progression. First, a number of cellular genes that encode proteins necessary for cell cycle progression, DNA synthesis in particular, contain E2F sites in their transcriptional control regions (see Table). They are thus, by implication, potential genes regulated by DRTF1/E2F. However, only a small number are known to contain functionally important E2F-binding sites. For example, the transcriptional activity of DHFR, which is required for the synthesis of purines and thymidylate prior to S phase, increases towards the end of G1 (Farnham and Schimke, 1985). Within the DHFR

Table. E2F Site-Containing Genes

Gene	Expression	Reference
DHFR	G1/S	Blake and Azizkhan, 1989
		Means et al, 1992
		Slansky et al, 1993
Thymidine kinase	G1/S	Kim and Lee, 1991
		Ogris et al, 1993
p34^{cdc2}	G1/S	Dalton, 1992
B-*myb*	G1/S	Lam and Watson, 1993
DNA polymerase α	G1/S	Pearson et al, 1991
Thymidylate synthase	G1/S	Joliff et al, 1991
c-*myc*	early G1*	Thalmeier et al, 1989
		Hiebert et al, 1989
N-*myc*	early G1*	Mudryj et al, 1990
Cyclin D1	Constitutive[†]	Sewing et al, 1993
Rb	Constitutive[†]	Zacksenhaus et al, 1993
Adenovirus E2A	Viral infection	Kovesdi et al, 1986

*in serum-stimulated cells
[†]in some cell types

promoter are two inverted and overlapping E2F sites which have been shown, by mutagenesis, to be important for the correct temporal increase in transcription at the G1/S border (Means et al, 1992; Slansky et al, 1993). Similarly, levels of DNA polymerase α mRNA increase as cells enter S phase, and E2F-binding sites within the promoter region of this gene are necessary and sufficient for G1/S phase regulation (Pearson et al, 1991). The mRNA level of p34^{cdc2}, a mitotic kinase involved in the regulation of progression through the cell cycle (Hunt, 1989), increases at the G1/S boundary and is highest at the S and G2 phases of the cell cycle. Deletion of a region containing an E2F site leads to growth-independent promoter activity (Dalton, 1992), implicating the E2F site in the transcriptional control of cdc2. The profile of cyclin A gene transcription in the cell cycle (Pines and Hunter, 1989) is similar to that of p34^{cdc2} and could therefore, be regulated in a similar way. Finally, the late G1 and S phase regulation of B-*myb* transcription requires the presence of an E2F-binding site since mutation of such a site results in a constitutively active promoter (Lam and Watson, 1993). In general, the genes that are known to be regulated by DRTF1/E2F play an important role in promoting cell cycle progression, arguing that DRTF1/E2F is critical in coordinating cell cycle events. Other promoters contain E2F sites, the importance of which have yet to be established (Li et al, 1994).

Secondly, DRTF1/E2F DNA-binding activity in synchronous cultures of 3T3 cells increases and peaks during S phase, correlating with the increase in transcriptional activity of some E2F site-containing genes

(Mudryj et al, 1990; Shirodkar et al, 1992). The regulation of the biochemical activities of DRTF1/E2F is therefore consistent with a role in the correct temporal regulation of transcription.

Finally, since the early studies it has become apparent that at the molecular level, DRTF1/E2F interacts with a variety of proteins that have an established role in regulating cell cycle progression both in a negative fashion, such as the retinoblastoma tumor suppressor gene product (pRb) (Bandara and La Thangue, 1991; Chellappan et al, 1991; Kaelin et al, 1991) and positively, such as certain cyclins and cyclin-dependent kinases (Mudryj et al, 1991; Devoto et al, 1992; Lees et al, 1992; Bandara et al, 1992). This further implicates DRTF1/E2F in the regulation of the cell cycle.

DRTF1/E2F is a Cellular Target for pRb

One of the first insights which suggested that DRTF1/E2F could integrate cell cycle events with the transcription apparatus came from the exciting discovery that pRb, a negative regulator of cellular proliferation (Weinberg, 1991), can bind to DRTF1/E2F (Bandara and La Thangue, 1991; Chellappan et al, 1991). Much of the excitement arose from the fact that *Rb* is a well characterised tumor suppressor gene. In contrast to oncogenes which actively promote cell division, pRb suppresses proliferation (Weinberg, 1991). In a range of human tumor cells (although not all), its mutation correlates with a loss of pRb's negative growth regulating activity (Weinberg, 1991). The Rb gene product is a 105kD nuclear protein constitutively expressed in many mammalian cells (Bernards et al, 1989), and mutational analysis of pRb has defined a region the integrity of which is essential for negative growth control and is required to bind viral oncoproteins (Hu et al, 1990; Huang et al, 1990; Kaelin et al, 1991). Referred to as the pocket (Kaelin et al, 1991), this region is approximately 400 amino acid residues long and resides towards the C-terminus of the protein (Figure 3.1).

Before the connection between pRb and DRTF1/E2F was made, it had been known for some time that viral oncoproteins, such as adenovirus E1A, SV40 large T antigen (large T), and the E7 proteins of certain human papilloma viruses, sequester pRb in a pocket-dependent manner (Dyson et al, 1989a, 1989b; Ewen et al, 1989, 1991), a mechanism widely believed to promote cellular proliferation. The pocket region can be subdivided into two segments, termed A and B, separated by a region referred to as the spacer (Figure 3.1). The spacer is not required for the binding of E1A or large T since mutation of the spacer, or substitution of

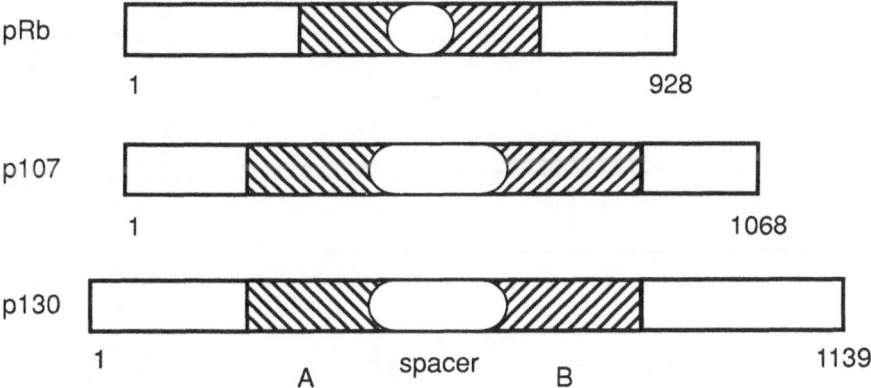

Figure 3.1 Pocket Proteins
Diagrammatic representation of pRb, p107 and p130. The pocket region is subdivided into
segments A and B, separated by a region referred to as the spacer.

unrelated sequences, does not affect the binding of these viral proteins
(Hu et al, 1990; Huang et al, 1990). Significantly, most naturally
occurring mutations of pRb, whether single amino acid substitutions,
large C-terminal truncations or chromosomal rearrangements, alter the
pocket region (Weinberg, 1991). Mutation of the pocket region or
sequestration of pRb by viral proteins have the common effect of
releasing cells from negative growth control (Figure 3.2). A possible
explanation is that these viral oncoproteins evolved structurally to mimic
cellular proteins recognised by the pRb pocket region. Thus, by sequestering pRb, these viral proteins divert pRb from regulating its normal
cellular substrates, thereby overcoming its negative growth regulating
effects.

One level through which the biological activity of pRb is believed to be
regulated is by phosphorylation since in some cells the level of protein
remains constant during the cell cycle while undergoing a cyclical
variation in phosphorylation state (DeCaprio et al, 1992). Thus, in
early G1 the predominant form of pRb is hypophosphorylated, but as
the cycle progresses to the G2/M transition, the overall level of pRb
phosphorylation increases (DeCaprio et al, 1992) until finally, late in M
phase, pRb is dephosphorylated (Ludlow et al, 1990, 1993). Since the
negative effects of pRb are exerted during early cell cycle progression, the
hypophosphorylated state may confer on pRb the ability to restrain cell
cycle progression (Goodrich et al, 1991). The precise phosphorylation
events that occur in vivo, their functional relevance and the specific
kinases and phosphatases involved have yet to be established. However, it
is known that many of the sites in pRb that are phosphorylated in a cell

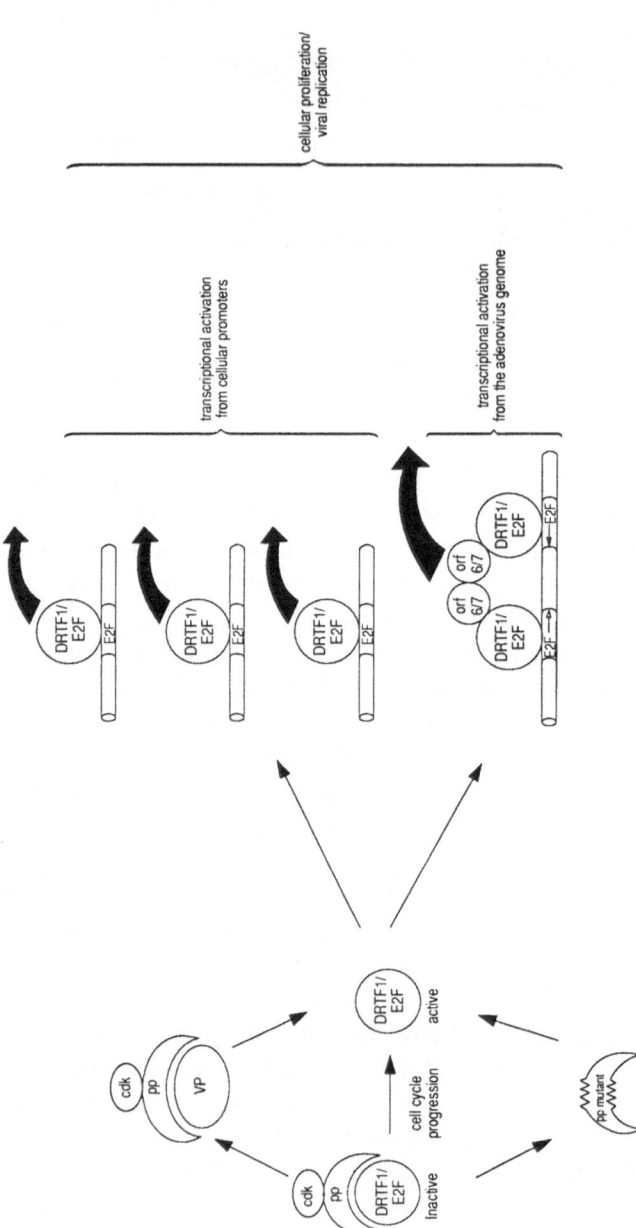

Figure 3.2 Regulation of DRTF1/E2F

The transcription factor DRTF1/E2F is inactive when complexed with a pocket protein (pp), such as pRb, p107 or p130. During cell cycle progression, transcriptionally active DRTF1/E2F is released from the pp complex, possibly due to the activity of cyclin-dependent kinases (cdks) which are recruited by cyclins to the complex. Viral oncoproteins (e.g. E1A, large T, E7) may also release active DRTF1/E2F by sequestering pps. Mutant pRbs, which occur in some tumor cells, fail to repress DRTF1/E2F transcription activity. Active DRTF1/E2F increases the transcription of cellular genes containing E2F sites (as indicated by the large arrows). In adenovirus infected cells an additional interaction occurs between the E4 orf 6/7 gene product and two molecules of DRTF1/E2F. This interaction enhances the binding of DRTF1/E2F or two appropriately spaced and oriented E2F-binding sites such as in the E2A promoter and therefore maintains the transcriptional activity of viral genes during infection.

cycle-dependent manner conform to the cyclin-dependent kinase (cdk) consensus site, thereby implicating members of the cdk group of kinases in regulating the activity of pRb. Furthermore, a kinase containing the conserved PSTAIRE amino acid motif, found in the cdc-2 class of cdks, is associated with pRb (Hu et al, 1992) and indeed cdc2-like kinases can phosphorylate pRb in vitro (Lees et al, 1991).

Cyclin-dependent kinases associate with DRTF1/E2F

Cdks are enzymatically active only if complexed with a member of a group of evolutionarily conserved proteins, called cyclins, which undergo periodic accumulation and destruction during the cell cycle (Hunt, 1989). Some cyclins are required for cell cycle progression (Giordano et al, 1989; Pines and Hunter, 1990; Hunter and Pines, 1991), and it is believed that each cyclin regulates a different transition point during cell cycle progression. For example, cyclin A is thought to regulate progression into S phase since inactivation of cyclin A, either through antibody or antisense approaches, prevents cells from completing S phase (Girard et al, 1991; Pagano et al, 1992); cyclin A also has a role later in the cell cycle in regulating M phase (Pagano et al, 1992). In support of the involvement of these cyclins in the phosphorylation and regulation of pRb, it has been shown that coexpression of cyclins A and E can override pRb-induced cell cycle arrest (Hinds et al, 1992; Ewen et al, 1993) in a fashion which correlates with increased pRb phosphorylation (Hinds et al, 1992). More directly, in vitro, cyclin A can recruit the catalytic kinase subunit $p33^{cdk2}$, a member of the cdc2-related family of protein kinases, to the DRTF1/E2F DNA-binding complex (Bandara et al, 1992) establishing that cyclins can target cdks to pRb, and thus activate kinase activity. In turn, this may influence the ability of pRb to interact with cellular targets such as DRTF1/E2F.

Of the increasing number of cyclins identified, several have been classified as G1-cyclins, that is, they form active cyclin-kinase complexes in G1 which promote entry into S phase (Sherr, 1993). Of particular interest have been the D-type cyclins which can interact with cdk2, cdk4, cdk5 and cdk6 (Matsushime et al, 1992; Xiong et al, 1992; Bates et al, 1994). All three D-type cyclins can activate cdk4 in vitro and coexpression with cdk4, or D2 and D3 with cdk2, in insect cells generates a kinase activity which efficiently phosphorylates pRb (Kato et al, 1993; Matsushime et al, 1992; Ewen et al, 1993). In vivo cdk4 may be the more relevant partner for D-type cyclins since in some cell types, cdk2 appears to be complexed with cyclins E and A (Koff et al,

1991; Dulic et al, 1992; Rosenblatt et al, 1992). Furthermore, in vitro pRb can interact efficiently with cyclins D2 and D3, an interaction dependent upon the intregrity of the pocket region (Kato et al, 1993; Ewen et al, 1993; Dowdy et al, 1993). Significantly, when D2 is expressed with *Rb* in *Rb-/-* cells (cells containing mutant allelles of *Rb*), pRb becomes hyperphosphorylated, and growth arrest is relieved (Hinds et al, 1992; Ewen et al, 1993). It is possible therefore, that in addition to cyclins A and E, D-type cyclins together with the appropriate catalytic subunits, are involved in regulating the phosphorylation level of Rb protein. As for the potential phosphatase(s) involved in dephosphorylation at M phase, a role for serine and threonine phosphoprotein phosphatase type 1 has been suggested, since a pRb-specific mitotic phosphatase activity can be inhibited by okadaic acid and protein phosphatase inhibitors 1 and 2 (Ludlow et al, 1993).

Until recently relatively little was known about the molecules that interact with pRb, and the mechanisms by which pRb exerts its biological effect of negative growth control. When pRb was found in a complex with DRTF1/E2F, a possible mechanism was implied: that the negative proliferative effects of pRb are exerted through its ability to regulate the transcriptional activity of DRTF1/E2F. This was indeed shown to be the case. Thus, by introducing the wild-type protein into *Rb-/-* tumor cells we, and others, found that the expression of wild-type pRb represses the transcriptional activity of DRTF1/E2F (Zamanian and La Thangue, 1992; Hiebert et al, 1992). Furthermore, adenoviral E1A can override the pRb-mediated repression and predictably this requires the pRb-binding regions. Thus, E1A and pRb have opposing effects on the activity of DRTF1/E2F; pRb represses, whereas E1A activates. Finally, since all the naturally occurring mutant Rb proteins studied so far fail to bind to and repress DRTF1/E2F activity (Zamanian and La Thangue, 1992; Hiebert et al, 1992), its normal control would seem likely to be deregulated in tumor cells.

DRTF1/E2F and viral replication

The interaction between pRb and DRTF1/E2F is disrupted by viral proteins E1A, large T and E7 proteins (Bagchi et al, 1990; Bandara et al, 1991). Two short but highly conserved regions, CR1 and CR2, in all these viral proteins, are necessary to disrupt pRb-DRTF1/E2F complexes (Bagchi et al, 1990; Bandara et al, 1991). Previous studies had shown these regions are also necessary for a wide range of biological effects including induction of DNA synthesis, cellular immortalisation and transformation (Moran and Mathews, 1987). It is possible that these viral proteins sequester pRb to overcome the repression of the transcription of cellular genes containing E2F sites in their promoter regions thus activating gene expression and hence creating an intracellular environment which allows efficient viral replication (Figure 3.2). Furthermore, given the nature of the genes regulated by E2F-binding sites, deregulation of their expression would also be expected to promote cell cycle progression and, in this respect, is consistent with the ability of E1A, large T antigen and E7 proteins to transform and immortalise cells.

Adenovirus is particularly adept at subverting cellular DRTF1/E2F for its own purposes and is the only virus known to contain functional E2F-binding sites in its genome. First, as mentioned above, the E1A protein sequesters pRb thus releasing transcriptionally active DRTF1/E2F, ensuring that the levels of cellular proteins are sufficiently high to allow replication. During infection a second viral protein, a product of the early E4 transcription unit referred to as orf 6/7, then enhances the binding of DRTF1/E2F to two appropriately positioned E2F sites within the adenovirus E2A promoter (Figure 3.2) (Hardy and Shenk, 1989; Huang and Hearing, 1989; Marton et al, 1990; Raychaudhuri et al, 1990). This mechanism maintains the transcriptional activity of the E2A promoter in infected cells.

DRTF1/E2F is a Cellular Target for Rb-Related Proteins

Two pRb-related cellular proteins, identified as binding to adenoviral E1A through similar domains as those required to bind pRb (Ewen et al, 1991; Cobrinik et al, 1993; Hannon et al, 1993; Li et al, 1993; Mayol et al, 1993), have been isolated and named according to their molecular weight, p107 and p130. Although they both show significant sequence similarity to pRb, p130 is more closely related to p107 (Li et al, 1993). Overall p130 and p107 show greater than 50% amino acid residue identity, whereas p130 and pRb show only 22% identity (Li et al, 1993). The pocket regions in particular, show a high degree of sequence similarity but differ in size (Figure 3.1). Thus, in pRb the spacer consists of approximately 60 amino acid residues while that of p107 and p130 is considerably larger, at approximately 200 residues. Sequence similarity between the spacers of p107 and p130 suggests some functional similarity to one another but possibly different to pRb. In this respect, the p107 spacer, but not that of pRb, can specifically bind cyclin A in vitro (Ewen et al, 1991, 1992). The overall sequence similarity of these Rb-related or pocket proteins suggests they share common biological effects. In support of this, both p107 and p130, like pRb, bind to DRTF1/E2F (Shirodkar et al, 1992; Cobrinik et al, 1993), an interaction modulated by viral oncoproteins (Shirodkar et al, 1992; Hannon et al, 1993). Furthermore p107, again like pRb, regulates the transcriptional activity of DRTF1/E2F, a process which correlates with arrest of cell cycle progression (Hinds et al, 1992; Schwarz et al, 1993; Zamanian and La Thangue, 1993; Zhu et al, 1993). Such negative growth control, mediated through the regulation of DRTF1/E2F, is likely to be a common property of Rb-related proteins. As yet there is also no clear evidence to indicate that the genes encoding p107 and p130, like pRb, are mutated in tumor cells although p130 has been broadly mapped to a region that undergoes allelic loss in a variety of human tumors (Li et al, 1993).

Whether p107 and/or p130 are, like pRb, also regulated by phosphorylation remains to be established, although this would seem a likely possibility given the presence of cdk consensus sites and their periodic association during the cell cycle with cyclin-kinase complexes (Lees et al, 1992; Shirodkar et al, 1992; Cobrinik et al, 1993; Schwarz et al, 1993). Thus, cyclins A and E can stably bind to p107-DRTF1/E2F complexes in vivo (Lees et al, 1992), recruit $p33^{cdk2}$ to the complexes and activate kinase activity (Cao et al, 1992; Devoto et al, 1992; Faha et al, 1993). As cells approach G1/S the p107-DRTF1/E2F complex associates with cyclin $E/p33^{cdk2}$. Later, as cells enter S phase this association decreases, and the p107-DRTF1/E2F complex associates with cyclin $A/p33^{cdk2}$ (Lees et al, 1992). Similarly the p130-DRTF1/E2F complex associates with cyclin $A/p33^{cdk2}$ and cyclin $E/p33^{cdk2}$ but in a temporally different manner, occurring predominantly in G1 (Cobrinik et al, 1993). The presence of $p33^{cdk2}$ in these complexes suggests a role for phosphorylation either in the regulation of the pocket protein (as discussed earlier), and possibly its interaction with DRTF1/E2F during the cell cycle, or in the direct phosphorylation of a substrate such as DRTF1/E2F.

Currently it is thought possible that the biological activities of p107 and pRb in growth control may be mediated through different mechanisms. For instance in certain cells, where both pRb and p107 can inhibit E2F-binding site-dependent transcription, only p107 blocks proliferation (Zhu et al, 1993). Furthermore, growth arrest by pRb, but not p107, is rescued by expression of cyclin A or E (Hinds et al, 1992; Zhu et al, 1993), and the growth suppressing activity of p107, unlike pRb, does not depend on its ability to associate with E1A (Zhu et al, 1993). Thus although p107, p130 and, perhaps, other pocket proteins are related to pRb, their mechanisms of action may be subtly different.

DRTF1/E2F-Pocket Protein Interactions and the Cell Cycle

During the cell cycle each pocket protein has its own characteristic profile of interactions with DRTF1/E2F, suggesting that transcriptional regulation by particular pocket proteins is temporally regulated. For instance p107-DRTF1/E2F complexes are prevalent throughout G1 and into S phase (Lees et al, 1992; Shirodkar et al, 1992; Schwarz et al, 1993) while p130-DRTF1/E2F complexes appear predominantly in G0 and G1 (Cobrinik et al, 1993). It is possible that each pocket protein controls a particular transition point during the cell cycle, perhaps acting as a checkpoint (that is, a control point through which the cell must pass before entering the next phase of the cell cycle).

Furthermore, the regulation of DRTF1/E2F by different pocket proteins suggests a mechanism by which distinct cell cycle signals can be relayed to the transcription apparatus and establishes a potential link between the cell cycle and cell cycle-regulated transcription. It is possible therefore that DRTF1/E2F plays a central role in integrating cell cycle signals with the transcription apparatus. The focus of recent research has been to elucidate the molecular composition of this important cell cycle-regulated and -regulating transcription factor, and recent exciting advances in this area are discussed below.

The Heterodimeric Nature of DRTF1/E2F

DRTF1/E2F remained a poorly characterised transcription factor, defined by its DNA-binding specificity in cell extracts until a DNA-binding polypeptide, named E2F-1, was isolated by virtue of its ability to bind pRb (Helin et al, 1992; Kaelin et al, 1992; Shan et al, 1992). E2F-1 was molecularly characterised and shown to have some of the properties expected for a component of DRTF1/E2F. For instance, E2F-1 binds pRb both in vitro and in vivo, the binding being modulated by viral oncoproteins (Helin et al, 1992). It also binds specifically to E2F DNA-binding sites and can trans-activate E2F site-dependent promoters (Helin et al, 1992; Kaelin et al, 1992; Shan et al, 1992). Further characterisation indicated that little or no E2F-1 mRNA is detectable in quiescent cells. However, levels reach a maximum at the G1/S boundary showing that E2F-1 RNA is influenced by the rate of cellular proliferation (Slansky et al, 1993). This may have functional significance for the regulation of certain E2F-responsive genes, such as the DHFR promoter (a promoter in which E2F sites are required for its correct temporal expression), since the increase in E2F-1 mRNA correlates with transcription of the DHFR

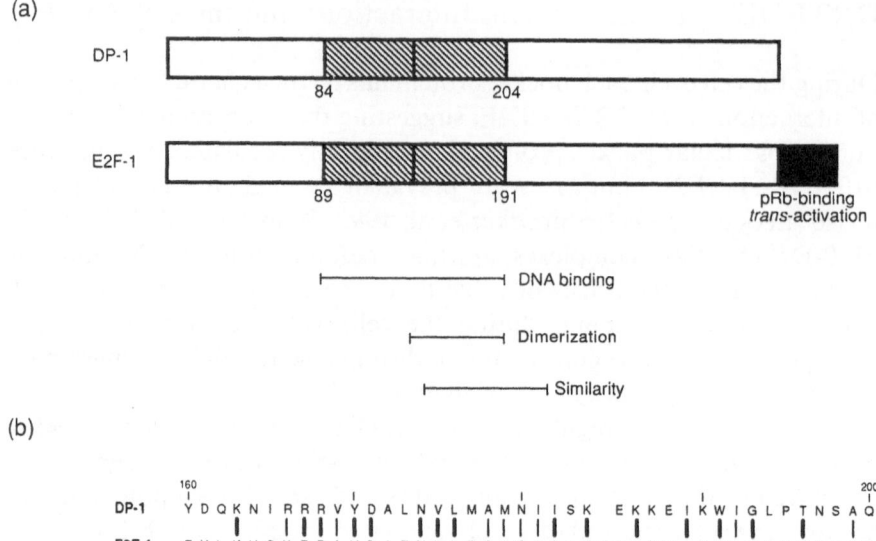

Figure 3.3 Comparison of DP-1 and E2F-1 Proteins
(a) Central to both proteins is a DNA-binding domain which can be subdivided to show the region required for dimerization of DP-1 and E2F-1. At the C-terminus of E2F-1 are overlapping pRb-binding and trans-activation domains.
(b) The region of similarity in DP-1 and E2F-1: heavy lines indicate identical and light lines similar amino acid residues. This region is conserved in all known DP and E2F proteins.

gene (Slansky et al, 1993; Li et al, 1994). In fact E2F-1 can trans-activate other known E2F site-responsive genes, but not all (Li et al, 1994). The expression pattern of *E2F-1* implies the presence of E2F-1 in DRTF1/E2F occurs only at certain points in the cell cycle.

Although E2F-1 is influenced by cell cycle progression, it does, nevertheless, appear to be able to influence cell cycle progression. For example, over-expression of E2F-1 causes cells to progress into S-phase, while micro-injection of two E2F-1 mutant cDNAs which are unable to trans-activate, does not induce S-phase entry (Johnson et al, 1993). The E2F-1 protein therefore has a dominant effect on cell cycle progression. However, these experiments do not address the regulation of endogenous E2F-1, about which little is known.

Organisation of E2F-1

An analysis of E2F-1 has defined a number of functional domains (Helin et al, 1992; Kaelin et al, 1992; Shan et al, 1992). A DNA-binding region in the N-terminal half of E2F-1 (Figure 3.3a) has a basic stretch of residues followed by a potential helix-helix motif, the latter being interrupted by a sequence which may form a loop. C-terminal to the DNA binding domain is a leucine zip-type motif, another motif which may be involved in protein-protein interactions but which, in this case, has no known function. The pRb-binding domain is present at the C-terminus (Figure 3.3a) (Helin et al, 1992; Kaelin et al, 1992; Shan et al, 1992) where 18 amino acid residues (409–426) are necessary to mediate the binding of pRb in vitro (Helin et al, 1992). Interestingly, a trans-activation domain was mapped to 69 amino acid residues at the C-terminus, overlapping the pRb-binding site (Figure 3.3a) (Kaelin et al, 1992). Such an overlap of functional domains suggested pRb could prevent transcriptional activation by physically excluding proteins which interact with the activation domain. Evidence to support this view came from muta-genic studies of E2F-1 where, by altering residues in the C-terminal region and abolishing the binding of pRb, it was shown that the activity of E2F-1 was no longer repressed by pRb; only when the pRb-binding domain was intact could E2F-1-mediated trans-activation be inhibited by coexpression of pRb (Flemington et al, 1993; Helin et al, 1993a). But how does E2F-1 trans-activate, and by what mechanism does pRb influence this activity?

A mechanism of trans-activation was proposed when it was noted that the C-terminal regions of TBP and TFIIB had significant sequence similarity to the pocket region of pRb (Hagemeier et al, 1993a; 1993b) and that certain mutations in the activation domain of E2F-1 had parallel effects on TBP and pRb binding (Hagemeier et al, 1993b). These observations suggested that TBP, TFIIB and pRb recognise similar structural features and therefore may interact with common targets such as E2F-1. A simple model to explain pRb repression of E2F-1 would be that the interaction between E2F-1 and TBP (and possibly TFIIB) is required for transcriptional activation, and subsequent binding of pRb could physically exclude TBP from contacting the activation domain (Figure 3.4). It is likely however that DRTF1/E2F can activate transcription through numerous pathways and the E2F-1/TBP interaction represents just one.

Although E2F-1 had some of the properties expected of DRTF1/E2F, several lines of evidence suggested the existence of multiple E2F site-binding activities. For example, in cell extracts DRTF1/E2F is found complexed with at least three different pocket proteins, yet E2F-1 can bind only pRb (Helin et al, 1992). One possible explanation for this would be the presence of distinct, but related activities, each possessing specificity for different pocket proteins. Furthermore, E2F-1 mRNA is

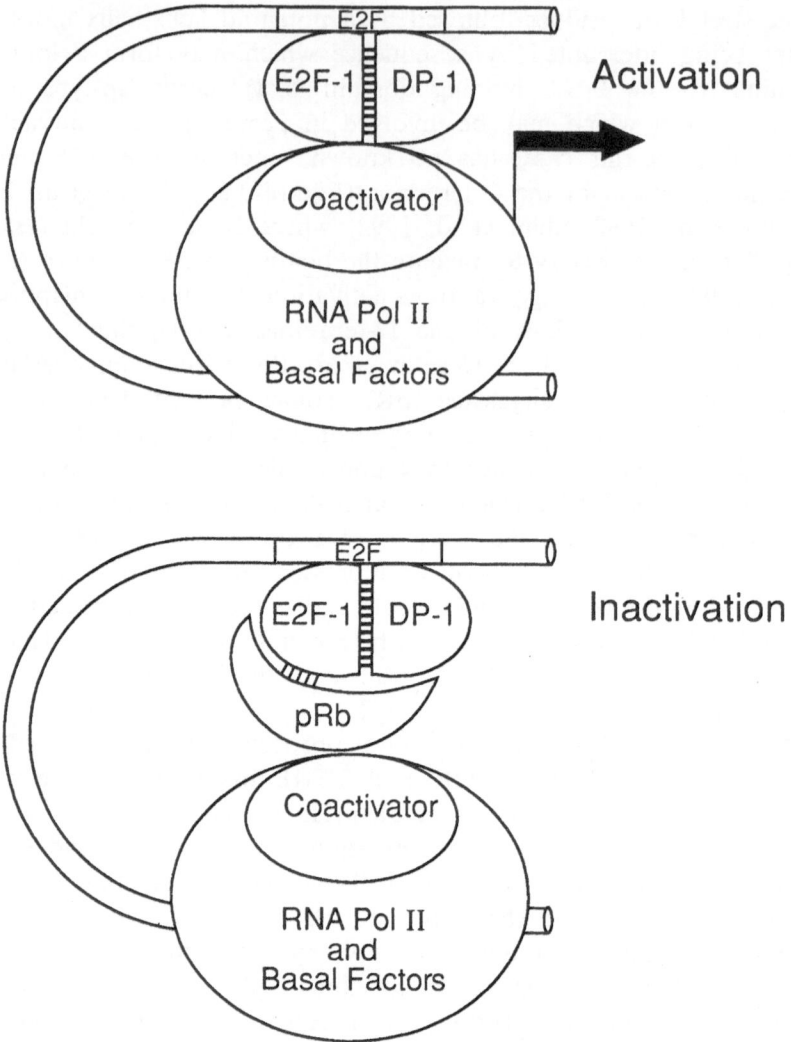

Figure 3.4 Possible Mechanism of Trans-activation by DRTF1/E2F and Regulation by pRb
The E2F-1/DP-1 heterodimer may trans-activate by interacting directly with the transcription apparatus. If E2F-1 contacts the transcriptional machinery via its trans-activation domain, which is overlapped by the pRb-binding domain, then pRb could interfere with this interaction and suppress transcriptional activity.

not detectable in quiescent cells yet these cells contain E2F site-binding activity (Kaelin et al, 1992; Shan et al, 1992), again suggesting that DRTF1/E2F consists of a number of DNA-binding species. Finally, antisera, specific for E2F-1, affect the mobility of only a proportion of E2F-binding site activities (Kaelin et al, 1992).

DP-1: a partner for E2F-1

Following the molecular characterisation of E2F-1 another DNA binding polypeptide, DP-1 (standing for DRTF1-Polypeptide-1), was isolated in our laboratory (Figure 3.3a) (Girling et al, 1993). In these studies purification of E2F site-binding activities from F9 EC cells revealed a number of polypeptides (Shivji and La Thangue, 1990), providing further proof that DRTF1/E2F is composed of multiple DNA-binding species. Micro-sequencing of one of these polypeptides gave information which led to the isolation of DP-1. This novel protein was subsequently shown to bind in vitro to different E2F-binding sites (Girling et al, 1993) as defined by binding-site selection (Chittenden et al, 1991), and is now known to be a component of DRTF1/E2F in many cell types.

During cell cycle progression in 3T3 cells, the DNA-binding activity of DRTF1/E2F undergoes a series of cell cycle-dependent changes (Bandara et al, 1993). The slower migrating form of DRTF1/E2F, in which proteins such as pRb, p107, cyclins A and E are present, is predominant in serum deprived cells (G0), but as the cell cycle progresses through G1, the uncomplexed, transcriptionally active form of DRTF1/E2F, increases and peaks at S phase (Mudryj et al, 1991; Shirodkar et al, 1992; Bandara et al, 1993; Schwartz et al, 1993). Recent studies indicate that DP-1 is a component of most E2F site DNA-binding activities that occur during the 3T3 cell cycle, suggesting that DP-1 is a frequent component of DRTF1/E2F in this cell type.

In 3T3 cells, DP-1 is a polypeptide of 55kD (p55) which can be detected throughout the cell cycle. However, during early cell cycle progression, a slightly faster migrating form of p55, p55L (lower) appears, suggesting that DP-1 is modified posttranslationally during the cell cycle. Since DP-1 has a number of potential phosphorylation sites and associates with cyclin-kinases, it seems likely that it is modified by phosphorylation, an idea confirmed in recent studies (Bandara et al, 1994). Since the changes in phosphorylation occur at a similar time to the increase in DNA-binding activity (Bandara et al, 1994), phosphorylation may regulate the DNA-binding activity of DP-1. Further functional consequences of regulating the phosphorylation level of DP-1 are likely but remain to be elucidated.

In contrast to its regulation during cell cycle progression, DP-1 is regulated in a different manner during the process of differentiation. As F9 EC cells differentiate, the level of p55 decreases correlating with down-regulation of DRTF1/E2F DNA-binding activity and the decrease in the rate of proliferation (La Thangue and Rigby, 1987; Partridge and La

Thangue, 1991). Thus the mechanisms which regulate DP-1 during the cell cycle and differentiation are likely to be mechanistically distinct.

E2F-1 and DP-1 functionally interact

Although DP-1 and E2F-1 are distinct proteins, they do contain a small region of similarity within their DNA-binding domains (Figure 3.3b) (Girling et al, 1993). Interestingly, this region is distantly related to the DNA-binding domains of some yeast cell cycle-regulated transcription factors (La Thangue and Taylor, 1993) and may therefore have a role in specific-sequence recognition. Secondary structure analysis predicts that this region forms an amphipathic α-helix (Girling et al, 1993) and hence has a possible role in dimerization.

Studies of the composition of biochemically purified DRTF1/E2F, suggest that DRTF1/E2F binds to DNA as a heterodimer (Huber et al, 1993), an idea supported by recent studies in which several groups have shown that DP-1 and E2F-1 heterodimerize to create a DNA-binding activity that efficiently binds to the E2F site (Bandara et al, 1993; Helin et al, 1993a; Krek et al, 1993). The interaction between DP-1 and E2F-1 requires the region of similarity (Bandara et al, 1993; Helin et al, 1993a), thus arguing that it is the interface involved in dimerization. Residues within this region may contact DNA, but structure/function experiments show that it must be extended N-terminally in both proteins, for efficient binding of the heterodimer to DNA (Bandara et al, 1993; Helin et al, 1993b; Krek et al, 1993). This additional N-terminal region is rich in α-helices, but shows no sequence similarity with other DNA-binding domains. Importantly, the efficiency of DNA-binding with the heterodimer is greater than with either homodimer indicating that the proteins interact synergistically. The interaction is also important for transcriptional activity because E2F site-dependent transcription is more efficiently activated in the presence of both proteins than with either protein alone. A further consequence of heterodimerization is that pRb binds more efficiently to DP-1/E2F-1 than to E2F-1 alone (Helin et al, 1993b; Krek et al, 1993; Bandara et al, 1994). This may be influenced by a direct interaction between DP-1 and pRb (Bandara et al, 1994). The DP-1/E2F-1 heterodimer also interacts with the adenovirus orf 6/7 protein, resulting in a DNA-binding activity that has the biochemical and functional properties of the adenovirus-infected cellular form of DRTF1/E2F (Bandara et al, 1994). All these properties of DP-1/E2F-1 are consistent with those expected of a physiologically relevant form of DRTF1/E2F, and in fact the DP-1/E2F-1 heterodimer does exist in HeLa cells (Bandara et al, 1993, 1994).

In summary, DP-1 is a frequent and cell cycle-regulated component of DRTF1/E2F DNA-binding complexes. In contrast, E2F-1 is influenced by cell cycle progression and may be limiting at certain points in the cell cycle. The two proteins heterodimerize to create an E2F site-binding activity which has all the hallmarks expected of a physiologically relevant and important form of DRTF1/E2F.

DRTF1/E2F: An Expanding Family of Heterodimeric Transcription Factors

It is now clear that E2F-1 and DP-1 are members of two distinct families of closely related proteins. Here proteins with similarity to E2F-1 are described as being members of the E2F family while those with similarity to DP-1 are described as members of the DP family. To date, the isolation of three E2F family members has been reported. E2F-2 and E2F-3 were identified by low stringency screening of cDNA libraries with the DNA-binding domain of E2F-1 (Ivey-Hoyle et al, 1993; Lees et al, 1993). All members of the E2F family share extensive sequence similarity, particularly across their DNA-binding domains (73% identity) and at the C-terminus of each E2F protein where pRb binds (56% identity). The sequence similarity of E2F-2 and E2F-3 with E2F-1 appears to extend to functional similarity since E2F-2 and E2F-3 can bind to E2F sites and to pRb in vitro and in vivo (Ivey-Hoyle et al, 1993; Lees et al, 1993). They also trans-activate in an E2F site-dependent fashion (Lees et al, 1993). Finally, in the majority of cell lines tested so far, E2F-1, E2F-2 and E2F-3 all appear to be constitutively expressed (Ivey-Hoyle et al, 1993).

A family of DP proteins also exists. Recent studies identified DP-2, by low stringency screening with murine DP-1 cDNA (Girling et al, 1994). Both DP-1 and DP-2 show a high degree of sequence similarity and structural organization although DP-2 is somewhat shorter than DP-1 at the N-terminus. It will be interesting to investigate the possible functional role of this N-terminal region. Heterodimerization of DP-1 and DP-2 with E2F-1, E2F-2 and E2F-3 has so far indicated that their properties are similar because, like DP-1 and E2F-1, the interactions are synergistic with respect to DNA-binding (Girling et al, 1994). However, there do appear to be differences in their expression during development which presumably reflects the functional importance of DP family members.

Like E2F family members, the DP proteins show remarkable sequence similarity. For example, DP-1 and DP-2 are 86% identical at the amino acid residue level (Girling et al, 1994). Sequence analysis of both families shows that the region of similarity between DP-1 and E2F-1, which is to our knowledge unique to E2F and DP proteins, is highly conserved in all members of both families suggesting that all E2F and DP proteins can heterodimerize. As mentioned above, this is indeed the case for all combinations examined so far. We are thus now led to the view that DRTF1/E2F is a group of heterodimeric transcription factors, whose DNA-binding activity arises when a DP protein and an E2F protein interact. The actual number of members within each family has yet to be established; however, the potential for many different heterodimers is already apparent. But what is the biological significance of having so many distinct DP/E2F dimers?

The number of heterodimers in itself suggests that each heterodimer may selectively bind to a specific E2F site-containing promoter. It is equally possible that each dimer binds to all E2F site-containing promoters, and regulation of transcriptional activity is determined by association with other proteins, e.g. surrounding transcription factors or pocket proteins. Pocket proteins are obvious candidates in this respect, since each one may bind preferentially to a given DP/E2F dimer. This is compatible with our earlier discussion in which E2F-1 binds preferentially to pRb, p107, p130 and other, as yet unidentified, pocket proteins may also bind preferentially to particular heterodimers. Cell cycle signals could then be relayed via the pocket protein to the transcription apparatus ensuring that transcription of certain genes is integrated with cell cycle events.

DRTF1/E2F and the Cell Cycle Response

A complex number of afferent signals converge on a cell nucleus leading to the specific transcriptional activation of genes required for cell cycle progression. The recent break-throughs in the characterization of DRTF1/E2F at the molecular level have advanced our understanding of this group of transcription factors in relation to periodic transcriptional activity during the cell cycle. Central to the transduction of signals to the transcription apparatus by DRTF1/E2F is the formation of a heterodimer comprising one member of the E2F family of proteins and one member of the DP family.

Trans-activation by DRTF1/E2F may be regulated in a number of ways. The composition of the heterodimer itself may influence sequence specificity which in turn, may be determined by availability of each E2F or DP protein, for instance in different tissues, where the population of heterodimers which exist in the nucleus may be limited. The temporal interaction of different heterodimers with different pocket proteins could theoretically have a very important role to play by regulating the activity of distinct

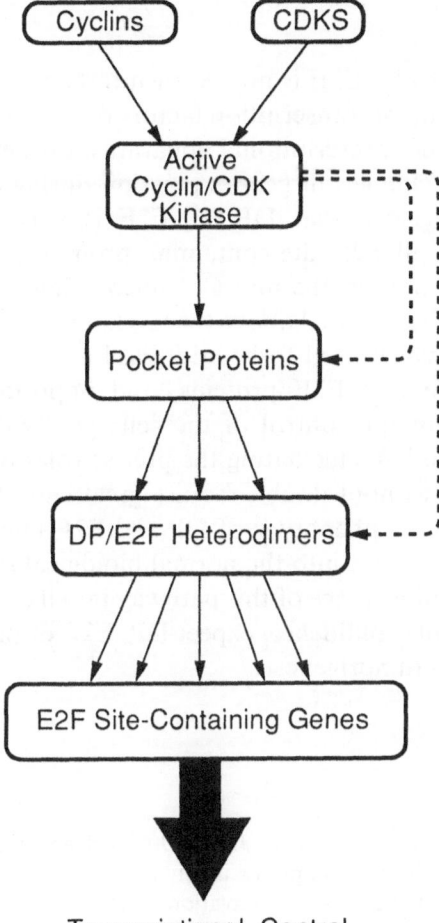

Figure 3.5 An Expanding Number of Regulatory Pathways Control Trans-activation by DRTF1/E2F

The solid arrows indicate the expanding number of possible interactions involved in regulating DRTF1/E2F. An active cyclin/kinase complex may interact with one of a variety of pocket proteins (pps) which, in turn, can interact with a number of DP/E2F dimers. Different heterodimers or different regulated forms of the same heterodimer could then bind to and trans-activate E2F site-containing genes. Broken arrows indicate possible regulatory mechanisms, such as the direct phosphorylation of pps or DP/E2F dimers, which may affect transcriptional activation

heterodimers. The involvement of cyclins and kinases with DRTF1/E2F further integrates the activity of DRTF1/E2F with the machinery of the cell cycle. Kinases associated with the transcription factor complexes could specifically modify either pocket proteins and hence their interaction with DRTF1/E2F or perhaps DRTF1/E2F itself (Figure 3.5).

Conclusions

To summarise, DRTF1/E2F is now known to be a complicated activity composed of a group of transcription factors functioning to integrate cell cycle events with the transcriptional apparatus, thus enabling the cell to make appropriate changes in gene-expression during cell cycle progression. The simplest idea that DRTF1/E2F is a single DNA-binding activity regulating all E2F site-containing promoters can no longer be sustained. We are now in the midst of uncovering its true complexity, which is likely to increase well above the level outlined in this review. The central questions which need to be addressed are: why does the animal need so many DP and E2F proteins, and importantly, how is this biologically relevant for control of the cell cycle? Given the apparent redundancy in function, elucidating the precise roles of these proteins in the cell cycle will, without doubt, require great care. We will obviously await with interest the phenotype of DP and E2F knockout mice which may provide some insight into the normal biological roles in the animal. Given the central importance of this pathway in cell cycle control, we can also, with reasonable confidence, expect DP, E2F or perhaps both, to be deregulated during tumorigenesis.

References

Bagchi S, Raychaudhuri P, Nevins JR (1990): Adenovirus E1A proteins can dissociate heterodimeric complexes involving the E2F transcription factor: A novel mechanism for E1A trans-activation. *Cell* 62: 659–669

Bandara LR, La Thangue NB (1991): Adenovirus E1a prevents the retinoblastoma gene product from complexing with a cellular transcription factor. *Nature* 351: 494–497

Bandara LR, Ademczewski JP, Hunt T, La Thangue NB (1991): Cyclin A and the retinoblastoma gene product complex with a common transcription factor. *Nature* 352: 249–251

Bandara LR, Ademczewski JP, Zamanian M, Hunt T, La Thangue NB (1992): Cyclin A recruits p33^{cdk2} to the cellular transcription factor DRTF1. *J Cell Sci* 16: 77–85

Bandara LR, Buck VM, Zamanian M, Johnston LH, La Thangue NB (1993): Functional synergy between DP-1 and E2F-1 in the cell cycle-regulating transcription factor DRTF1/E2F. *EMBO J* 12: 4317–4324

Bandara LR, Lam EW-F, Sorensen TS, Zamanian M, Girling R, La Thangue NB (1994): DP-1: a cell cycle-regulated and phosphorylated component of transcription factor DRTF1/E2F which is functionally important for recognition by pRb and the adenovirus E4 orf 6/7 protein. *EMBO J* 13: 3104–3114

Bates S, Bonetta L, MacAllan D, Parry D, Holder A, Dickenson C, Peters G (1994): CDK6 (PLSTIRE) and CDK4 (PSK-J3) are a distinct subset of the cyclin-dependent kinases that associate with cyclin D1. *Oncogene* 9: 71–79

Bernards R, Schackleford GM, Gerber MR, Horowitz JM, Friend SH, Schartl M, Bogenmann E, Rapaport JM, McGee T, Dryja TP, Weinberg RA (1989): Structure and expression of the murine retinoblastoma gene and characterisation of its encoded protein. *Proc Natl Acad Sci USA* 86: 6474–6478

Blake MC, Azizkhan J (1989): Transcription factor E2F is required for efficient expression of hamster dihydrofolate reductase gene in vitro and in vivo. *Mol Cell Biol* 9: 4994–5002

Cao L, Faha B, Dembski M, Tsai L-H, Harlow E, Dyson N (1992): Independent binding of the retinoblastoma protein and p107 to the transcription factor E2F. *Nature* 355: 176–179

Chellappan SP, Hiebert S, Mudryj M, Horowitz JM, Nevins JR (1991): The E2F transcription factor is a cellular target for the RB protein. *Cell* 65: 1053–1061

Chittenden T, Livingston DM, Kaelin WG (1991): The T/E1A-binding domain of the retinoblastoma product can interact selectively with a sequence-specific DNA-binding protein. *Cell* 65: 1073–1082

Cobrinik D, Whyte P, Peeper DS, Jacks T, Weinberg RA (1993): Cell cycle-specific association of E2F with the p130 E1A-binding protein. *Genes Dev* 7: 2392–2404

Dalton S (1992): Cell cycle regulation of the human cdc2 gene. *EMBO J* 11: 1797–1804

DeCaprio JA, Furukawa Y, Ajchenbaum F, Griffen JD, Livingston DM (1992): The retinoblastoma susceptibility gene product becomes phosphorylated in multiple stages during cell cycle entry and progression. *Proc Natl Acad Sci USA* 89: 1795–1798

Devoto SH, Mudryj M, Pines J, Hunter T, Nevins JR (1992): A cyclin A-protein kinase complex possesses sequence-specific DNA-binding activity: P33[cdk2] is a component of the E2F-cyclin A complex. *Cell* 68: 167–176

Dowdy SF, Hinds PW, Louie K, Reed SI, Arnold A, Weinberg RA (1993): Physical interaction of the retinoblastoma protein with human D cyclins. *Cell* 73: 499–511

Drapkin R, Merino A, Reinberg D (1993): Regulation of RNA polymerase II transcription. *Curr Opin Cell Biol* 5: 469–476

Dulic V, Lees E, Reed SI (1992): Association of human cyclin E with a periodic G1-S phase protein kinase. *Science* 257: 1958–1961

Dyson N, Buchkovich K, Whyte P, Harlow E (1989a): The cellular p107 protein that binds to adenovirus E1A also associates with the large T antigens of SV40 and JC virus. *Cell* 58: 249–255

Dyson N, Howley PM, Munger K, Harlow E (1989b): The human papilloma virus-16 E7 oncoprotein is able to bind to the retinoblastoma gene product. *Science* 242: 934–937

Ewen ME, Faha B, Harlow E, Livingston DM (1992): Interaction of p107 with cyclin A independent of complex formation of viral oncoproteins. *Science* 255: 85–87

Ewen ME, Ludlow JW, Marsilio E, DeCaprio JA, Millikan RC, Cheng S-H, Paucha E, Livingston DM (1989): An N-terminal transformation-governing sequence of SV40 large T antigen contributes to the binding of both p110[Rb] and a second cellular protein, p120. *Cell* 58: 257–267

Ewen ME, Sluss HS, Sherr CJ, Matsushime H, Kato J, Livingston DM (1993): Functional interactions of the retinoblastoma protein with mammalian D-type cyclins. *Cell* 73:487–497

Ewen ME, Xing Y, Lawrence JB, Livingston DM (1991): Molecular cloning, chromosomal mapping and expression of the cDNA for p107, a retinoblastoma gene product-related protein. *Cell* 66: 1155–1164

Faha B, Harlow E, Lees E (1993): The adenovirus E1A-associated kinase consists of cyclin E-p33^{cdk2} and cyclin A-p33^{cdk2}. *J Virol* 67: 2456–2465

Farnham PJ, Schimke RT (1985): Transcriptional regulation of mouse dihyrofolate reductase in the cell cycle. *J Biol Chem* 260: 7675–7680

Flemington EK, Speck SH, Kaelin WG (1993): E2F-1 mediated trans-activation is inhibited by complex formation with the retinoblastoma susceptibility gene product. *Proc Natl Acad Sci USA* 90: 6914–6918

Giordano A, Whyte P, Harlow E, Franza BJ, Beach D, Draetta G (1989): A 60kD cdc2-associated polypeptide complexes with the E1A proteins in adenovirus-infected cells. *Cell* 58: 981–990

Girard F, Strausfeld U, Fernandez A, Lamb NJC (1991): Cyclin A is required for the onset of DNA replication in mammalian fibroblasts. *Cell* 67: 1169–1179

Girling R, Bandara LR, Ormondroyd E, Lam EW-F, Surendra K, Mohun T, La Thangue NB (1994): Molecular characterisation of *Xenopus leavis* DP proteins. *Mol Biol Cell* 5: 1081–1092

Girling R, Partridge JF, Bandara LR, Burden N, Totty NF, Hsuan JJ, La Thangue NB (1993): A new component of the transcription factor DRTF1/E2F. *Nature* 362: 83–87

Goodrich DW, Wang NP, Qian Y-W, Lee EY-HP, Lee W-H (1991): The retino-blastoma gene product regulates progression through the G1 phase of cell cycle. *Cell* 67: 293–302

Hagemeier C, Bannister AJ, Cook A, Kouzarides T (1993a): The activation domain of transcription factor PU.1 binds the retinoblastoma (RB) protein and the transcription factor TFIID in vitro: RB shows sequence similarity to TFIID and TFIIB. *Proc Natl Acad Sci USA* 90: 1580–1584

Hagemeier C, Cook A, Kouzarides T (1993b): The retinoblastoma protein binds E2F residues required for activation in vivo and TBP binding in vitro. *Nuc Acid Res* 21: 4998–5004

Hannon GJ, Demetrick D, Beach D (1993): Isolation of the Rb-related p130 through its interaction with CDK2 and cyclins. *Genes Dev* 7: 2378–2391

Hardy S, Shenk T (1989): E2F from adenoviral-infected cells binds cooperatively to DNA containing two properly oriented and spaced recognition sites. *Mol Cell Biol* 9: 4495–4506

Helin K, Harlow E, Fattaey A (1993a): Inhibition of E2F-1 trans-activation by direct binding of the retinoblastoma protein. *Mol Cell Biol* 13: 6501–6508

Helin K, Lees JA, Vidal M, Dyson N, Harlow E, Fattaey A (1992): A cDNA encoding a Rb-binding protein with properties of the transcription factor E2F. *Cell* 70: 337–350

Helin K, Wu C-L, Fattaey AR, Lees JA, Dynalacht BD, Ngwu C, Harlow E (1993b): Heterodimerization of the transcription factors E2F-1 and DP-1 leads to cooperative trans-activation. *Genes Dev* 7: 1850–1861

Hiebert SW, Chellappan SP, Horowitz JM, Nevins JR (1992): The interaction of Rb with E2F coincides with an inhibition of the transcriptional activity of E2F. *Genes Dev* 6: 177–185

Hiebert SW, Lipp M, Nevins JR (1989): E1A-dependent trans-activation of the human MYC promoter is mediated by the E2F factor. *Proc Natl Acad Sci USA* 86: 3594–3598

Hinds PW, Mittnacht S, Dulic V, Arnold A, Reed SI, Weinberg RA (1992): Regulation of retinoblastoma protein functions by ectopic expression of human cyclins. *Cell* 70: 993–1006

Hu Q, Dyson N, Harlow E (1990): The regions of the retinoblastoma protein needed for binding to adenovirus E1A or SV40 large T antigen are common sites for mutations. *EMBO J* 9: 1147–1155

Hu Q, Lees JA, Buchkovich KJ, Harlow E (1992): The retinoblastoma protein physically associates with the human cdc2 kinase. *Mol Cell Biol* 12: 971–980

Huang MM, Hearing P (1989): The adenovirus early region 4 open reading frame 6/7 protein regulates the DNA-binding activity of the cellular transcription factor, E2F, through a direct complex. *Genes Dev* 3: 1699–1710

Huang S, Wang N, Tseng BY, Lee W-H, Lee EY-HP (1990): Two distinct and frequently mutated regions of retinoblastoma protein are required for binding to SV40 T antigen. *EMBO J* 9: 1815–1822

Huber HE, Edwards GM, Goodhart PJ, Patrick DR, Huang PS, Ivey-Hoyle M, Banett SF, Oliff A, Heimbrook DC (1993): Transcription factor E2F binds DNA as a heterodimer. *Proc Natl Acad Sci USA* 90: 3525–3529

Hunt T (1989): Maturation promoting factor, cyclins and the control of M-phase. *Curr Opin Cell Biol* 1: 268–274

Hunter T, Pines J (1991): Cyclins and cancer. *Cell* 66: 1071–1074

Ivey-Hoyle M, Conroy R, Huber HE, Goodhart PJ, Oliff A, Heimbrook DC (1993): Cloning and characterization of E2F-2, a novel protein with the biochemical properties of transcription factor E2F. *Mol Cell Biol* 13: 7802–7812

Johnson DG, Schwarz JK, Cress WD, Nevins JR (1993): Expression of transcription factor E2F1 induces quiescent cells to enter S phase. *Nature* 365: 349–352

Jolliff K, Li Y, Johnson LF (1991): Multiple protein-DNA interactions in the TATA-less mouse thymidylate synthase promoter. *Nuc Acid Res* 19: 2267–2274

Kaelin WG, Krek W, Sellers WR, DeCaprio JA, Ajchenbaum F, Fuchs CS, Chittenden T, Li Y, Farnham PJ, Blanar MA, Livingston DM, Flemington EK (1992): Expression cloning of a cDNA encoding a retinoblastoma-binding protein with E2F-like properties. *Cell* 70: 351–364

Kaelin WG, Pallas DC, DeCaprio JA, Kaye FJ, Livingston DM (1991): Identification of cellular proteins that can interact specifically with the T/E1A-binding regions of the retinoblastoma gene product. *Cell* 64: 521–532

Kato J, Matsushime H, Hiebert SW, Ewen ME, Sherr CJ (1993): Direct binding of cyclin D to the retinoblastoma gene protein (pRb) and pRB phosphorylation by the cyclin-dependent kinase CDK4. *Genes Dev* 7: 331–342

Kim YK, Lee AS (1991): Identification of a 70-base-pair cell cycle regulatory unit within the promoter of the human thymidine kinase gene and its interaction with cellular factors. *Mol Cell Biol* 11: 2296–2302

Koff A, Cross F, Fisher A, Schumacher J, Leguellec K, Phillipe M, Roberts JM (1991): Human cyclin E, a new cyclin that interacts with two members of the CDC2 gene family. *Cell* 66: 1217–1228

Kovesdi I, Reichel R, Nevins JR (1986): Identification of a cellular transcription factor involved in E1A trans-activation. *Cell* 45: 219–228

Krek W, Livingston DM, Shirodkar S (1993): Binding to DNA and the retinoblastoma gene product promoted by complex formation of different E2F family members. *Science* 262: 1557–1560

La Thangue NB, Rigby PWJ (1987): An adenovirus E1A-like transcription factor is regulated during the differentiation of murine embryonal carcinoma stem cells. *Cell* 49: 507–513

La Thangue NB, Taylor W (1993): A structural similarity between mammalian and yeast transcription factors for cell-cycle regulated genes. *Trends Cell Biol* 3: 75–76

Lam EW-F, Watson RJ (1993): An E2F binding site mediated cell cycle repression of mouse B-*myb* transcription. *EMBO J* 7: 2705–2713

Lees JA, Buchkovich KJ, Marshak DR, Anderson WW, Harlow E (1991): The retinoblastoma protein is phosphorylated on multiple sites by human cdc2. *EMBO J* 10: 4279–4290

Lees E, Faha B, Dulic V, Reed SI, Harlow E (1992): Cyclin E/cdk2 and cyclin A/cdk2 kinases associate with p107 and E2F in a temporally distinct manner. *Genes Dev* 6: 1874–1885

Lees JA, Saito M, Vidal M, Valentine M, Look T, Harlow E, Dyson N, Helin K (1993): The retinoblastoma protein binds to a family of E2F transcription factors. *Mol Cell Biol* 13: 7813–7825

Li Y, Graham C, Lacy S, Duncan AMV, Whyte P (1993): The adenovirus E1A-associated 130-kD protein is encoded by a member of the retinoblastoma gene family and physically interacts with cyclins A and E. *Genes Dev* 7: 2366–2377

Li Y, Slansky JE, Myers DJ, Drinkwater NR, Kaelin WG, Farnham PJ (1994): Cloning, chromosomal location and characterisation of mouse E2F1. *Mol Cell Biol* 14: 1861–1869

Ludlow JW, Glendening CL, Livingston DM, DeCaprio JA (1993): Specific enzymatic dephosphorylation of the retinoblastoma protein. *Mol Cell Biol* 13: 367–372

Ludlow JW, Shan J, Pipas JM, Livingston DM, DeCaprio JA (1990): The retinoblastoma susceptibility gene product undergoes cell cycle-dependent dephosphorylation and binding to and release from SV40 large T. *Cell* 60: 387–396

Marton MJ, Baim SB, Ornelles DA, Shenk T (1990): The adenovirus E4 17-kilodalton protein complexes with the cellular transcription factor E2F, altering its DNA-binding properties and stimulating E1A induced accumulation of E2 mRNA. *J Virol* 64: 2345–2359

Matsushime H, Ewen ME, Strom DK, Kato JY, Hanks SK, Roussel MF, Sherr CJ (1992): Identification and properties of an atypical catalytic subunit (p34PSK-J3/cdk4) for mammalian D type G1 cyclins. *Cell* 71: 323–334

Mayol X, Grana X, Baldi A, Sang N, Hu Q, Giordano A (1993): Cloning of a new member of the retinoblastoma gene family (pRb) which binds to the E1A transforming domain. *Oncogene* 8: 2561–2566

Means AL, Slansky JE, McMahon SL, Knuth MW, Farnham PJ (1992): The HIP1 binding site is required for growth regulation of the dihydrofolate reductase gene promoter. *Mol Cell Biol* 12: 1054–1063

Moran E, Matthews MB (1987): Multiple functional domains in the adenovirus E1A gene. *Cell* 48: 177–178

Mudryj M, Devoto S, Hiebert SW, Hunter T, Pines J, Nevins JR (1991): Cell cycle regulation of the E2F transcription factor involves an interaction with cyclin A. *Cell* 65: 1243–1253

Mudryj M, Hiebert SW, Nevins JR (1990): A role for the adenovirus inducible E2F transcription factor in a proliferative dependent signal transduction pathway. *EMBO J* 9: 2179–2184

Ogris E, Rotheneder H, Mudrak I, Pilcher A, Wintersberger E (1993): A binding site transcription factor E2F is a target for trans-activation of murine thymidine kinase by polyomavirus large T antigen and plays an important role in growth regulation of the gene. *J Virol* 67: 1765–1771

Pagano M, Pepperkok R, Verde F, Ansorge W, Draetta G (1992): Cyclin A is required at two points in the human cell cycle. *EMBO J* 11: 961–971

Partridge JF, La Thangue NB (1993): A developmentally regulated and tissue-dependent transcription factor complexes with the retinoblastoma gene product. *EMBO J* 10: 3819–3827

Pearson BE, Nasheuer H-P, Wang TS-F (1991): Human DNA polymerase a gene: Sequences controlling expression in cycling and serum-stimulated cells. *Mol Cell Biol* 11: 2081–2095

Pines J, Hunter T (1989): Isolation of a human cyclin cDNA: evidence for cyclin mRNA and protein regulation in the cell cycle and for interaction with p34cdc2. *Cell* 58: 833–846

Pines J, Hunter T (1990): Human cyclin A is adenovirus E1A associated protein p60 and behaves differently from cyclin B. *Nature* 346: 760–763

Raychaudhuri P, Bagchi S, Neill SD, Nevins JR (1990): Activation of the E2F transcription factor in an adenovirus-infected cell involves E1A-dependent stimulation of DNA-binding activity and induction of cooperative binding mediated by an E4 gene product. *J Virol* 64: 2702–2710

Rosenblatt J, Gu Y, Morgan DO (1992): Human cyclin-dependent kinase 2 is activated during the S phase and G2 phase of the cell cycle and associated with cyclin A. *Proc Natl Acad Sci USA* 89: 2824–2828

Schwarz JK, Devoto SH, Smith ES, Chellappan SP, Jakoi L, Nevins JR (1993): Interactions of the p107 and Rb proteins with E2F during the cell proliferation response. *EMBO J* 12: 1013–1020

Sewing A, Burger C, Brusselbach S, Schalk C, Lucibello FC, Muller R (1993): Human cyclin D1 encodes a labile nuclear protein whose synthesis is directly induced by growth factors and suppressed by cyclic AMP. *J Cell Sci* 104: 545–555

Shan B, Zhu X, Chen P-L, Durfee T, Yang Y, Sharp D, Lee W-H (1992): Molecular cloning of cellular genes encoding retinoblastoma-associated proteins: Identification of a gene with properties of the transcription factor E2F. *Mol Cell Biol* 12: 5620–5631

Sherr CJ (1993): Mammalian G1 cyclins. *Cell* 73: 1059–1065

Shirodkar S, Ewen M, DeCaprio JA, Morgan J, Livingston DM, Chittenden T (1992): The transcription factor E2F interacts with the retinoblastoma product and a p107-cyclin A complex in a cell cycle-regulated manner. *Cell* 68: 157–166

Shivji MK, La Thangue NB (1991): Multicomponent differentiation-regulated transcription factors in F9 embryonal carcinoma stem cells. *Mol Cell Biol* 11: 1686–1695

Slansky JE, Li Y, Kaelin WG, Farnham PJ (1993): A protein synthesis-dependent increase in E2F1 mRNA correlates with growth regulation of the dihydrofolate reductase promoter. *Mol Cell Biol* 13: 1610–1618

Thalmeier K, Synovzik H, Mertz R, Winnacker E-L, Lipp M (1991): Nuclear factor E2F mediates basic transcription and trans-activation by E1A of the human MYC promoter. *Genes Dev* 3: 527–536

Wahl AF, Geis AM, Spain BH, Wong SW, Korn D, Wang TS-F (1988): Gene expression of human DNA polymerase α during cell proliferation and the cell cycle. *Mol Cell Biol* 8: 5016–5025

Weinberg RA (1991): Tumor suppressor genes. *Science* 254: 1138–1146

Xiong Y, Zhang H, Beach D (1992): D type cyclins associate with multiple protein kinases and the DNA replication and repair factor PCNA. *Cell* 71: 505–514

Zacksenhaus E, Gill RM, Phillips RA, Gallie BL (1993): Molecular cloning and characterisation of the mouse RB1 promoter. *Oncogene* 8: 2343–2351

Zamanian M, La Thangue NB (1992): Adenovirus E1A prevents the retinoblastoma gene product from repressing the activity of a cellular transcription factor. *EMBO J* 11: 2603–2610

Zamanian M, La Thangue NB (1993): Transcriptional repression by the Rb-related protein p107. *Mol Biol Cell* 4: 389–396

Zhu L, van den Heuvel S, Helin K, Fattaey A, Ewen M, Livingston D, Dyson N, Harlow E (1993): Inhibition of cell proliferation by p107, a relative of the retinoblastoma protein. *Genes Dev* 7: 1111–1125

4

Direct Signal Transduction by Tyrosine Phosphorylation of Transcription Factors with SH2 Domains

XIN-YUAN FU

Introduction

A living cell is constantly receiving and responding to a variety of extracellular signals. Many of these signals are polypeptides such as hormones, growth factors, cytokines, neural transmitters, antigens and cell matrix proteins. The state of cell metabolism, proliferation, differentiation, and other phenotypes are controlled by these polypeptide ligands. The signal transduction of these polypeptide ligands depends on their specific cell surface receptors. A ligand exerts its action by binding to its receptor on the cell surface and then initiating a chain of reactions within the cell through the intracellular domain of the receptor, which often contains or associates with certain enzymatic activities.

There are two major conceptions of mechanism of signal transduction. The first is the second messenger theory. In this theory, a ligand is the first messenger. The ligand activated receptor and its associated enzymes, produce a number of small molecules, such as cAMP, diacylglycerol etc., within the cell as the second messengers. These second messengers further induce pleiotropic cellular effects (Majerus et al, 1990; Borrelli et al, 1992). One question still to be answered by this theory is how specificity of a signal is maintained; many polypeptide ligands can affect levels of the same pool of second messengers, although these ligands may produce different effects in a cell. The other major theory is that signals are transduced through a cascade of kinase reactions. For example, many receptors are associated with, or have intrinsic tyrosine kinase activities, and through some mediators, the tyrosine kinases can activate a series of downstream serine/threonine kinases. These serine/threonine kinases in turn can stimulate effectors such as transcription factors in the cytoplasm

INDUCIBLE GENE EXPRESSION, VOLUME 2
P.A. Baeuerle, Editor
© 1995 Birkhäuser Boston

and nucleus (Hunter and Karin, 1992; Schlessinger and Ullrich, 1993). A vast amount of evidence indicates that the second messenger and kinase cascade pathways play important roles in mediating intracellular signal transduction. Many nuclear transcription factors, such as Fos, Jun, Myc, and SRF (serum responsive factor) have been shown to be stimulated by these downstream thr/ser kinases (Hunter and Karin, 1992).

However, recent studies on the mechanism of interferon-induced gene expression reveal a novel signalling pathway from cell surface receptors to the transcription factors (Hunter, 1993). In this pathway, the signal is mediated directly by a new class of transcription factors containing Src homology region 2 and 3 (SH2 and SH3) domains, and which are phosphorylated and activated in the cytoplasm by tyrosine kinases (Fu, 1992; Schindler et al, 1992b). The activated transcription factors then translocate to the nucleus, joined by a nuclear DNA binding factor, to form an active transcriptional complex (Fu et al, 1990; Kessler et al, 1990). This signalling mechanism seems to require no second messenger involvement, and tyrosine kinases, instead of thr/ser kinases, directly activate transcription factors. During the past year, it has been shown that many other cytokins and growth factors may use this direct signalling pathway to control gene expression in the nucleus. This article reviews the key experiments leading to the discovery of this signalling pathway, and discusses recent progress and possible future directions in this field.

Interferon-α/β, ISGs, ISRE and ISGF3

One physiological consequence of the binding of a polypeptide ligand to its cell surface receptor is the induction of primary response genes in the nucleus. During the late 1970s and early 1980s, several groups of researchers began studies on the nature of this induction. New mRNAs and proteins were shown to be induced in cells treated with serum or specific mitogens, such as PDGF or EGF (Riddle et al, 1979; Pledger et al, 1981). Cochran et al used the technique of differential screening cDNA library to identify clones that are induced by PDGF (Cochran et al, 1982). Greenberg and Ziff analyzed expression of some proto-oncogenes in response to serum and mitogens (Greenberg and Ziff, 1984). They showed that c-*fos*, c-*myc* and other important genes are immediately induced by mitogens. Interestingly, the induction of these genes requires no protein synthesis, indicating that preexisting, latent transcription factors are activated in this induction process. These early experiments provided a base for further studies on mechanisms of signal transduction from receptors to transcription factors.

Interferons are cellular antiviral factors and potential cell growth inhibitors (De Maeyer et al, 1988). There are two distinct types of interferons. Type I interferon comprises interferon-α (IFN-α), which is a family of related proteins, and interferon-β (IFN-β). Type II interferon is also called interferon-γ (IFN-γ), which is produced primarily by lymphocytes, and has a wide effect on cellular immune responses, besides its anti-viral activities. Like other polypeptide ligands, these two types of interferons mediate their effects by binding to specific cell surface receptors, and induce many specific cellular effects.

The interferon-α/β induced gene expression was analyzed with approaches similar to those in PDGF studies. A number of interferon-α/β stimulated genes (ISGs) were isolated using differential screening procedures (Friedman et al, 1984; Larner et al, 1984; Williams, 1991). These ISGs were induced within a few minutes after interferon treatment, and de novo protein synthesis was not required, indicating activation of latent transcription factors. Characterization of 5' flank regions of these ISGs revealed a consensus sequence, ISRE (Interferon Stimulated Response Element), which is conserved in the promoters of nearly all interferon-α/β regulated-genes (Reich et al, 1987; Levy et al, 1988; Porter et al, 1988; Sen and Ransohoff, 1993). Functional analyses of this ISRE sequence by extensive mutagenesis demonstrated that ISRE is essential for induction of ISGs by interferon-α (Kessler et al, 1988; Dale et al, 1989). Gene transfer experiments with this ISRE sequence, inserted around the basic promote of TK gene, showed that ISRE is also sufficient for ISG induction (Reich and Darnell, 1989). One conserved core sequence GAAA is tandem repeated in most ISREs found in interferon α/β induced genes, implying that ISRE may be bound by a dimer (see below for more discussion). It is interesting that this GAAA sequence is also found in many other regulatory elements, including PRDI, PRDII and PRDIII in the interferon-β gene, HIV LTR, SV40 early promoter, and NF-AT binding site in IL-2 gene promoter suggesting that this GAAA motif may be recognized by similar or related DNA binding factors.

A number of ISRE binding proteins were identified by a number of groups (Cohen et al, 1988; Levy et al, 1988; Porter et al, 1988; Rutherford et al, 1988; Williams, 1991). At least three specific ISRE-binding complexes (termed ISGFs: Interferon Stimulated Gene Factors) were characterized by gel retardation assay (Levy et al, 1988; Dale et al, 1989). ISGF1 activity is observed in untreated cells. ISGF2 activity is induced by interferon treatment, but this induction requires protein synthesis. Therefore, neither ISGF1 and ISGF2 is likely to be the primary activator of ISGs. ISGF3 is the only ISRE binding factor whose activity profile parallels the process of ISG activation; induction of ISGF3 activity does not require protein synthesis, and purified ISGF3 stimulates specific ISRE dependent transcription in vitro (Fu et al, 1990).

Dale et al first presented evidence suggesting that ISGF3 preexists in a latent form in the cytoplasm (Dale et al, 1989). They showed that ISGF3 activity appears first in the cytoplasm, then translocates to the nucleus. More convincingly, they demonstrated that ISGF3 is induced in the denucleated cells. These observations were confirmed, and it was found that sodium fluoride can efficiently block nuclear translocation of ISGF3 . (Levy et al, 1989). Furthermore, ISGF3 activity can be primed by pretreatment of cells with IFN-γ, suggesting accumulation of ISGF3 precursors. Formation of active ISGF3 complex was achieved in vitro by mixing cytosols from IFN-α treated cells and IFN-γ treated cell extracts which contained enriched latent factors (Levy et al, 1989). These experiments indicated that ISGF3 activity is composed of two components: one is induced by IFN-α treatment (ISGF3-α component), the other enriched by IFN-γ treatment (ISGF3-γ component). Protein alkylation reagent N-ethyl maleimide (NEM) selectively inhibits ISGF3-γ, but not ISGF3-α, thus allowing discrimination between these two components (Levy et al, 1989).

The key step revealing the composition of ISGF3 complex has been the purification of ISGF3 to near homogeneity (Fu et al, 1990). The success of the ISGF3 purification is based on alternative use of the ISRE-affinity chromatography and a point mutant ISRE-affinity chromatography: ISGF3 activity, but not the other nonspecific ISRE binding proteins, flowed through the mutant ISRE affinity column and then was collected on the following ISRE affinity column. The most purified fraction of ISGF3 contains four distinct polypeptides with sizes of 113, 91, 84, and 48 kD (they are therefore termed p113, p91, p84, and p48 respectively, see Figure 4.1). This result has been confirmed by an experiment in which the proteins in ISGF3 complex were excised from the mobility shifted gel, eluted, and then resolved in a SDS-denaturing protein gel. In this analysis, the same four polypeptides were recovered from the ISGF3 complex, but not from control or nonspecific complex, indicating ISGF3 is composed of these four different sized proteins. The ISRE-binding subunit of ISGF3 was first determined by a gel renaturation assay. In this assay, a protein with a size between 45–50 kD can be renatured from the SDS-denaturing gel, and forms an ISRE binding complex with a faster mobility than ISGF3 complex. The same faster moving complex can be formed by purified p48 (see Figure 4.1), showing that p48 is the DNA binding subunit of ISGF3. Since this ISRE-binding activity can be specifically enhanced by IFN-γ pretreatment, p48 is also termed ISGF3-g components (Fu et al, 1990). A photoaffinity UV-crosslinking

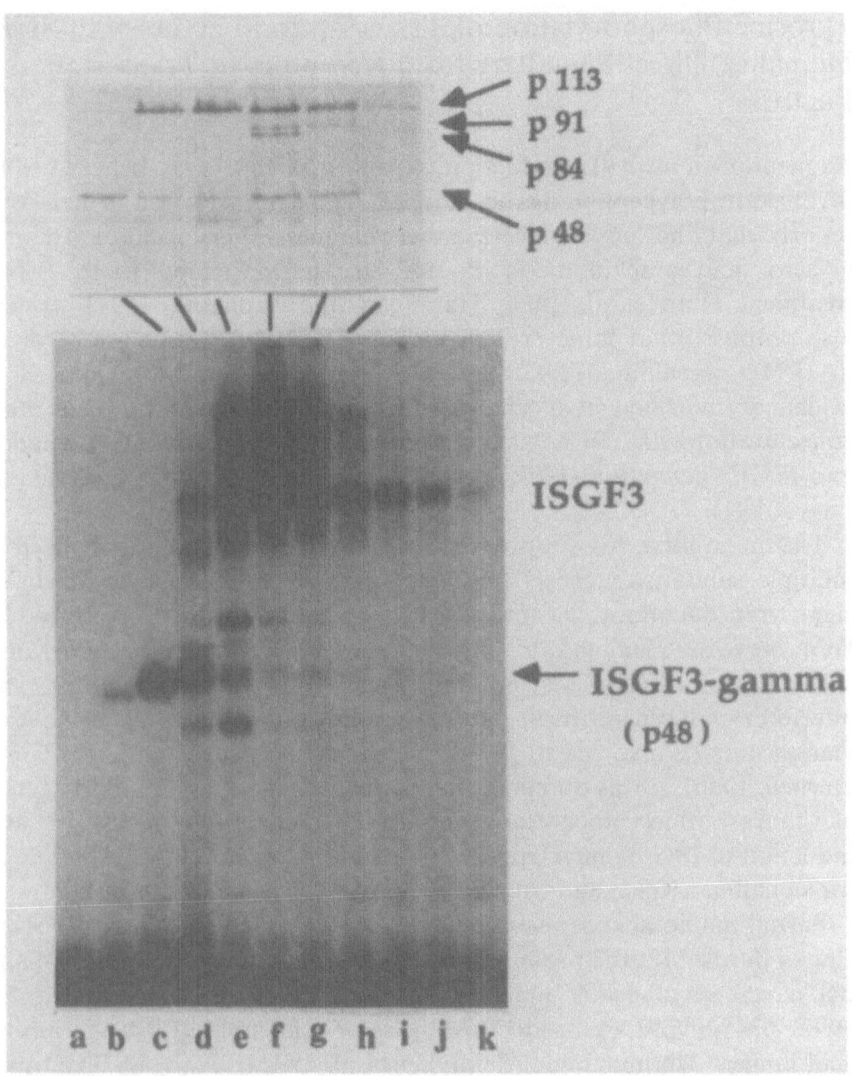

Figure 4.1 Final purification of ISGF3 revealed four peptides (p113, p91, p84 and p48, according to their relative sizes) in the peak fraction (lane f). p48, which is also termed a gamma component, can itself bind to the ISRE probe to form a faster moving complex (lane c).

experiment and a glycerol gradient sedimentation analysis with the highly purified ISGF3 fraction confirmed that p48 binds to ISRE, and ISGF3 is a multimeric complex (Kessler et al, 1990). However, p48 seems to bind to ISRE with intrinsic, lower affinity, and complexing p48 with the other components (p113, 91, and 84, i.e. ISGF3-α components) substantially increases the ISRE binding affinity (Kessler et al, 1990; Banyopadhyay et al, 1990).

Tyrosine Phosphorylation of Transcription Factors with SH2
Domains: Direct Signalling from Receptors to Transcription
Factors

As mentioned in the Introduction, second messengers are believed to be
involved in polypeptide ligand induced signal transduction. There were
reports that the levels of certain second messengers including diacylglycerol and arachidonic acid, are changed in response to IFN-α
treatment (Yap et al, 1986; Hannigan and Williams, 1991). It was
also proposed that some protein kinase C (PKC) isoforms are involved
in IFN-α signalling (Pfeffer et al, 1991). However, more and more
evidences indicated that changes in intracellular second messenger
concentrations and PKC activities have little effect on ISG induction
and ISGF3 activation (Larner et al, 1986; Lew et al, 1989; Kessler and
Levy, 1991).

The finding that ISGF3 preexists in the cytoplasm and is composed of
multiple subunits, suggests that IFN-α uses a different mechanism of
signal transduction. It was proposed that signal transduction induced by
IFN may be mediated through direct protein/protein interactions:
subunit(s) of ISGF3α may interact specifically with the receptor, and
interferon-induced posttranslational modifications including phosphorylation may be involved in ISGF3 activation (Fu et al, 1990; Levy and
Darnell, 1990). It was observed that some protein kinase inhibitors, such
as staurosporine, block the cytoplasm accumulation of ISGF3 and
induction of ISG by interferon-α, indicating phosphorylation is required
for signalling (Reich and Pfeffer, 1990; Kessler and Levy, 1991).

Partial amino acid sequences have been obtained from microsequen-
cing of purified ISGF3 proteins, and cDNAs encoding p113, p91, p84 and
p48 have been cloned (Fu et al, 1992; Schindler et al, 1992a; Veals et al,
1992). p84 and p91 are products of the same gene, but p91 has a 38 amino
acid longer C-terminal end (Schindler et al, 1992a). A detailed comparison between the p113 and p91 sequences reveals that there are stretches
of amino acid identity. These similar areas are scattered throughout
almost the entire 700 amino acid length, and the entire proteins are more
than 42% identical, but are apparently derived from different members of
a new gene family (Fu et al, 1992). When compared at moderate or high
stringency to the Genbank and EMBL data bases, there are no sequences
related to either p113 or p91.

Subunits p113 and p91/84 contain several conserved structural motifs,
such as basic region and heptad leucine repeats, and an acidic region in
the C-terminal end of p113 (Fu et al, 1992) (Figure 4.2A). These motifs
are commonly found in transcription factors (Landschultz et al, 1988;

Ptashne, 1988). Notably, some conserved tyrosine residues also have been observed in p113 and p91/84 (Fu et al, 1992). The cDNA clones for p48 have been isolated and the primary sequence of p48 has been shown to be related to the IRF (Interferon Regulatory Factor) (Miyamoto et al, 1988) and Myb families of DNA-binding proteins (Veals et al, 1992).

The most interesting structural feature of this new class of transcription factors is that all these ISGF3-α factors (p113, p91 and p84) contain a conserved SH2 domain and a SH3-like domain (Fu, 1992). As shown in Figure 4.2B, the longest region of amino acid identity is a stretch of 11 amino acids (GTFLLRFSESS) near the C-terminal ends. This sequence resembles the most conserved motif (GTFLVRESETTK in c-*src*) in the SH2 domains (Figure 4.2B). The other conserved motifs and structural features, α-helices, β-sheets, and loops such as those recently identified in the crystal structure of SH2 domains of Src and Abl (Waxman et al, 1992; Overduin et al, 1992), are also mostly conserved in the p113 and p91/84 (Figure 4.2B). The overall homology to c-*src* is 60% in this region of about 100 amino acid residues of p113 and p91/84. Similar to several members of the c-*src* protein family, conserved SH3 domains (Mayer and Hanafusa, 1990) have been identified at the immediate N-end of the SH2 domains in p113 and p91/84 (Figure 4.2A). Another notable feature of the primary sequences of p113 and p91/84 is that several conserved tyrosine residues are located separately at the further C-terminal ends of SH2 domains, but the sequences around them show some variation from protein tyrosine phosphorylation sites previously characterized (Cantley et al, 1991).

The SH2 domain has been shown to mediate the interactions of signalling proteins with the phosphorylated receptor tyrosine kinases (Koch et al, 1991; Mayer and Baltimore, 1993). The SH2 domains bind specifically to phosphotyrosine residues. When a tyrosine kinase is activated by autophosphorylation, this kinase may become a high affinity receptor for the SH2 domains. The binding of a SH2-containing protein to the tyrosine kinase usually results in tyrosine phosphorylation of the SH2-containing protein. Therefore, the fact that ISGF3α proteins contain SH2 domains immediately implies that these proteins are associated with tyrosine kinases during activation, or in other words, IFN-α signalling is involved with tyrosine phosphorylation of these SH2-containing transcription factors. As anticipated, ISGF3α proteins are transiently associated with the tyrosine kinase(s) after interferon-α treatment (Fu, 1992). Immunoprecipitated ISGF3-α proteins from [32]p labeled cells after IFN-α treatment have been shown to be phosphorylated on tyrosine residues (Schindler et al, 1992a; Fu, 1991). The tyrosine phosphorylation of ISGF3α is required for ISGF3 activation and can be

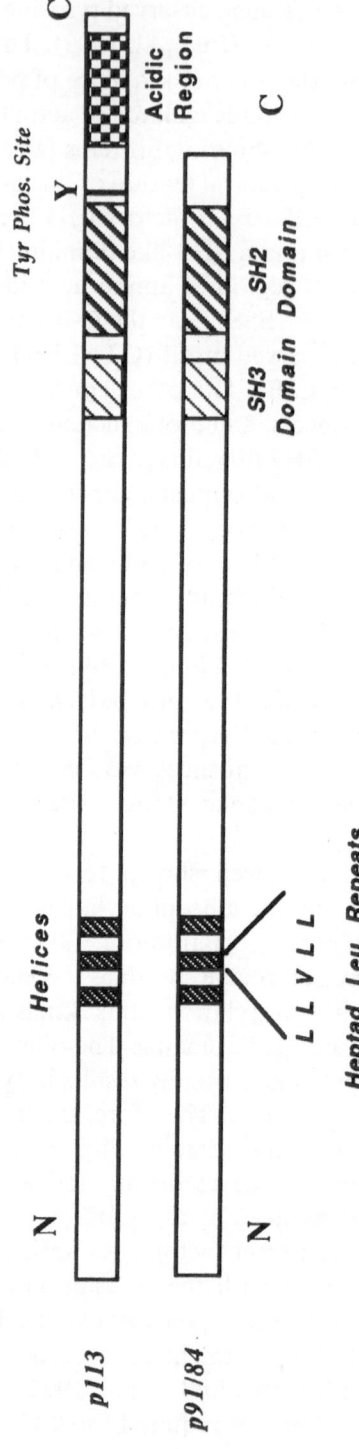

Figure 4.2A Schematic representation of structural motifs in p91/84 and p113 proteins.

	βA1		αA2			βB5			
p91/84	WND	GCIMGFI	SKERERALLK	DQQP	G	TFLLRFS	ESSRE	GAITFTWVER	608
p113	WND	GRIMGFV	SRSQERRLLK	KTMS	G	TFLLRFS	ESSEG	ITCSWVEH	617
c-src	WYF	GKI	TRRESERLLL	NPENPR	G	TFLVRES	ETTK	GAYCLSVSD	194
c-abl	WYH	GPV	SRNAAEY-LL	SSGIN	G	SFLVRES	DRRP	GQRSISLRY	·167

		(p91 & p113 Specific Loop)		βD6			
p91/84	S QN	GGEPDFHAVEPYTKKELSAVTF PD		IIRNYKV	MAAENIPENPL		664
p113	QD	DDKVLIYSVQPYT KEVLQSLPLTE		IIRHYQL	LTEENIPENPL		661
src	FFDNAK	GL		NVKHYKI	RKLDS	G	210
abl	E E	G		RVYHYRI	NTASD	G	200

				(p91 & p113 Phosph. Site)							
		βB9									
p91/84	KYLY	P NID K	DHAFGKYYSR	PK	EA	PEP	MELD	GPKGTG	Y	IKTE	704
p113	RFLY	P RIP R	DEAFGCYYQ	EK	VNLQE			RRK	Y	LKHR	694
c-src	GFYI	TSR TQF S	SLQQLVAYYSKH	AD	GL	CH	RLT	NVCPTS			248
c-abl	KLYV	SSE SRF N	TLAELVHHHSTV	AD	GL	IT	TLH	YPAPKR			238

Figure 4.2B Comparison of SH2 domains of p91, p113, Src and Abl proteins. A p91 specific loop and the tyrosine phosphorylation site are highlighted in boxes. (Adapted with permission from Fu, 1992; Waksman et al, 1992; Overduin et al, 1992)

inhibited by protein kinase inhibitors genistein and staurosporine (Fu, 1992; Gutch et al, 1992; Schindler et al, 1992).

These findings suggest that a novel signalling mechanism is used by interferon-α. A direct effector model for signal transduction has been proposed (Fu, 1992). In contrast to the conventional second messenger mechanism for signal transduction by many extracellular stimuli, signal transduction from the cell membrane receptor of interferon-α to the nucleus is mediated directly by tyrosine phosphorylation of an effector. In this model, the direct effector is the regulatory (p113 and p91/84) subunits of the transcription factor ISGF3 which are directly activated by an interferon-α-induced protein tyrosine kinase and translocated to the nucleus to regulate specific gene expression. It has been hypothesized that this interferon-α-induced protein tyrosine kinase(s) may also be with interferon-α-receptor during stimulation. The proper interactions between the ligand and receptor may induce allosteric changes of the intracellular domain of the receptor, somehow resulting in activation of associated kinase(s) by autophosphorylation. The conserved SH2 domains of p113 p91/84 may direct or facilitate the formation of a soluble complex with the activated tyrosine kinase and the receptor. The binding of these transcription factors to the activated kinase might further result in tyrosine phosphorylation of these proteins. This covalent modification of p113 and p91/84 could in turn change their protein conformations leading to dissociation of these proteins from the kinase/receptor complex, translocation to the nucleus, and activation of their ability to form a functional ISGF3 complex with the 48 kD protein. This active ISGF3 complex binds to the ISRE site, and initiates transcription of a number of interferon-α inducible genes. Recent data further support this direct effector model for signal transduction of interferon and other cytokines (see below).

Interferon-γ, GAS and GAF

Lew et al first identified an interferon-γ activated sequence (GAS) in the 5′ flanking region of the guanylate binding protein (GBP) gene (Lew et al, 1991). This GAS site is partially overlapped with an ISRE site at the 5′ end. Mutations that change either site seem to reduce response to interferon-γ. A GAS binding factor, GAF (interferon-γ activated factor) is activated within minutes in cells treated with interferon-γ (Decker et al, 1991). The GAF activity was first detected by a sensitive exonuclease protection assay which showed that this GAF activity shares many similarities to ISGF3. GAF activation occurs in the cytoplasm, and activated GAF translocates to the nucleus; the profile of GAF activation correlates with interferon-γ induced transcription (Decker et al, 1991). These observations raise the possibility that GAF and ISGF3 are related factors.

Using GAS containing oligonucleotide as a probe, several laboratories observed interferon-γ induced GAF activities by gel retardation assay (Shuai et al, 1992; Igarashi et al, 1993a, b; Pearse et al, 1993; Pine et al, 1994). With the availability of specific antibodies to components of ISGF3 (Schindler et al, 1992a), Shuai et al convincingly demonstrated that the GAF complex contains p91, but not p113 and p48 of ISGF3 (Shuai et al, 1992). The GAF complex detected by the gel retardation assay is supershifted by an anti-p91 antiserum but not by anti-p113 antiserum. A novel two-dimensional gel mobility shift-SDS PAGE analysis confirmed that a 91 kD protein in the GAF complex is recognized by the anti-p91 antiserum. A UV cross-linking experiment with partially purified GAF indicated that p91 can directly interact with the GAS sequence. Phospho-amino acid analysis of ^{32}p labeled p91 demonstrated that p91 is tyrosine phosphorylated in response to interferon-γ (Schindler et al, 1992a, b; Shuai et al, 1992). A single tyrosine residue immediately following the SH2 domain (amino acid number 701 in p91) in the p91 family of proteins is the phosphorylation site in response to either interferon-α or -γ (Shuai 1993a). As shown previously in ISGF3 activation (Fu, 1992), this tyrosine phosphorylation is required for the specific GAS binding activity; phosphatase treatment of the GAF complex will disrupt this specific DNA-protein complex (Shuai et al, 1992; Igarashi et al, 1993) (see discussion below).

By comparison of the sequences in the responsive elements of several interferon-γ inducible genes including Ly-6E/A antigen (Khan et al, 1990), IFN-γ responsive region (GRR) of the high affinity Fc receptor for IgG (FcgRI) (Pearce et al, 1991; Perez et al, 1993), guanylate binding protein (GBP) (Lew et al, 1991), and macrophage-induced gene (MIB) (Wright and Farber, 1991), a GAS core consensus sequence **TTC/ACNNNAA** has been identified (Pearce et al, 1993). Point mutations at any of these conserved (in bold) nucleotides result in a diminished response to interferon-γ (Lew et al, 1991; Pearce et al, 1993). Recently this GAS core sequence has been found in many primary response gene promoters that may respond to a number of different cytokines (see below).

One Possible Mechanism of Activation of Transcription Factor by Tyrosine Phosphorylation

One of the most distinguished structural features of p91 is the presence of a well-conserved SH2 domain near the C-terminal end of the protein and this SH2 domain is essential for EGF or interferon induced tyrosine phosphorylation and p91 complex formation (see below). The key question is how tyrosine phosphorylation of p91 or p91-like factors leads to functional activation with regard to unblocking domains or signals for formation of p91 complex, nuclear translocation and DNA binding.

As shown in Figure 4.2A, a motif of heptad leucine repeats are identified in p91 which may be involved in p91 complex formation. These leucine repeats may be blocked for a role in the possible intermolecular association before p91 is tyrosine phosphorylated, since unphosphorylated p91 is not in a complex with its potential partners (such as p113) before interferon treatment (Schindler et al, 1992a, b). Recently, it has been shown that the leucine repeats are involved in complex formation since a short polypeptide containing this region can compete for the normal ISGF3 formation in vitro (Fu, 1994). Additionally, by careful examining sequences of p91, a putative and conserved nuclear localization signal (NLS) in these proteins (RKSKR starts at amino acid 83–87 in p91) is identified. However, the function of the NLS in p91 must be masked before interferon or EGF treatment since p91 is only present in the cytoplasm in untreated cells.

The direct consequence of tyrosine phosphorylation of a protein is that this protein may become a target for binding SH2 domains. One model of activation is that the newly created phosphotyrosine site in p91 or p91-like protein may provide a site for binding of SH2 domain of p91 itself or another protein. This intra- or intermolecular binding or interacting may initiate and help homo- or heterodimer formation through heptad leucine repeats. This dimer formation will further induce a certain conformation change to unmask or activate the NLS. The affinity of this intra- or intermolecular binding to p91 itself or another protein, such as p113, may be higher than the affinity of this SH2 domain to the phosphorylated tyrosine kinase; thus the phosphorylated p91 can be released from the kinase to form a homo- or heterodimer with each other.

There are several lines of evidence to support this hypothesis. First, it has been demonstrated that tyrosine phosphorylation is crucial for ISGF3 formation; phosphatase treatment of tyrosine phosphorylated ISGF3 proteins will abolish complex formation and releasing out of p48 from the complex (Fu, 1992). The antibody which was against the SH2 domain of p91 can also disrupt the p91-SIF complex, releasing a smaller DNA-binding protein (Fu and Zhang, 1993). The GAF complex formation can be prevented with either monoclonal antiphosphotyrosine antibody or treatment of protein tyrosine phosphatase (Igarashi et al, 1993a, b). Second, a competition assay with either phenylphosphate or phosphotyrosine can compete for p91-GAF or p91-SIF complex formation (David et al, 1993a, b; Sadowski et al, 1993). More convincingly, Shuai et al showed that the phosphopeptides that resemble the p91 and p113 tyrosine phosphorylation sites, can specifically dissociate the p91 complex (Shuai et al, 1994). These data may indicate that both the phosphorylated tyrosine site and the SH2 domains are involved in complex formation.

Mutant Cell Lines Defective in Interferon Response

The essential function of a given signalling protein can be clearly demonstrated if cells defective in expression of the signalling protein are also defective in the signalling. For example, proteins like p91, p113, and p48 have been demonstrated biochemically to be the necessary components of interferon induced transcription factors, their essential roles in the interferon induced gene expression can be further determined in cell lines in which they are defective.

Laboratories of George Stark, Ian Kerr, and Richard Flavell pioneered a genetic approach to isolated mutant cell lines which are unresponsive to interferons (Stark and Kerr, 1992). In this approach, the bacterial guanine-phosphoribosyl transferase (gpt) gene is regulated by an interferon-inducible promoter. Then 6-thioguanine is used to select interferon unresponsive mutants after cells are mutated by a frame-shift mutagen ICR-191. Furthermore, the back-selection can be done after genomic DNA transfection using hypoxanthine-aminopterin-thymidine (HAT) medium. The isolated unresponsive mutant cell lines are divided in several complementation groups (U1 and U6 cell lines) (Pellegrini et al, 1989; John et al, 1991; McKendry et al, 1991; Muller et al, 1993a, b; G Stark, 1994) (see Table). This genetic approach proves to be very fruitful, which not only confirms the biochemistry studies of the essential

Table. Mutant Cell Lines in Interferon Signaling Pathway

Protein	Mutant cell lines	IFN-α	IFN-γ	Reference
p48	U2	+/−	+/−	*
p91/84	U3	−	−	Muller et al, 1993b
p113	U6	−	+	*
Tyk2	U1	−	+	Velazquez et al, 1992
Jak1	U4	−	−	Muller et al, 1993a
Jak2	γ2	+	−	Watling et al, 1993

*G. Stark, 1994

signalling proteins in the pathway, but also reveals additional missing signalling components and possible mechanisms of activation.

Mutant cell line U2A, which expresses a truncated p48, fails to respond to interferon-α/β, showing p48 is a necessary component for this pathway (John et al 1991). Interestingly, although U2A cells are responsive to interferon-γ, a few of interferon-γ inducible genes are unresponsive in these cells, indicating p48 may be partially required for interferon-γ-induced gene expression, or interferon-γ may use more than one (p48 dependent or independent) pathway. The U3A cell line was selected based on unresponsiveness to interferon-α/β, but this cell line is not responsive to interferon-γ either, suggesting a signalling component shared by both interferon-α and interferon-γ is defective (McKendry et al, 1991). Molecular characterization of the defect revealed that the transcriptional complexes ISGF3 and GAF are not formed, due to lack of p91 and p84 expression in this cell line (Muller et al, 1993a). Introducing p91 into these U3A cells restores the responsiveness to both interferon-α/β, and interferon-γ, whereas p84 only restores the responsiveness to interferon-α. These results indicate that p91, but not p84, is required for the interferon-γ pathway, and the C-terminal end of p91, which is truncated in p84, may contain an activation domain. However, p84 can function in the interferon-α pathway, possibly because p113 has a longer acidic C-terminal end which may act as an activation domain (Fu et al, 1992). Recently, the cell line defective in p113 (U6) also has been isolated. As anticipated, this cell line is defective in interferon-α/β response, but not in interferon-γ response (Stark G, 1994). Mutant cell lines that may affect the expression of secondary responsive genes also have been isolated (Mao et al, 1993). These genetic characterizations are strikingly consistent with biochemical studies of ISGF3 and GAF transcriptional complexes (see previous sections).

Although the receptors for type I and Type II interferons have been cloned (Aguet et al, 1988; Uze et al, 1990), how these receptors may transduce signals was unknown until recently. Cloned interferon receptors, like most cytokine receptors, do not contain any recognizable catalytic activities. One breakthrough was achieved using the above genetic approach: a class of tyrosine kinases were shown to be essential mediators of the interferon signalling pathway. Pellegrini and colleagues first presented evidence showing a potential tyrosine kinase may be a crucial component in interferon-α signal transduction (Velazquez et al, 1992). They demonstrated the gene encoding Tyk2 tyrosine kinase seems to be able to restore the interferon-α responsiveness in a mutant cell line 11,1 (U1A) using a genomic DNA transfection method. Tyk2 was previously isolated by a degenerate PCR method, but its function was not known (Firmbach-Kraft et al, 1990). Tyk2 belongs to a new class of tyrosine kinase, the Jak family of kinases (Wilks et al, 1991). These nonreceptor tyrosine kinases do not contain the SH2 and SH3 domains, but have two tandem repeated tyrosine kinase domains (Bernards, 1991). Intriguingly, Tyk2 kinase also seems to be required for the high affinity binding of interferon-α to its receptors, since these U1A cells lack the high affinity receptors (Pellegrini et al, 1989). If Tyk2 kinase is the only defective protein in the U1A cells for interferon-α signalling, then these data suggest either that Tyk2, although it does not contain the trans-membrane domain, may be a necessary component in the high affinity receptor, or that Tyk2 may stimulate the formation of high affinity receptors. Although Tyk2 is assumed to directly phosphorylate transcription factors such as ISGF3α, there is yet no solid biochemical data showing ISGF3α proteins are substrates of Tyk2 or other members of Jak family kinases. Therefore, the biochemical relationship of ISGF3α proteins and the Jak family of kinases needs to be further established (see below).

Besides Tyk2, two other known members of this tyrosine kinase family, Jak1 and Jak2, have recently been shown to play an essential role in the interferon signalling. Mutant cell line U4A, which is completely defective in both interferon-α and -γ signalling (McKendry et al, 1991), expresses no Jak1 protein, and introducing Jak1 expressing vector into these cells restores interferon response (Muller et al, 1993a, b). Possibly as a result, p91 is not tyrosine phosphorylated in U4A cells unless these cells express introduced Jak1 kinase. Consistent with these mutant cell line studies, Shuai et al show that Jak1 becomes tyrosine phosphorylated in response to both interferon-α and -γ (Shuai et al, 1993). These results indicating that Jak1 kinase is required in both interferon-α and -γ signalling pathways. In contrast to Jak1, Jak2 kinase appears to be required only for interferon-γ response (Watling et al, 1993). Surprisingly, Tyk2 kinase, which is present normally in U4A cells, is not tyrosine phosphorylated in response to interferon-α in these cells. Similarly, Jak1 kinase fails to be phosphorylated in response to interferon-α in U1 cells that do not express Tyk2 kinase. Therefore, it is concluded that there exists an interdependence, but not a linear activating cascade, between Tyk2 and Jak1 kinases (Muller et al, 1993a, b). This dual kinase relationship is also observed in the interferon-γ pathway; however, in this case, Jak1 and Jak2 kinases are involved (Watling et al, 1993). It would be worthwhile to determine whether these two kinases may form a heterodimer in the receptor-kinase complex in response to interferon, a process which may be similar to the receptor dimer formation and activation during growth factor (such as EGF) signal transduction (Schlessinger and Ullrich, 1992).

A Common Signalling Pathway from Ligand-bound Receptors to Transcription Factors

At least two signal pathways linking cell surface receptors to transcription factors have been proposed. In the first, a number of growth factors, such as epidermal growth factor (EGF), platelet-derived growth factor (PDGF) and nerve growth factor (NGF), may activate a kinase cascade linking protein tyrosine kinases to serine/threonine kinases, such as RAF kinase, MAP/ERK kinase and p90^{rsk} kinase. These serine/threonine kinases in turn lead to activation of transcription factors localized in the nucleus (Cantley et al, 1991; Schlessinger and Ullrich, 1992; Blenis, 1993).

The other signalling pathway, as discussed above in the interferon system, involves activation of cytoplasmic transcription factors by tyrosine phosphorylation. Several questions about these two pathways remain to be answered. Do different growth factors and cytokins, such as EGF and interferons, use different and independent pathways (for example, kinase cascade or direct tyrosine phosphorylation of transcription factors) to transduce their signals to the nucleus, or is there a merging point, and cross-interaction between these two pathways? A more intriguing question is whether these two pathways may simultaneously function in response to a certain ligand and modulate a concerted regulation of the induced nuclear event. While one pathway activates the cytoplasmic regulatory subunits of transcription factor(s), the other pathway modulates activity of the nuclear subunits of transcription factor(s). The combined action of these two kinds of subunits would lead to proper expression of a specific set of genes.

Although the receptors for EGF and interferons are different, the initial activation of both pathways requires tyrosine kinase activities. There are also many similarities between interferon and EGF or PDGF induced gene expression. For example, a transcription factor SIF (c-sis-inducible factor) is rapidly induced by PDGF and EGF in the absence of de novo protein synthesis (Wagner et al, 1990). SIF specifically binds to a regulatory element, SIE (c-sis-inducible element), in the c-*fos* gene promoter. This SIE element is sufficient to confer PDGF responsiveness to the c-*fos* gene promoter (Wagner et al, 1990). Although the component(s) of SIF have not been determined, Sadowski and Gilman first presented evidence suggesting that SIF may be activated in a similar manner to ISGF3: i.e., activation of SIF might be mediated by a SH2 domain and involve a protein tyrosine kinase(s) (Sadowski and Gilman, 1993).

It is interesting to note that there is a striking similarity between the DNA sequence of SIE (AGTTCCCGTCAAATCCT) (Wagner et al, 1990) and the consensus sequence of GAS (TTC/(A)CNNNAA) (see above). This finding suggests the possibility that related transcription factors may be involved in binding to these sites in response to EGF and interferons.

Studies in several laboratories have shown that the transcription factor p91 is involved in EGF and PDGF induced gene expression. First, tyrosine phosphorylation of p91 is observed in the cells treated with EGF or PDGF, and p91 is a component in the induced SIF complex (Ruff-Jamison et al, 1993; Fu and Zhang, 1993; Sadowski et al, 1993; Shuai et al, 1993b; Levy et al, 1993). During the activation, p91 is associated directly with the EGF receptor complex (Fu and Zhang, 1993). This is the first demonstration that a transcription factor can interact directly with the receptor complex through its SH2 domains. This SH2 domain-mediated transcription factor-receptor interaction is essential for activation of p91, since a point mutation in the FLLRSESS motif (Arg to Gln) in the SH2 domain of p91 abolishes its interaction with the receptor, and as a consequence, this mutant p91 fails to be tyrosine phosphorylated and remains in the cytoplasm in response to EGF (Fu and Zhang, 1993). Besides p91, other factors are apparently involved in SIF complex. The SIF complex can be resolved into at least three different migrating complexes in a gel retardation assay (Sadowski et al, 1993; Ruff-Jamison et al, 1993). The lower and the middle bands, which are specifically interrupted by anti-p91 antiserum, contain p91. These results may suggest that p91 can either form a homodimer itself, or a heterodimer with another protein; the lower band is probably the homodimer of p91, and the middle band is the heterodimer of p91 with the protein in the higher band. In a p91-deficient cell line (U3), these two lower and middle complexes with p91 are not formed unless this cell line is stably transfected with a p91 expression vector (Sadowski et al, 1993). Furthermore, epitope-tagged p91, which can be specifically analyzed by anti-tag monoclonal antibody, has been shown to be tyrosine phosphorylated just as the endogenous p91 (Fu and Zhang, 1993). These results indicate that the same p91, but not an antigenic-related factor, is involved in both the interferon and EGF signalling pathways. As discussed in the direct effector model, signal transduction with these SH2 domain-containing transcription factors does not involve a kinase cascade or need another intermediator. Silvennoinen et al have shown that growth factor-induced p91 activation is independent of the Ras-Raf pathway, because a Ras dominant negative mutant which blocks the Ras-Raf pathway, has no effect on tyrosine phosphorylation of p91 (Silvennoinen et al, 1993a, b).

In the past year, a growing number of cytokines have been shown to use this novel direct signalling pathway to activate p91-like transcription factors. One well-studied example is interleukine 6 (IL-6) activated transcription factors. The high affinity receptor for IL-6 is composed of the ligand binding subunit gp80 and a signal-transducing subunit, gp130 (Taga et al, 1989; Hirano and Kishimoto, 1990). The ligand-induced dimerization of gp130 is associated with a tyrosine kinase (Murakami et al, 1993). On the other hand, Horn and colleagues have shown that a transcription factor, APRF (acute-phase response factor), is rapidly activated posttranslationally from a latent form by IL-6 treatment (Wegenka et al, 1993). Similarly to the response of p91 to interferons, a 89 kD protein has been shown to be tyrosine phosphorylated, binding to an ARPE (acute-phase response element) site which contains the core sequences of the GAS site (Wegenka et al, 1994; Yuan et al, 1994). Interestingly, it has been observed that IL-6 can activate a p91-related transcriptional complex which binds to the GAS core sequence (Sadowski et al, 1993), and this p89-APRF complex can be recognized by an antiserum against the N-terminal part of p91, suggesting this p89 is related to p91 (Wegenka et al, 1994). Recent cloning of this IL-6 stimulated transcription factor has indicated that this factor(s) is a new member of the p91 (termed STAT-3 or APRF) protein family: it shares about 50% identical amino acid sequence with p91 (Zhong et al, 1994; Akira et al, 1994). It is known that the gp130 signalling protein is shared by several other cytokine receptors, such as receptors for LIF (leukemia inhibitory factor), CNTF (ciliary neurotropic factor), and OSM (oncostatin M) (Ip et al, 1992; Stahl and Yancopoulos, 1993). As anticipated, all these related cytokine receptors appear to use this direct signalling pathway; Jak family of kinases and p91-related transcription complexes are activated in response to these cytokines (Bonni et al, 1993; Lutticken et al, 1994; Stahl et al, 1994). In contrast to interferon systems, in which a specific pair of Jak kinases are activated, gp130 mediated signalling pathways seem to activate all known members of Jak kinases (Stahl et al, 1994), possibly indicating gp130 may act as a specific docking protein for Jak kinases.

Other cytokines using this direct signalling pathway include another subfamily of hemopoiesis cytokines, interleukin-3 (IL-3), IL-5 and granulocyte macrophage colony-stimulating factor (GM-CSF). These cytokines share a common signal-transducing β-chain of the receptors (Kastelein and Shanafelt, 1993). It has been demonstrated that Jak2 kinase is activated in the IL-3 signal transduction (Silvennoinen et al, 1993a, b), and all three of these cytokines can induce specific DNA binding complexes involving p91-like proteins (Larner et al, 1993). Jak2 kinase is also implicated in signal transduction of growth hormone and erythropoietin (Argetsinger et al, 1993; Witthuhn et al, 1993). Moreover, it has been shown recently that different p91-like transcription factors, which bind to a DNA probe with the GAS core sequence, may be differently activated by growth hormone and erythropoietin (Finbloom et al, 1994). IL-4, another lymphocyte growth factor, also appears to activate p91-like transcription factors; however, no known members of Jak kinases are involved (Kotanides and Reich, 1993). IL-2, which recently has been shown to share a common receptor γ chain with the receptors of IL-4 and IL-7 (Kondo et al, 1994; Noguchi et al, 1994; Russell et al, 1994), appears to activate two different sized tyrosine kinases, which are antigenically related to Jak kinases, but are not the same as any known Jak kinase members (D'Andrea, 1994).

In summary, this tyrosine kinase/transcription factor pathway is widely used by many, if not all, tyrosine kinase initiated signalling processes. Different ligands seem to use different tyrosine kinases to phosphorylate the specific members of a family of transcription factors. Classical second messengers are not required for the signalling in this pathway. Although more than one tyrosine kinase may be required by this pathway, they function interdependently of each other. In other words, there is no kinase cascade involved in the activation of transcription factors in this pathway. This interdependent dual tyrosine kinase requirement may suggest that a heterodimer of these two kinases is formed during activation.

The Constitutive Nuclear Specific DNA binding Proteins: ISGF3γ, IRF-1/ISGF2, and Related Factors

It has been shown that p48 of ISGF3 is a DNA binding subunit of ISGF3, and binds to the ISRE site with a similar specificity but a lower affinity than ISGF3 (Fu et al, 1990; Kessler et al, 1990). The protein level of the p48 subunit is increased in response to interferon-γ treatment; thus p48 is also called a ISGF3γ component. This ISGF3γ originally was suggested as a cytoplasmic component in an ISGF3 reconstitution experiment (Levy et al, 1989; Bandyopadhyay et al, 1990). However, unlike p113, p91 and p84 of ISGF3, which are present in the cytoplasm in a latent form before stimulation, this p48 is essentially a nuclear protein with the constitutive ISRE binding activity, and may be leaked out to the cytosol fraction (Fu, 1994). The sequence of p48 indicates a significant similarity to the IRF-1 (interferon regulatory factor) family of DNA binding proteins, and also to some weak conserved imperfect tryptophan repeats similar to the DNA binding domain of Myb oncoprotein (Veals et al, 1992). Therefore, p48 may represent another family of transcription factors possessing constitutively the specific DNA binding activity, and forming an active transcriptional complex with nuclear translocated, and tyrosine phosphorylated p91-like proteins after stimulation.

The IRF protein previously had been identified (Fujita et al, 1988; Miyamoto et al, 1988) as a binding protein to the virus inducible element PRDI in the interferon-β gene promoter and cloned (Goodburn and Maniatis, 1988). IRF-1 is identical to the ISGF2, a factor originally characterized as a newly synthesized ISRE-binding protein in response to interferon-α (Levy et al, 1988; Pine et al, 1990). IRF-2 is thought to play as antagonistic a role as IRF-1 (Harada et al, 1989). Another member of IRF family proteins is ICSBP (interferon consensus sequence binding protein), which is primarily expressed in cells of lymphocyte or macrophage lineage (Shirayoshi et al, 1988; Driggers et al, 1990). IRF factors had been believed to be involved in interferon and interferon-inducible genes expression (Harada et al, 1989). However, recent results demonstrate that IRF proteins might not be required for activation of interferon-inducible genes, since induction of interferon-inducible genes seems normal in the cells which are deficient in IRF-1 or 2 (Matsuyama et al, 1993; Ruffner et al, 1993). Similarly, induction of type 1 interferon genes appears unaffected in IRF-1 deficient cells in response to infections of some virus (Ruffner et al, 1993; Matsuyama et al, 1993). What are the real functions of these IRF factors? It has been observed that mRNA for IRF1 is also immediately induced by prolactin and Il-6 (Yu-Lee et al, 1990; Abdollahi et al, 1991). Therefore, it is possible that IRF-1 may participate in the secondary responses induced by cytokines other than interferons. Another possibility is that IRF factors, similar to p48 of ISGF3, may be joined by other unidentified p91-like factors in response to these cytokines. Although their real functions need to be further explained, the importance of these IRF factors in cytokine-mediated responses, such as in lymphocyte development, is supported by analysis of IRF-1 or -2 deficient mice. IRF-1 deficient mice are defective in thymocyte development, especially in production of CD8+ T cells, and IRF-2 deficient mice appear to have somewhat suppressed B lymphopoiesis (Matsuyama et al, 1993). However, it is not clear how these IRF deficiencies might cause these altered phenotypes.

In interferon-γ signal transduction, it originally had been proposed that p91 alone binds to the GAS site, and only a single interferon-γ induced specific GAS/p91 complex (GAF) was observed (Shuai et al, 1992). As mentioned earlier, the p91 family of proteins contains a basic-leucine zipper-like structure at its N-terminal (Fu et al, 1992); therefore it is possible that this class of proteins may contact DNA. However, several laboratories have shown that other factor(s) or partners in addition to p91 must be coactivated with p91 in response to interferon-γ, since at least three distinct GAS binding complexes can be induced (Wilson et al, 1992; Pearse et al, 1993; Igarashi et al, 1993; Sadowski et al, 1993). Moreover, a UV cross-linking experiment has indicated a 43 kD DNA binding protein is involved in GAF complexes (Igarashi et al, 1993). Similarly, a smaller specific DNA binding protein, which has distinct SIE binding activity, also has been detected in the EGF-induced SIF-p91 complex (Fu and Zhang, 1993). Therefore, the probable general mechanism is that ligand-activated p91-like factors, translocated to the nucleus, join a specific nuclear DNA binding protein to form a transcriptionally active complex (see below for more discussion).

The Puzzle on Specificity: a Divergence and Reconvergence Hypothesis for Signal Transduction

Several more questions have been raised by recent studies on this direct signalling pathway. For example, p91 is clearly a necessary component of the transcription factor ISGF3, and is tyrosine phosphorylated and activated by an interferon-induced kinase (Fu, 1992; Schindler et al, 1992b; Velazquez et al, 1992). However, p91 also plays a crucial role in the SIF complex induced by EGF and PDGF (Fu and Zhang, 1993; Sadowski et al, 1993). An intriguing problem is how specificity is maintained. One possibility is that additional p91-related protein(s) may also be involved. It appears that a 75 kD protein is tyrosine phosphorylated and cotranslocated with p91 to the nucleus in response to EGF treatment (Fu and Zhang, 1993). This p75 is recognized by an anti-P91 antibody, suggesting p75 may be another member of p91-like transcription factors. Recently, Darnell and colleagues have isolated and identified a 92 kD protein (p92 or STAT-3) as a new member of p91 protein family and it contains a protein sequence about 50% identical to original p91 (Darnell, 1994). The antibodies raised against this p92 protein seemingly can recognize a protein in the SIF complex that is not the original p91, suggesting this protein may play a role in EGF induced transcription. As previously shown, in the interferon-α induced signalling pathway, another p91-related factor, p113, which seems not to be involved in the EGF signalling, is tyrosine phosphorylated after stimulation, and together with p91 could then form a specific transcriptional complex in the nucleus. Although p113 or p92 are related to p91, they are differently induced by different peptide ligands such as EGF and interferon-α. Therefore, it is

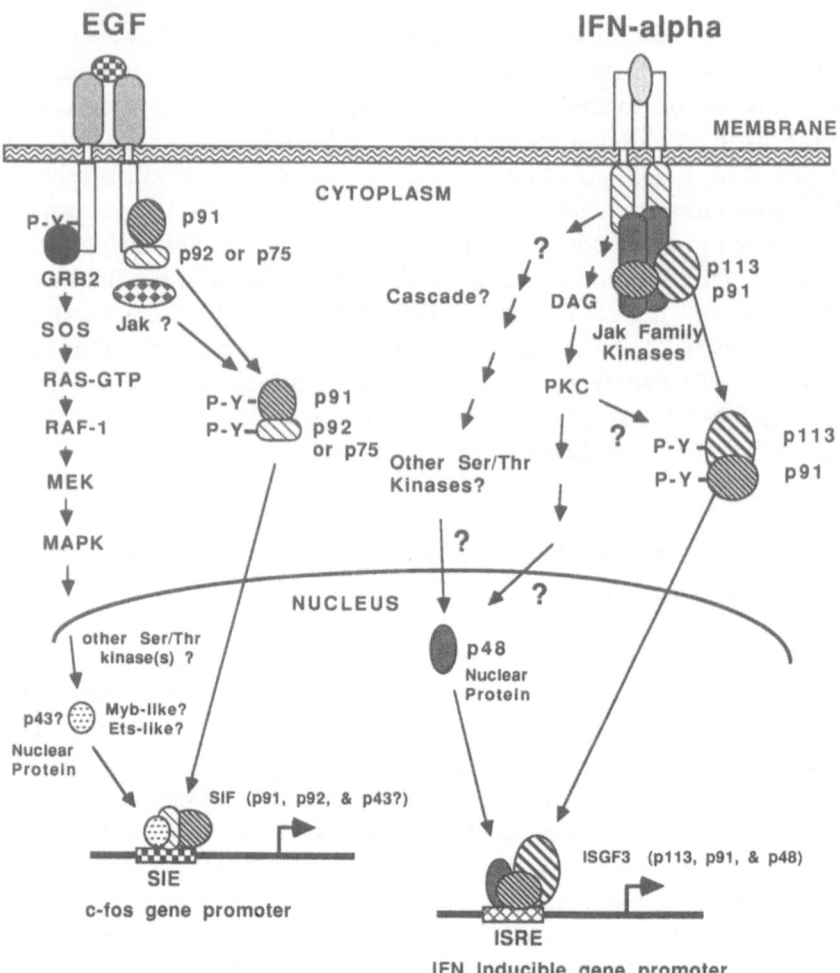

Figure 4.3 Divergence & reconvergence of two signalling pathways examplified by EGF and interferon.

likely that a different combination of p91 and other related protein(s), such as p113 or p92, might be a way to differentiate and maintain specificity for each signal induced by a certain ligand (Figure 4.3).

Another way to maintain signalling specificity might be through nuclear DNA binding factors which could form specific transcriptional complexes with p91 or related proteins. As mentioned above, the partial disruption of the SIF complex by the anti-p91 antibody produces a faster moving SIE/protein complex (Fu and Zhang, 1993), a phenomenon that also has been observed when ISGF3 complex is partially disassociated; p48 of ISGF3 can form a smaller but still specific complex (Fu et al, 1990,

1992) (see Figure 4.3). The faster SIE-binding complex derived from SIF is clearly distinguishable from the complex formed by p48 of ISGF3, as it is competed by a SIE probe but not by an ISRE probe. Whether this smaller DNA binding protein is present in SIF activity and belongs to the gene family of the DNA binding subunit p48 of ISGF3 (Veals et al, 1992) can be determined when the SIF complex is highly purified and its composition is analyzed.

An intriguing question is whether p91-like proteins may form specific transcriptional complexes with any previously identified nuclear DNA binding proteins, such as the Egr, Ets, and Fos/Jun families of proteins (Hunter and Karin, 1992). These nuclear proteins, like the p48 and IRF family of proteins, are immediately induced by a variety of polypeptide ligands, suggesting they play important roles in control of gene expression required for cellular responses. Furthermore, many of these DNA binding proteins may be constitutively expressed, and their DNA binding activities can be regulated by ser/thr kinases, such as MAP kinases (Hunter and Karin, 1992). As mentioned in the Introduction, tyrosine kinases mediated signalling may also control transcription factors through the RAS, RAF, and MAP kinase cascade pathway. One working model proposes that the signals from the ligand-bound receptor diverge into at least two pathways (Figure 4.3). One is the cytoplasmic transcription factors, which are directly tyrosine phosphorylated by receptor associated kinases and then translocated to the nucleus; the other pathway may eventually activate ser/thr kinases through a cascade of reactions. These ser/thr kinases, such as MAP kinase, may stimulate nuclear DNA binding proteins which themselves can induce transcription of response genes to a certain degree. However, if these two kinds of factors, cytoplasmic activated p91-like factors and the nuclear DNA binding proteins, can reconverge to form a complex, a higher affinity and more active complex may be generated, resulting in maximal transcriptional responses. This hypothetical divergence and reconvergence of signals during transcriptional responses to polypeptide ligands (see Figure 4.3) needs to be further tested. However, there are demonstrations that nuclear DNA binding proteins, such as IRF-1, are posttranslationally modified, which seems to be required for their transcriptional activity (Pine, 1992; Watanabe et al, 1991). Recent results also show that p48 of ISGF3, which is probably phosphorylated by ser/thr kinase, fails to bind DNA after phosphatase treatment (Levy, 1994), indicating that the posttranslational ser/thr phosphorylation is required for the function of p48 and ISGF3, although it is not clear yet whether this phosphorylation is regulated by a specific ligand.

Questions and Future Perspectives

Two kinds of protein kinases are involved phosphorylation of p91-like transcription factors: the receptor tyrosine kinases (such as EGF and PDGF receptor tyrosine kinases) and the cytoplasmic tyrosine kinases (such as Jak family kinases). Deficiency of Jak1 kinase in U3A cells results in the failure of tyrosine phosphorylation of p91 in response to either interferon-α and -γ, suggesting Jak1 kinase may phosphorylate p91 (Muller et al, 1993a, b). Similarly, Jak1 kinase is tyrosine phosphorylated in response to EGF; therefore, it is believed that phosphorylation of p91 induced by EGF may also employ Jak1 kinase (Shuai et al, 1993b). The question here is whether Jak family kinase(s) are the only kinases that can phosphorylate p91 or p91-like factors. It is likely that other protein tyrosine kinases may also interact directly with p91. For example, it has been shown that p91 can directly interact with EGF receptor through its SH2 domain (Fu and Zhang, 1993). Therefore, it is possible that EGF receptor kinase may directly phosphorylate p91. Although it does not rule out the possibility that a member of the Jak family of kinases is involved in this p91-EGF receptor complex, it can be determined whether Jak kinase is involved in EGF induced tyrosine phosphorylation of p91 by using cells that are deficient in Jak kinase, such as U4 cells which do not express functional Jak1 kinase. If EGF still can induce tyrosine phosphorylation of p91 in U4 cells, Jak1 is not required for p91 phosphorylation induced by EGF. Actually, there is no direct evidence yet that any member of the Jak family of kinases or a dual Jak kinase may actually phosphorylate p91 or p91-like factors. More experiments are needed to determine whether p91 can be phosphorylated by tyrosine kinases other than Jak family members.

The second question is whether nuclear DNA binding factors other than p48 of ISGF3 can form a complex with activated and translocated p91, or the p91 family of factors, and similarly, whether other leucine zipper containing transcription factors may form a heterodimer with p91 or its related proteins through leucine repeat region in the N-terminal end of p91 family of proteins. A two hybrid system (Fields and Song, 1989) may be useful in finding possible partners of p91 family of proteins. It has been observed that a proline rich region in p48 of ISGF3 may interact with p91 or p113 in the complex (Veals et al, 1993); it will be interesting to determine whether the SH3 domains (Mayer and Baltimore, 1993) in p91 and p113 mediate interactions with this proline rich region.

An important, but less studied, question is how active transcriptional complexes like ISGF3 and GAF interact with the basic transcriptional machinery to initiate transcription. As shown in Figure 4.2A, an acidic region is present at the C-terminal end of p113, which is suspected of being involved in activation (Fu et al, 1992). Similarly, the C-terminal end of p91 seems important for gene activation by interferon-γ, because p84, which is missing the last 38 amino acids at the C-terminal end, fails to complementate the mutant cell line U3 (Muller et al, 1993a, b). Other interesting questions are how do activated transcription complexes become inactivated, and what is the mechanism of transcriptional down-regulation of the responsive genes in the nucleus? One possibility is that some nuclear tyrosine phosphatases may dephosphorylate p91-like factors (David et al, 1993a, b), or these proteins may have a short half-life in the nucleus.

Some growth factors and cytokins, such as EGF, PDGF or IL-3, have been shown to play critical roles in development and early cell differentiation and tumorigenesis. It will be interesting to see whether this tyrosine kinase/p91 signalling pathway has a decisive function in these important biological systems. For example, it will be interesting to determine whether this pathway is essential for cell growth, and whether the high expression of activated p91-like factor(s) will result in cellular transformation. Indeed, constitutive activation of p91 has been implicated in a number of breast cancer cells, and p91 is expressed very early in the development (Chin, Zhang and Fu, 1994), suggesting that this tyrosine kinase/p91 pathway must be widely used in these biological processes.

Acknowledgements

The author was a Markey Research Fellow. I thank Dr. R. Pine for comments on the manuscript.

References

Abdollahi A, Lord KA, Hoffman-Liebermann B, Liebermann DA (1991): Interferon regulatory factor 1 is a myeloid differentiation primary response gene induced by interleukin 6 and leukemia inhibitory factory: role in growth inhibition. *Cell Growth & Diff* 2: 401–407

Aguet JM, Dembic Z, Merlin G (1988): Molecular cloning and expression of the human interferon-γ receptor. *Cell* 55: 273–280

Akira S, Nishio Y, Inoue M, Wang X-J, Wei S, Matsusaka T, Yoshida K, Sudo T, Naruto M, Kishimoto T (1994): Molecular cloning of APRF, a novel IFN-stimulated gene factor 3 p91-related transcription factor involved in the gp130-mediated signaling pathway. *Cell* 77: 63–71

Argetsinger L, Campbell GS, Yang X, Witthuhun BA, Silvennoinen O, Ihle JN, Carter-Su C (1993): Identification of JAK2 as a growth hormone receptor-associated tyrosine kinase. *Cell* 74: 237–244

Bandyopadhyay SK, Kalvakolanu DVR, Sen GC (1990): Gene induction by interferons: Functional complementation between trans-acting factors induced by alpha interferon and gamma interferon. *Mol Cell Biol* 10: 5055–5063

Bernards A (1991). Predicted tyk2 protein contains two tandem protein kinase domains. *Oncogene* 6(7): 1185

Blenis J (1993): Signal transduction via the MAP kinase: proceed at your own risk. *Proc Natl Acad Sci USA* 90: 5889–5892

Bonni A, Frank DA, Schindler C, Greenberg ME (1993): Characterization of a pathway for ciliary neurotrophic factor signaling to the nucleus. *Science* 262: 1575–1579

Borrelli E, Montmayeur J-P, Foulkes NS, Sassone-Corsi P (1992): Signal transduction and gene control: The cAMP pathway. *Crit Rev in Oncogenesis* 3: 321–338

Cochran BH, Reffel AC, Stiles CD (1983): Molecular cloning of gene sequences regulated by platelet-derived growth factor. *Cell* 33: 939–947

Cantley LC, Auger KR, Carpenter C, Duckworth B, Graziani A, Kapeller R, Soltoff S (1991): Oncogenes and signal transduction. *Cell* 64: 939–947

Carter-Su C, Stubbart JR, Wang X, Stred SE, Argetsinger LS, Shafer JA (1989): Phosphorylation of highly purified growth hormone receptors by growth hormone receptor-associated tyrosine kinase. *J Biol Chem* 264: 18654–18661

Chin Y, Zhang J, Fu X-Y (1994): Unpublished results

Cohen B, Peretz D, Vaiman D, Benech P, Chebath J (1988): Enhancer-like interferon responsive sequences of the human and murine (2′–5′) oligoadenylate synthetase gene promoters. *EMBO J* 7: 1411–1419

Dale TC, Imam AMA, Kerr IM, Stark GR (1989): Rapid activation by interferon-α of a latent DNA-binding protein present in the cytoplasm of untreated cells. *Proc Natl Acad Sci USA* 86: 1203–1207

D'Andrea A (1994): Personal communication

Darnell JE (1994): Personal communication

David M, Larner AC (1992): Activation of transcription factors by interferon alpha in a cell free system. *Science* 257: 813–815

David M, Crimley PM, Finbloom DS, Larner AC (1993): A nuclear tyrosine phosphatase downregulates interferon-induced gene expression. *Mol Cell Biol* 13: 7515–7521

David M, Romero G, Zhang Z-Y, Dixon JE, Larner AC (1993b): in vitro activation of transcription factor ISGF3 by IFNα involved a membrane associated tyrosine phosphatase and tyrosine kinase. *J Biol Chem* 268: 6593–6599

Decker T, Lew DJ, Mirkovitch J, Darnell JE Jr (1991): Cytoplasmic activation of GAF, an IFN-α-regulated DNA-binding factor. *EMBO J* 10: 927–932

De Maeyer E, Maeyer-Guignard J (1988): *Interferons and Other Regulatory Cytokines.* New York: John Wiley & Sons

Driggers PH, Ennist DL, Gleason SL, Mak W-H, Marks B-Z, Levi J, Flanagan R, Appella E, Ozato K (1990): An interferon-β-regulated protein that binds the interferon-inducible enhancer element of major histocompatibility complex class I genes. *Proc Natl Acad Sci USA* 87: 3743–3747

Fields S, Song O (1989): A novel genetic system to detect protein-protein interactions. *Nature* 340: 245–246

Finbloom DS, Petricoin EF, Hackett RH, David M, Feldman GM, Igarashi K-I, Fibach E, Weber MJ, Thorner MO, Silva CM, Larner AC (1994): Growth hormone and erythropoietin differentially activate DNA-binding proteins by tyrosine phosphorylation. *Mol Cell Biol* 14: 2113–2118

Friedman RL, Manly SP, McMahon M, Kerr IM, Stark GR (1984): Transcriptional and posttranscriptional regulation of interferon-induced gene expression in human cells. *Cell* 38: 745–755

Fu X-Y (1994): Unpublished results

Fu X-Y (1992): A transcription factor with SH2 and SH3 domains is directly activated by an interferon α-induced protein tyrosine kinase(s) *Cell* 70: 323–335

Fu X-Y, Zhang J-J (1993): Transcription factor p91 interacts with the EGF receptor and mediates activation of the c-fos gene promoter. *Cell* 74: 1135–1145

Fu X-Y, Kessler DS, Veals SA, Levy DE, Darnell JE (1990): ISGF3, the transcriptional activator induced by interferon-α, consists of multiple interacting polypeptide chains. *Proc Natl Acad Sci USA* 87: 8555–8559

Fu X-Y, Schindler C, Improta T, Aebersold R, Darnell JE (1992): The proteins of ISGF-3, the interferon α-induced transcriptional activator, define a gene family involved in signal transduction. *Proc Natl Acad Sci USA* 89: 7840–7843

Goodbourn SEY, Maniatis T (1988): Overlapping positive and negative regulatory domains of the human β-interferon gene. *Proc Natl Acad Sci USA* 85: 1135–1451

Greenberg ME, Ziff EB (1984): Stimulation of 3T3 cells induces transcription of c-fos proto-oncogene. *Nature* 311: 433–438

Gutch MJ, Daly C, Reich NC (1992): Tyrosine phosphorylation is required for activation of an α-interferon-stimulated transcription factor. *Proc Natl Acad Sci USA* 89: 11411–11415

Hannigan GE, Williams BRG (1991): Signal transduction by interferon-α through arachidonic acid metabolism. *Science* 251: 204–207

Harada H, Fujita T, Miyamoto M, Kimura Y, Maruyama M, Furia A, Miyata T, Taniguchi T (1989): Structurally similar but functionally distinct factors, IRF-1 and IRF-2, bind to the same regulatory elements of IFN and IFN-inducibgenes. *Cell* 58: 729–739

Hirano T, Kishimoto T (1990): Interleukin-6. In: *Peptide Growth Factors and Their Receptors Vol 1*. Sporn MB, Roberts AB, eds. Berlin: Springer-Verlag

Hunter T (1993): Cytokine connections. *Nature* 366: 114–116

Hunter T, Karin M (1992): The regulation of transcription by phosphorylation. *Cell* 70: 375–387

Igarashi K, David M, Finbloom DS, Larner AC (1993a): In vitro activation of the transcription factor gamma interferon-activating factor by gamma interferon: evidence for a tyrosine phosphatase/kinase signaling cascade. *Mol Cell Biol* 13: 1634–1640

Igarashi K-I, David M, Larner AC, Finbloom DS (1993b): In vitro activation of a transcription factor by gamma interferon requires a membrane-associated tyrosine kinase and is mimicked by vanadate. *Mol Cell Biol* 13: 3984–3989

Ip NY, Nye SH, Boulton TG, David S, Taga T, Li Y, Birren SJ, Yasukawa K, Kishimoto T, Anderson DJ, Stahl N, Yancopoulos GD (1992): CNTF and LIF act on neuronal cells via shared signaling pathways that involve the IL-6 signal transducing receptor component gp130. *Cell* 69: 1121–1132

John J, McKendry R, Pellegrini S, Flavell D, Kerr IM, Stark GR (1991): Isolation and characterization of a new mutant human cell line unresponsive to alpha and beta interferons. *Mol Cell Biol* 11: 4189–4195

Kastelein RA, Shanafelt AB (1993): GM-CSF receptor: interactions and activation. *Oncogene* 8: 231–236

Kessler DS, Levy DE (1991): Protein kinase activity required for an early step in interferon-α signalling. *J Biol Chem* 266: 23471–23476

Kessler DS, Veals SA, Fu X-Y, Levy DE (1990): IFN-alpha regulates nuclear translocation and DNA-binding affinity of ISGF3, a multimeric transcriptional activator. *Genes Dev* 4: 1753–1765

Khan KD, Lindwall G, Maher SE, Bothwell ALM (1990): Characterization of promote elements of an interferon-inducible Ly-6E/A differentiation antigen, which is expressed on activated T cells and hematopoietic stem cells. *Mol Cell Biol* 10: 5150–5159

Koch CA, Anderson D, Moran MF, Ellis C, Pawson T (1991): SH2 and SH3 domains: elements that control interactions of cytoplasmic signaling proteins. *Science* 252: 668–674

Kondo M, Takeshita T, Ishii N, Nakamura M, Watanabe S, Arai K-I, Sugamura K (1993): Sharing of the interleukin-2 (IL-2) receptor-chain between receptors for IL-2 and IL-4 *Science* 262: 1874–1877

Kotanides H, Reich NC (1993): Requirement of tyrosine phosphorylation for rapid activation of a DNA binding factor by IL-4. *Science* 262: 1265–1267

Landschulz WH, Johnson PF, McKnight SL (1988): The leucine zipper: a hypothetical structure common to a new class of DNA binding proteins. *Science* 240: 1759–1764

Larner AC (1993): Tyrosine phosphorylation of DNA binding proteins by multiple cytokines. *Science* 261: 1730–1733

Larner AC, Jonak G, Cheng Y-SE, Korant B, Knight E, Darnell JE Jr (1984): Transcriptional induction of two genes in human cells by α-interferon. *Proc Natl Acad Sci USA* 81: 6733–6737

Levy DE (1994): Personal communication

Levy DE, Kessler DS, Pine R, Reich N, Darnell JE Jr (1988): Interferon-induced nuclear factors that bind a shared promoter element correlate with positive and negative control. *Genes Dev* 2: 383–393

Levy DE, Kessler DS, Pine R, Darnell JE Jr (1989): Cytoplasmic activation of ISGF3, the positive regulator of interferon-α-stimulated transcription, reconstituted in vitro. *Genes Dev* 3: 1362–1371

Lew DE, Decker T, Strehlow I, Darnell JE Jr (1991): Overlapping elements in the guanylate-binding protein gene promoter mediate transcriptional induction by alpha and gamma interferons. *Mol Cell Biol* 11: 182–191

Lutticken C, Wegenka UM, Yuan J, Buschmann J, Schindler C, Ziemiecki A, Harpur AG, Wilks AF, Yasukawa K, Taga T, Kishimoto T, Barbieri G, Pellegrini S, Sendtner M, Heinrich PC, Horn F (1994): Association of transcription factor APRF and protein kinase Jak1 with the interleukin-6 signal transducer gp130. *Science* 263: 89–92

Majerus PW, Ross TS, Cunningham TW, Caldwell KK, Jefferson AB, Bansal VS (1990): Recent insights in phosphatidylinositol signaling. *Cell* 63: 59–465

Mao C, Davies D, Kerr IM, Stark GR (1993): Mutant human cells defective in induction of major histocompatibility complex class II genes by interferon-γ. *Proc Natl Acad Sci USA* 90: 2880–2884

Matsuyama T, Kimura T, Kitagawa M, Pfeffer K, Kawakami T, Watanabe N, Kundig TM, Amakawa R, Kishihara K, Wakeham A, Potter J, Furlonger CL, Narendran A, Suzuki H, Ohashi PS, Paige CJ, Taniguchi T, Mak TW (1993): Targeted disruption of IRF-1 or IRF-2 results in abnormal type I IFN gene induction and aberrant lymphocyte development. *Cell* 75: 83–97

Mayer BM, Baltimore D (1993): Signaling through SH2 and SH3 domains. *Trends Cell Biol* 3: 8–13

Mayer BJ, Hanafusa H (1990): Association of the v-crk oncogene product with phosphotyrosine-containing proteins and protein kinase activity. *Proc Natl Acad Sci USA* 87: 2638–2642

McKendry R, John J, Flavell D, Muller M, Kerr IM, Stark GR (1991): High-frequency mutagenesis of human cells an characterization of a mutant unresponsive to both a and b interferons. *Proc Natl Acad Sci USA* 88: 11455–11459

Miyamoto M, Fujita T, Kimura Y, Maruyama M, Harada H, Sudo Y, Miyata T, Taniguchi T (1988): Regulated expression of a gene encoding a nuclear factor, IRF-1, that specifically binds to IFN-b gene regulatory elements. *Cell* 54: 903–913

Muller M, Lexton C, Briscoe J, Schindler C, Importa T, Darnell JE, Stark GR, Kerr IM (1993a): Complementation of a mutant cell line: central role of the 91 kD polypeptide of ISGF3 in the interferon-α and -γ signal transduction pathway. *EMBO J* 12: 4221–4228

Muller M, Briscoe J, Laxton C, Guschin D, Ziemiecki A, Silvennoinen O, Harpur AG, Barbieri G, Witthuhn BA, Schindler C, Pellegrini S, Wilks AF, Ihle JN, Stark GR, Kerr IM (1993): The protein tyrosine kinase JAK1 complements defects in interferon-α/β and -γ signal transduction. *Nature* 366: 129–135

Murakami M, Hibi M, Nakagawa N, Nakagawa T, Yasukawa K, Yamanishi K, Taga T, Kishimoto T (1993): IL-6-induced homodimerization of gp130 and associated activation of a tyrosine kinase. *Science* 260: 1808–1810

Noguchi M, Nakamura Y, Russell SM, Ziegler SF, Tsang M, Cao X, Leonard WJ (1993): Interleukin-2 receptor γ chain: A functional component of the interleukin-7 receptor. *Science* 262: 1877–1880

Overduin M, Rios CB, Mayer BJ, Baltimore D, Cowburn D (1992): Three dimensional solution structure of the scr homology 2 domain of c-abl. *Cell* 70: 697–704

Pearse RN, Feinman R, Ravetch JV (1991): Characterization of the promoter of the human gene encoding the high-affinity IgG receptor: transcriptional induction by γ-interferon is mediated through common DNA response elements. *Proc Natl Acad Sci USA* 88: 11305–11309

Pearse RN, Feinman R, Shuai K, Darnell JE Jr, Ravetch JV (1993): Interferon γ-induced transcription of the high-affinity Fc receptor for IgG requires assembly of a complex that includes the 91-kDa subunit of transcription factor ISGF3. *Proc Natl Acad Sci USA* 90: 4314–4318

Pellegrini S, John J, Shearer M, Kerr IM, Stark GR (1989): Use of a selectable marker regulated by alpha interferon to obtain mutations in the signaling pathway. *Mol Cell Biol* 9: 4605–4612

Perez C, Wietzerbin J, Benech PD (1993): Two cis-DNA elements involved in myeloid-cell-specific expression and gamma interferon (IFN-γ) activation of the human high-affinity Fcγ receptor gene: a novel IFN regulatory mechanism. *Mol Cell Biol* 13: 2182–2192

Pfeffer LM, Eisenkraft BL, Reich NC, Improta T, Baxter G, Daniel-Issakani S, Strulovici B (1991): Transmembrane signaling by interferon-α involves diacylglycerol production and activation of the {ge} isoform of protein kinase C in daudi cells. *Proc Natl Acad Sci USA* 88: 7988–7992

Pine R (1992): Constitutive expression of an ISGF2/IRF1 transgene leads to interferon-independent activation of interferon-inducible genes and resistance to virus infection. *J Virology* 66: 4470–4478

Pine R, Canova A, Schindler C (1994): Tyrosine phosphorylated p91 binds to a single element in the ISGF2/IRF-1 promoter to mediate induction by IFNa and IFNg, and is likely to antoregulate the p91 gene. *EMBO J* 13: 158–167

Pine R, Decker T, Kessler DS, Levy DE, Darnell JE Jr (1990): Purification and cloning of interferon-stimulated-gene factor 2: ISGF2 (IRF-1) can bind to the promoters of both beta interferon and interferon- stimulated genes but is not a primary transcriptional activator of either. *Mol Cell Biol* 10: 2448–2457

Pledger WJ, Hart CA, Locatell KL, Scher CD (1981): Platelet-derived growth factor-modulated proteins: Constitutive synthesis by a transformed cell line. *Proc Natl Acad Sci USA* 78: 4358–4362

Porter ACG, Chernajovsky Y, Dale TC, Gilbert CS, Stark GR, Kerr IM (1988): Interferon response element of the human gene 6–16. *EMBO J* 7: 85–92

Ptashne M (1988): How eukaryotic transcriptional activators work. *Nature* 355: 683–689

Reich N, Darnell Jr JE(1989): Differential binding of interferon-induced factors to an oligonucleotide that mediates transcriptional activation. *Nucleic Acids Res* 17: 3415–3424

Reich N, Evans B, Levy DE, Fahey D, Knight E, Darnell JE Jr (1989): Interferon-induced transcription of a gene encoding a 15 kDa protein depends on an upstream enhancer element. *Proc Natl Acad Sci USA* 84: 6394–6398

Riddle VGH, Dubrow R, Pardee AB (1979): Changes in the synthesis of actin and other cell proteins after stimulation of serum-arrested cells. *Proc Natl Acad Sci USA* 76: 1298–1302

Ruff-Jamison S, Chen K, Cohen S (1993): Induction by interferon-γ of tyrosine phosphorylated DNA binding proteins in mouse liver nuclei. *Science* 261: 1733–1736

Ruffner H, Reis LFL, N f D, Weissmann C (1993): Induction of type I interferon genes and interferon-inducible genes in embryonal stem cells devoid of interferon regulatory factor 1. *Proc Natl Acad Sci USA* 90: 11503–11507

Russell SM, Keegan AD, Harada N, Nakamura Y, Noguchi M, Leland P, Friedmann MC, Miyajima A, Puri RK, Paul WE, Leonard WJ (1993): Interleukin-2 receptor γ chain: A functional component of the interleukin-4 receptor. *Scien* 1880–1883

Rutherford MN, Hannigan GE, Williams BRG (1988):0 Interferon-induced binding of nuclear factors to promoter elements of the 2–5A synthetase gene. *EMBO J* 7: 751–759

Sadowski HB, Gilman MZ (1993): Cell-free activation of a DNA binding protein by epidermal growth factor. *Nature* 362: 79–83

Sadowski HB, Shuai K, Darnell JE, Gilman MZ (1993): A common nuclear signal transduction pathway activated by growth factor and cytokine receptors. *Science* 261: 1739–1744

Schindler C, Fu X-Y, Importa T, Aebersold R, Darnlee JE Jr (1922a): Proteins of transcription factor ISGF-3: one gene encodes the 91- and 84-kDa ISGF-3 proteins that are activated by interferon α. *Proc Natl Acad Sci USA* 89: 7836–7839

Schindler C, Shuai K, Prezioso VR, Darnell JE Jr (1992b): Interferon-dependent tyrosine phosphorylation of a latent cytoplasmic transcription factor. *Science* 257: 809–813

Schlessinger J, Ullrich A (1992): Growth factor signaling by receptor tyrosine kinases. *Neuron* 9: 383–391

Sen GC, Ransohoff RM (1993): Interferon-induced antiviral actions and their regulation. In: *Advances in Virus Research*, 42, Academic Press, Inc

Shirayoshi Y, Burke PA, Appella E, Ozato K (1988): Interferon-induced transcription of a major histocompatibility class I gene accompanies binding of inducible nuclear factors to the interferon consensus sequence. *Proc Natl Acad Sci USA* 85: 5884–5888

Shuai K, Hovath CM, Huang LHT, Qureshi SA, Cowburn D, Darnell JE (1994): Interferon activation of the transcription factor STAT91 involves dimerition through SH2-phosphotyosyl peptide interactions. *Cell* 76: 821–828

Shuai K, Schindler C, Presiozo VR, Darnell JE Jr (1992): Activation of transcription by IFNγ: tyrosine phosphorylation of a 91 *kDa DNA binding protein*. *Science* 258: 1808–1812

Shuai K, Stark GR, Kerr IM, Darnell JE (1993a): A single phosphotyrosine residue of stat91 required for gene activation by interferon-γ. *Science* 261: 1744–1746

Shuai K, Ziemiecki A, Wilks AF, Harpur AG, Sadowski HB, Gilman MZ, Darnell JE (1993b): Polypeptide signaling to the nucleus through tyrosine phosphorylation of Jak and STAT proteins. *Nature* 366: 580–583

Silvennoinen O, Schindler C, Schlessinger J, Levy DE (1993a): Ras-independent growth factor signalling by transcription factor tyrosine phosphorylation. *Science* 261: 1736–1739

Silvennoinen O, Witthuhn B, Quelle FW, Cleveland JL, Yi T, Ihle JN (1993b): Structure of the JAK2 protein tyrosine kinase and its role in IL-3 signal transduction. *Proc Natl Acad Sci USA* 90: 8429–8433

Stahl N, Boulton TG, Farruggella T, Ip NY, Davis S, Witthuhn BA, Quelle FW, Silvennoinen O, Barbieri G, Pellegrini S, Ihle JN, Yancopoulos JD (1994): Association and activation of Jak-Tyk kinases by CNTF-LIF-OSM-IL-6 γ receptor components. *Science* 263: 92–95

Stahl N, Yancopoulos GD (1993): The alphas, betas, and kinases of cytokine receptor complexes. *Cell* 74: 587–590

Stark GR (1994): Personal communication

Stark GR, Kerr IM (1992): Interferon-dependent signaling pathways: DNA elements, transcriptional factors, mutations, and effects of viral proteins. *J Interferon Res* 12: 147–151

Taga T, Hibi M, Hirata Y, Yamasaki K, Yasukawa K, Matsuda T, Hirano T, Kishimoto T (1989): Interleukin-6 triggers the association of its receptor with a possible signal transducer, gp130. *Cell* 58: 573–581

Uze G, Lutfalla G, Gresser I (1990): Genetic transfer of a functional human interferon-α receptor into mouse cells. Cloning and expression of its cDNA. *Cell* 60: 225–234

Veals SA, Santa Maria T, Levy DE (1993): Two domains of ISGF3-γ that mediate protein-DNA and protein-protein interactions during transcription factor assembly contribute to DNA-binding specificity. *Mol Cell Biol* 13: 196–206

Veals SA, Schindler C, Leonard D, Fu X-Y, Aebersold R, Darnell JE Jr, Levy DE (1992): Subunit of an alpha-responsive transcription factor is related to interferon regulatory factor and Myb families of DNA-binding proteins. *Mol Cell Biol* 12: 3315–3324

Velazquez L, Fellous M, Stark RF, Pellegrini S (1992): A protein tyrosine kinase in the interferon α/γ signaling pathway. *Cell* 313–322

Wagner BJ, Hayes TE, Hoban CJ, Cochran BH (1990): The SIF binding element confers sis/PDGF inducibility onto the c-fos promoter. *EMBO J* 9: 4477–4484

Waksman G, Kominos D, Robertson SC, Pant N, Baltimore D, Birge RB, Cowburn D, Hanafusa H, Mayer BJ, Overduin M, Resh MD, Rios CB, Silverman L, Kuriyan J (1992): Crystal structure of the phosphotyrosine recognition domain SH2 of v-src complexed with tyrosine-phosphorylated peptides. *Nature* 358: 646–653

Waksman G, Shoelson SE, Pant N, Cowburn D, Kuriyan J (1993): Binding of a high affinity phosphotyrosyl peptide to the Src domain: crystal structures of the complexed and peptide-free forms. *Cell* 72: 779–790

Watanabe N, Sakakibara J, Hovanessian AG, Taniguchi T, Fujita T (1991): Activation of IFN-β element by IRF-1 requires a posttranslational event in addition to IRF-1 synthesis. *Nucleic Acids Res* 19: 4421–4428

Watling D, Guschin D, Muller M, Silvennoinen O, Witthuhn BA, Quelle FW, Rogers NC, Schindler C, Stark GR, Ihle JN, Kerr IM (1993): Complementation by the protein tyrosine kinase JAK2 of a mutant cell line defective in the interferon-γ signal transduction pathway. *Nature* 11: 166–170

Wegenka UM, Buschmann J, Lutticken C, Heinrich PC, Horn F (1993): Acute-phase response factor, a nuclear factor binding to acute-phase response elements, is rapidly activated by interleukin-6 at the posttranslational level. *Mol Cell Biol* 13: 276–288

Wilks AF, Harpur A, Kurban RR, Ralph SJ, Zurcher G, Ziemiecki A (1991): Two novel protein-tyrosine kinases, each with a second phosphotransferase-related catalytic domain, define a new class of protein kinase. *Mol Cell Biol* 11: 205–2065

Williams BRG (1991): Transcriptional regulation of interferon- stimulated genes. *Eur J Biochem* 200: 1–11

Wilson KC, Finbloom DS (1992): Interferon-γ rapidly induces in human monocytes a DNA-binding factor that recognizes the γ response region within the promoter of the gene for the high-affinity Fcγ receptor. *Proc Natl Acad Sci USA* 89: 11964–11968

Witthuhn BA, Quelle FW, Silvennoinen O, Yi T, Tang B, Miura O, Ihle JN (1993): JAK2 associates with the erythropoietin receptor and is tyrosine phosphorylated and activated following stimulation with erythropoietin. *Cell* 74: 227–236

Wright TM, Farber JM (1991): 5′ Regulatory region of a novel cytokine gene mediates selective activation by interferon gamma. *J Exp Med* 173: 417–422

Yap WH, Teo TS, Tan YH (1986): *Science* 234: 355–358

Yuan J, Wegenka UM, Lutticken C, Buschmann J, Decker T, Schindler C, Heinrich PC, Horn F (1994): The signalling pathways of interleukin-6 and gamma interferon converge by the activation of different transcription factors which bind to common responsive DNA elements *Mol Cell Biol* 14: 1657–1668

Yu-Lee L-Y, Hrachovy JA, Stevens AM, Schwarz LA (1990): Interferon-regulatory factor 1 is an immediate-early gene under transcriptional regulation by prolactin in Nb2 T cells. *Molec Cell Biol* 10: 3087–3094

Zhong Z, Wen Z, Darnell JE (1994): A STAT family member activated by tyrosine phosphorylation in response to epidermal growth factor and interleukin-6. *Science* 264: 95–98

5

The Glucocorticoid Hormone Receptor

Martin Eggert, Marc Muller and Rainer Renkawitz

Introduction

The glucocorticoid receptor (GR) is present in nearly all mammalian tissues and mediates the cellular response to glucocorticoids. The GR belongs to the superfamily of steroid hormone receptors (Evans, 1988) and was the first transcription factor to be isolated and studied in detail. It is structured in three domains: the N-terminal domain, the DNA binding domain (DBD) and the C-terminal hormone binding domain (HBD).

The N-terminus is required for full transcriptional activity, and the DBD, encompassing approximately 80 amino acids (aa), contains two zinc fingers which mediate DNA binding. The HBD confers ligand binding to the receptor and promotes DNA binding and transcriptional activation. In the absence of hormone, the HBD represses receptor function and maintains the inactive state of the GR. The inactive GR, associated with heat shock proteins (hsp) with molecular weights of 90 and 59 kDa, is localized in the cytoplasm. In the presence of hormone, this cytoplasmic complex dissociates, and the hormone loaded receptor dimerizes and translocates to the nucleus. Subsequently, the GR binds to specific DNA sequences, named glucocorticoid response elements (GRE), resulting in activation or inhibition of gene transcription.

The glucocorticoid receptor is autoregulated as we know because it has been shown in most cells that prolonged glucocorticoid induction reduces GR-levels by a process termed homologous down regulation (Bellingham et al, 1992; Oakley and Cidlowski, 1993). Possible mechanisms for this down regulation include an alteration in GR mRNA stability or a down regulation of GR transcription by GR and/or other transcription factors. An exception for this hormone mediated autoregulation is found in T cells which show an increase in GR mRNA levels after continuous hormone induction (Denton et al, 1993). Nevertheless, the GR autoregulation appears to be a highly conserved mechanism for attenuating cellular responsiveness to hormone.

INDUCIBLE GENE EXPRESSION, VOLUME 2
P.A. Baeuerle, Editor
© 1995 Birkhäuser Boston

After characterization of the receptor molecule and of the response elements, research was focused on the mechanisms of transcriptional activation. In this review, we will summarize the knowledge about structure and function of GR, and focus on synergizing and modulating factors that influence GR driven gene expression. Furthermore, we will discuss putative bridging factors which may mediate an interaction of the DNA bound GR and the transcription initiation complex.

Structure of the GR

The GR was characterized in several organisms including man (hGR), mouse (mGR) and rat (rGR) in the 1980s. The primary structure was determined (Hollenberg et al, 1985; Danielsen et al, 1986; Miesfeld et al, 1986) and revealed a receptor consisting of 777 aa (hGR), 783 aa (mGR) and 795 aa (rGR) with a molecular mass of about 86–94 kDa as shown by SDS-PAGE followed by GR specific immunoblotting. Characterization of domains was achieved by functionally analyzing receptor mutants, such as point mutations, insertions or deletions as well as chimeric proteins containing defined regions of the GR (Giguere et al, 1986; Danielsen et al, 1987; Hollenberg and Evans, 1988). The results of these studies led to a GR model containing at least four main functional domains, the DBD, the HBD and two transactivation domains (Figure 5.1). These and other characteristics such as nuclear translocation signals and phosphorylation will be discussed below.

The DNA binding domain

In several functional assays, the DBD of the GR was assigned to a region between aa 421 to aa 488 (Hollenberg et al, 1987; Hollenberg and Evans, 1988) in man, and studies in mouse and rat revealed a similar region responsible for DNA binding (Danielsen et al, 1987; Godowski et al,

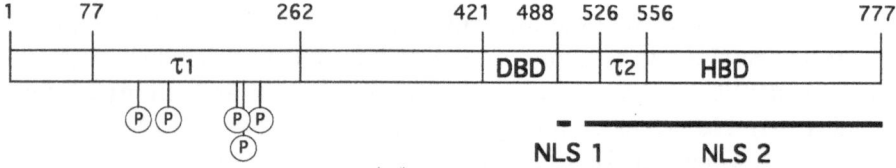

Figure 5.1 Linear structure of the human GR. Indicated are the transactivating domains (τ1 and τ2), the DNA binding domain (DBD), the hormone binding domain (HBD), and the nuclear translocation signals (NLS1 and NLS2). Serines conserved in hGR and rGR, which were found to be phosphorylated in mGR, are indicated (P).

1988). The DBD is positioned in the central basic part of the GR, which is rich in lysine, arginine and cysteine residues (Danielsen et al, 1987). The structural motif of the DBD is two zinc fingers formed by coordination of four cysteines to one zinc atom (Freedman et al, 1988). Site directed mutagenesis revealed that seven out of the eight cysteines are absolutely required for receptor function (Severne et al, 1988), which strongly supports the model of zinc finger mediated DNA binding of the GR. The major groove of the DNA double helix has been defined as the contact area (Miesfeld et al, 1987). More precisely, it was shown that the amino terminal zinc finger is involved in the specificity of the binding to a GRE (Danielsen et al, 1989; Umesono and Evans, 1989). In this experiment the Gly-Ser sequence between the distal pair of cysteines in the amino terminal zinc finger was changed to Glu-Gly residues, typical for the zinc finger of the estrogen receptor (ER). This led to an activated transcription through an estrogen response element (ERE) rather than through a GRE. The GRE specificity seemed to be achieved by 3 amino acids positioned adjacent to the first zinc finger, which has been confirmed by recent data (Zilliacus et al, 1992). Especially the Val residue 443 (hGR) appears to form both positive contacts with specific basepairs in the cognate binding site and negative contacts in the noncognate site. The second carboxy terminal zinc finger is also required for DNA binding, since several types of mutations in this area generate an inactive receptor (Hollenberg et al, 1987; Danielsen et al, 1987; Hollenberg and Evans, 1988).

The three-dimensional structure of the GR-DBD was solved by two-dimensional nuclear magnetic resonance (Härd et al, 1990). In this study a DBD fragment produced in *Escherichia coli* was investigated. The resulting model confirmed the occurrence of two zinc fingers and determined an α helical structure between both fingers contacting the DNA. In addition, the model predicted that the amino acids positioned at the amino terminal base of the second finger are in close contact in the homodimeric receptor complex bound to a palindromic GRE. Five amino acids comprise the D-box, which is involved in homodimerization of the GR (Dahlmann-Wright et al, 1991) by interacting with the equivalent part of the other DBD in a GR homodimer (Dahlmann-Wright et al, 1993). Taken together, this indicates that the DBD of GR is important for both DNA binding and dimerization. A second important dimerization domain resides in the HBD (see below). Although the GR normally seems to bind as a homodimer to a GRE, one exception should be noted. Drouin et al were able to show that the GR appears to bind as a homotrimer to a negative GRE (nGRE) of the pro-opiomelanocortin (POMC) gene They suggested that first a GR homodimer binds to this nGRE, followed by the binding of a GR monomer to this complex on the opposite side of the double helix (Drouin et al, 1993).

The hormone binding domain

The hormone binding domain (HBD) is located in the carboxy terminal part of the GR (Hollenberg et al, 1987; Miesfeld et al, 1987; Danielsen et al, 1987). More precise studies have revealed that amino acids involved in hormone binding cover a rather large domain (Rusconi and Yamamoto, 1987; Pratt et al, 1988). Only a short region termed hinge is excluded from ligand binding (Hollenberg et al, 1987), since a receptor lacking residues 490–515 retains binding activity. Therefore, the N-terminal boundary of the HBD seems to be located near aa 515. The C-terminal end of the HBD is positioned near to the C-terminus of the receptor, since deletions cause a decrease in hormone binding affinity (Rusconi and Yamamoto, 1987). In addition, the hGR β which is homologous to hGR α except for the last 50 amino acids of hGR α is unable to bind hormone (Hollenberg et al, 1985). Covalent affinity labelling of the GR by a steroid led to the identification of amino acid residues Met 622, Cys 656 and Cys 754 of rGR (Simons et al, 1987; Carstedt-Duke et al, 1988). These residues are located in hydrophobic segments within the HBD, providing evidence for a three dimensional structure where the ligand is bound in hydrophobic pocket.

Deletion of the HBD produces a C-terminal truncated receptor which is constitutively active (Hollenberg et al, 1987; Miesfeld et al, 1987), and which exhibits nearly wild-type transactivation capacity independent of ligand. The hormone-responsiveness of the HBD was shown to be transferrable since several chimeric proteins containing the HBD, fused to an unrelated DNA binding domain, confer hormone inducible gene activity (Hollenberg and Evans, 1988; Godowski et al, 1988). Furthermore, the transactivation potency of the Ela protein fused to the HBD of GR is repressed in the absence of hormone (Picard et al, 1988). Taken together, these results support the idea that the HBD might function as a repressor of transactivation in the absence of hormone.

Additionally, as judged from experiments with other steroid receptors, the HBD seems to be involved in dimerization. Studies on the estrogen and the thyroid hormone receptor revealed a region important for dimerization within the ligand binding domain (Forman et al, 1989; Glass et al, 1989; Fawell et al, 1990). Sequence comparison showed that all members belonging to the nuclear receptor gene superfamily contain conserved residues within the ligand binding domain which may be organized as α helical heptad repeats (Laudet et al, 1992).

The transactivation domains

The human glucocorticoid receptor contains two transactivation domains, termed τ1 and τ2. The τ1-domain corresponds to the enh 2 domain of rGR, and in mGR the homologous area is also involved in enhancement (Hollenberg and Evans, 1988; Godowski et al, 1988; Danielsen et al, 1987). The τ1 domain and the enh 2 domain mediate transcriptional activation, since they confer constitutive activation when fused to unrelated DNA binding domains (e.g. GAL4-DBD). Analysis of deletion mutants revealed that the τ1 domain is located between aa 77 and aa 262 of the hGR (Hollenberg and Evans, 1988). The τ2 domain of hGR has no exact counterpart in mouse and rat, since it is located adjacent to the hinge region of hGR and not the enh 1 domain of rGR which overlaps with the DBD (Hollenberg and Evans, 1988: Godowski et al, 1988). The τ2 domain mediates transactivation, independent from the position within the recombinant protein, as shown by hGR constructs containing two τ2 domains in the N-terminal part of the receptor (Hollenberg and Evans, 1988). The position of the 30 amino acid long τ2 domain has been situated between amino acid residues 526 and 556 of the hGR. Taken together the τ1 and τ2 domains are essential for transactivation and act in a position independent manner. Both protein sequences are structurally unrelated but share an acidic character.

Conclusions

The GR contains 3 functional types of domains: (1) the DBD which mediates correct and specific DNA binding of the GR to a GRE and is important for receptor dimerization on the palandromic response element; (2) the HBD which confers ligand binding to the receptor, suppresses receptor activity in the absence of a ligand, and mediates dimerization in solution in the presence of ligand; and (3) the transactivation domains which mediate transcriptional activity. All of these domains act independent of their position and, as recently reported, they do not need to be present in a cis configuration, but can act in trans as well (Spanjaard and Chin, 1993). This supports the hypothesis that these functional domains might have evolved from separate genes.

GR-Phosphorylation

Posttranslational modifications that may also influence receptor function have been investigated. In contrast to glycosylation, which has not been detected (Yen and Simons, 1991), GR phosphorylation has been found. During the last five years, two phosphorylated forms of the GR have been characterized (Orti et al, 1989). One, described as the basal phosphory-lated form of the GR, is found when inactive GR is complexed with the heatshock proteins in the cytoplasm; the other, called the hyperpho-sphorylated form of the GR, is found when GR is activated by hormone and becomes translocated to the nucleus. The role of this phosphoryla-tion still remains unclear, but there may be evidence that phosphorylation plays a role in GR-induced transactivation.

Recently, seven main phosphorylation sites in the hyperphosphory-lated mGR have been identified at serines 122, 150, 212, 220, 234, and 315 and threonine 159 by solid phase sequencing (Bodwell et al, 1991). Except for Ser 315 and Thr 159 these sites are conserved in both rat and man. All these phosphorylation sites are in the amino terminal part of the GR adjacent to the DBD in agreement with earlier results (van der Weyden et al, 1990; Hoeck and Groner, 1990), and, except for the nonconserved Ser 315, are part of the transactivation domain ($\tau 1$) of the GR. However, these seven sites account for only 80 per cent of the total phosphate incorporation in the hyperphosphorylated GR and, therefore, other yet unidentified phosphorylation site(s) may exist (Bodwell et al, 1991). Some of the seven phosphorylation sites found in hyperphosphorylated GR match consensus sequences for known kinases. Since there are only Ser/Thr-phosphorylations found in the hyperphosphorylated GR, the kinases involved belong to the Ser/Thr specific kinases. The serine residues 212, 220 and 234 are all phosphoacceptor-sites in putative consensus sequences for the proline directed kinase and/or the cdc2 kinase. A similar motif predominance has been found within the progesterone receptor (Denner et al, 1990). Moreover, Ser 122 is the phosphorylation site of a putative CKII motif (Bodwell et al, 1991). In several studies, a correlation between phosphorylation state and cellular localization has been investigated (Housley and Pratt, 1983; Bodwell et al, 1991; Orti et al, 1993). It can be shown that GR hyperphosphorylation is hormone-dependent. The phosphorylation state of GR when induced by hormone increases up to fivefold in NIH3T3 cells (Hoeck et al, 1989). Therefore, hyperphosphorylation is found in activated GR, both when translocated to the nucleus or when already bound in the nucleus. The inactive cytoplasmic complexed GR appears to be only basally phosphorylated (Orti et al, 1989; Tashima et al, 1990).

It has been predicted that GR phosphorylation alters the GR transac-tivation potency, but the mechanism remains unclear. Phosphorylation therefore might be just a prerequisite for efficient specific DNA-binding (Bodwell et al, 1991) followed by transactivation, and/or GR-hyperphosphorylation might be important for GR interaction with other transcription factors. The observation that hyperphosphorylated GR, when bound to the DNA, becomes dephosphorylated in at least two sites (Orti et al, 1989) suggests that GR hyperphosphorylation is only required for the initial steps in receptor dependent transactivation. Furthermore, it can be shown that PKA and PKC activators increase the glucocorticoid induced gene expression (Moyer et al, 1993), without altering the state of receptor phosphorylation. Possibly the binding of putative partners to the GR, essential for glucocorticoid-induced gene expression, is modulated via the phosphorylation state of these partners. However, site directed mutagenesis of the seven hyperphosphorylation sites in mouse GR does not reveal major differences in receptor dependent transactivation. The transactivation potency of single point mutated GR shows no difference to wt-GR. Even multiple substitutions show only weak effects and not a significant decrease in transactivation as expected (Mason et al, 1993). This is particularly surprising since the mutation affected the three serines in the acidic region of $\tau 1$, which is necessary for full GR-transactivation.

The Inactive GR

In the absence of hormone, the GR is present in the cytoplasm; only after activation by hormone does the GR translocate to the nucleus. The inactive GR is associated with other factors in a multiprotein complex. Cytosolic extracts of cell homogenates reveal a receptor complex that sediments at 9S in density gradients (Vedeckis, 1983) corresponding to a molecular mass of 300–330 kDa (Vedeckis, 1983; Rexin et al, 1988a; Gehring, 1993). This 9S form can bind glucocorticoids and then change to a 4S (100 kDa) form which is now able to bind DNA in vitro after activation by increasing the temperature to 25°C (Sanchez et al, 1987; Denis et al, 1988). Copurifications showed that the 9S form consisted of 3 main components, the 100 kDa receptor protein, a 90 kDa protein and a 59 kDa protein. The 90 kDa and 59 kDa proteins are known to belong to the family of heat shock proteins (hsp). In crosslinking studies, the high molecular weight complex has been further characterized, and the in vivo association of hsp 90 to GR has been demonstrated (Rexin et al, 1988a; Gehring, 1993). To determine the stoichiometry of the different proteins in this 9S form, the in vitro crosslinked, nonactivated, high molecular weight complex was isolated from S49 and from S49.1 mouse lymphoma cells (Rexin et al, 1988b). The S49.1 cells contain a mutated N-terminally truncated GR (50 kDa). The comparison of these cross-linked complexes including either wild-type GR or the mutated GR revealed that one receptor molecule is associated with two hsp 90 molecules (Figure 5.2).

The 56/59 kDa protein is the third partner of the tetrameric high molecular weight complex and belongs to the hsp-family as well. Immunoprecipitation studies have shown that it is directly bound to the hsp 90 (Renoir et al, 1990). Rexin et al has shown that a 50 kDa protein (which is likely to be identical to the 56/59 kDa protein) can be crosslinked to GR as well (Rexin et al, 1988b). Consequently, the tetrameric complex containing inactive GR consists of one GR, two hsp 90 and one hsp 56/59 molecule (Gehring, 1993). However, in vitro incubation of immunopurified hormone-free mGR with rabbit reticulocyte lysate results in ATP and monovalent cation-dependent assembly of the GR into a heterocomplex with hsp 90, hsp 59 and hsp 70 (Hutchison et al, 1993a). Such an in vitro assembly is accompanied by conversion of the GR from a form that exhibits only low affinity to glucocorticoids to a high affinity steroid-binding conformation (Bresnick et al, 1989). Therefore it became important to define the hsp 90 binding domain within the receptor. Deletion mutants, missing aa 568–616 of rGR, completely abolished the binding of GR to hsp 90 (Howard et al, 1990). This region is part of the HBD so therefore, binding of hsp 90 might result in a conformational change in the HBD, yielding efficient hormone binding of the receptor. The interaction of the HBD with hsp 90 has

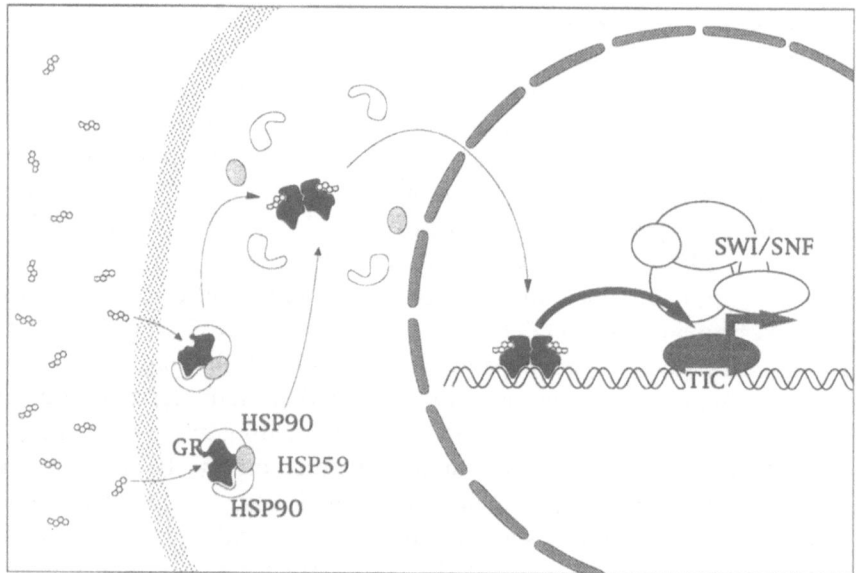

Figure 5.2 Signalling pathway of glucocorticoid hormones. For details and nomenclature see text.

been confirmed by showing that the HBD confers hormonal control to fusion proteins which bind to hsp 90 in the absence of hormone (Scherrer et al, 1993). The hsp 90 domain responsible for interaction with the HBD of the GR is not very well-characterized. Recent investigations revealed biochemical and immunological evidence that an acidic domain of hsp 90 is involved in binding to and stabilization of untransformed GR-complexes (Tbarka et al, 1993). A hypothesis for the GR/hsp90 interaction, based on computer structure predictions, proposes a metal-linked gapped zipper model (Schwartz et al, 1993). The model predicts two widely spaced leucine zipper-like heptads on either side of the HBD which might interact with similar regions of hsp 90.

Recent results indicate that the hsp 56/59 component of the GR heterocomplex is an immunophiline of the FK 506 binding class (Yen et al, 1992; Tai et al, 1992). Contradicting results show that FK 506 has no effect on receptor folding and function (Hutchison et al, 1993b) and that FK 506 can affect GR mediated expression (Ning and Sanchez, 1993). In this study FK 506 enhances the GR driven CAT expression at low (10^{-8} M and 10^{-7} M), but not at higher concentrations of dexamethasone (10^{-6} M). This FK 506 potentiation effect can be blocked with glucocorticoid antagonist RU 486. There is evidence, that FK 506 activation is due to increased nuclear translocation of GR.

Nuclear Translocation

The GR contains two nuclear localization signals (NLS) which mediate the nuclear import of the GR in a hormone dependent manner. One NLS (NLS1) is positioned adjacent to the second zinc finger and consists of a basic amino acid stretch between aa 497-524 in rGR (Picard and Yamamoto, 1987). The region fused to other proteins leads to constitutive nuclear localization of these hybrid proteins. In the GR this region is repressed by the HBD and only functions in the presence of hormone. A monoclonal antibody raised against this region only recognizes the hormone bound activated receptor (Urda et al, 1989). The second NLS (NLS2) is closely associated with the HBD and its function is also controlled by ligand (Picard and Yamamoto, 1987). Furthermore, the significance of NLS1 and NLS2 has been tested by analyzing nuclear translocation of recombinant proteins. Further studies revealed that nuclear import of a specific recombinant GR can be mediated even in the absence of hormone by these NLS, when NLS1 and HBD are repositioned to the N-terminus of GR. However, the transcriptional activation remains hormone dependent (Picard et al, 1988).

The nuclear translocation of the GR is thought to be mediated by carrier proteins that recognize and bind to the NLS of GR. Recently, two NLS binding proteins that recognize GR and the thyroid hormone receptor (TR) have been identified (LaCasse et al, 1993). Both proteins with a MW of 60 kDa and 76 kDa bind specifically to the NLS in the hinge region of GR and TR. Competition studies have revealed a higher specificity in NLS binding of the 60 kDa protein. The 60 kDa protein has been found in the nucleus and in the cytoplasm and might be identical with a 57 kDa protein isolated from rat liver cytosol which increases the binding of activated GR to nuclei in the presence of ATP (Okamoto et al, 1993). Whether this 57 kDa protein, named ASTP (ATP-stimulated GR translocation promoter), mediates the GR transport by the assistance of the cytoskeleton is not known. In fact, there is evidence for interaction of the activated GR with a cytoskeletal complex consisting of tubulin, actin, and vimentin (Scherrer and Pratt, 1992). The GR region binding to this complex includes the C-terminal part, beginning at aa 445 of mGR, which contains domains homologous to NLS 1 and 2 of the rGR. The DNA binding activity is not required for this partially energy dependent association of the GR with the cytoskeletal complex even if the receptor is already activated. Furthermore, the GR has the ability to reversibly transverse the nuclear envelope (Madan and DeFranco, 1993). The ability of various GR mutants to shuttle between nuclei of heterokaryons excluded transcriptional activation and DNA binding as prerequisite for nucleocytoplasmic shuttling of the GR. These results might be related to the localization of the 60 kDa NLS binding protein in the cytoplasm and nucleus (La Casse et al, 1993).

Interacting Partners of the GR

The GR, when activated by hormone, is translocated to the nucleus. The active nuclear form is then able to bind DNA (Figure 5.2). In various experiments, it has been shown that the GR is able to enhance transcription of genes with promotors containing a glucocorticoid response element (GRE). This enhancement is not only the result of the GR dimer binding to a GRE, but the interaction of GR with other transcription factors also seems to play a crucial role. These interactions modulate the glucocorticoid response. Therefore, the molecular details of these interactions are of major interest. Several transcription factors interacting with GR have been described including the AP1 complex. For some factors, synergizing effects on GR induced gene expression have been described; for others, a direct interaction with the GR can be shown. An interaction of the GR homodimer with the transcription-initiation complex (TIC) via direct contact or by bridging proteins has been postulated but is still not proven. In case of the progesterone and the estrogen receptors a direct interaction with TFIIB has been found (Ing et al, 1992)

Synergism

A detailed analysis of the long terminal repeat (LTR) of the mouse mammary tumor virus (MMTV) revealed the presence of several glucocorticoid receptor binding sites (Payvar et al, 1983; Scheidereit et al, 1983; Buetti and Diggelmann, 1983; Ponta et al, 1985; Strähle et al, 1988; Schüle et al, 1988a), the progressive deletion of which resulted in a gradual loss of inducibility (Hynes et al, 1983; Kühnel et al, 1986). Two or more GR binding sites were also shown to be necessary for full inducibility of other glucocorticoid regulated genes (Cato et al, 1984; Renkawitz et al, 1984; von der Ahe et al, 1985; Danesch et al, 1987; Jantzen et al, 1987). The presence of different GREs and estrogen responsive elements in the chicken vitellogenin II gene allowed the demonstration of synergistic induction in the presence of both steroids (Ankenbauer et al, 1988). Synergism between a thyroid hormone responsive element and a very weak GR binding site recently has been observed (Leers et al, 1994). This cryptic GRE does not mediate glucocorticoid response unless a TRE loaded with ligand-bound TR is present in the vicinity. All these observations indicate that several hormone receptor binding sites can cooperate to yield the strong induction observed in natural genes.

Synergism of the GR with unrelated transcription factors is now well documented. In the rat tryptophan oxygenase (TO) gene, a CACCC-sequence 5' of the receptor binding site is required for full inducibility (Schüle et al, 1988b). Similarly, the TAT gene contains, in addition to the cooperating GREs, a CAAT-box and a CACCC-element which are essential for inducibility (Becker et al, 1986; Strähle et al, 1988; De Vack et al, 1993). In the MMTV-LTR promoter, deletion and mutation analysis identified, among others, a NFI binding site in the vicinity of a GRE that is required for the inducibility by glucocorticoids and progestins (Buetti and Kühnel, 1986; Miksicek et al, 1987; Cato et al, 1988). Different mutations had varying effects on inducibility, depending on the type of receptor involved. This indicates that requirements for neighboring sequences are receptor specific (Chalepakis et al, 1988). These observations support the idea that the strength of a hormone responsive unit (HRU) (Klein-Hitpaβ et al, 1988) is determined by both receptor and nonreceptor binding sites. Combinations of a receptor binding site with sites for other transcription factors (e.g. CACCC-box factor, NF1, CAAT-box factor, Sp1, Oct1) show strong synergistic effects on steroid induction (Schüle et al, 1988a, b; Wieland et al, 1991). The best cooperation is seen between two GREs, while the other sequences cooperate to varying extents.

Interestingly, the effects of the different transcription factors have been shown to be strongly dependent on the cell line, probably reflecting the relative abundance of the various factors in these cell lines (Strähle et al, 1988). This observation offers a possible explanation for the variable inducibility of natural genes in different cell types, which does not always correlate with corresponding amounts of receptor in the cell (Tora et al, 1988; Bocquel et al, 1989). An example for cell-specific synergism with GR controlled gene expression is the glutamine synthetase (GS) expression in the chicken embryo retina (Ben-Dror et al, 1993). Although there is enough functional GR present in these cells on days six and seven of embryonic development, no induction of GS expression has been detected. At later stages of development, between days ten and twelve of embryonic development, induction of GS can be detected, although there is no change in the level of functional GR. GS promoter analysis reveals the presence of a binding sequence for a transcription factor, similar to the CCAAT enhancer binding protein (C/EBP), that seems to be the limiting factor for GS expression (Ben-Or and Okret, 1993). This retinal C/EBP like protein is not detectable on day seven but it is expressed on day twelve of embryonic development. The GR and NF-IL6, a cell-specific factor required for IL-6 expression, synergistically activate transcription of the rat α1-acid glycoprotein via direct protein-protein interaction (Nishio et al, 1993). Thus, it appears that in many cases a hormone responsive unit is composed of one or several receptor binding sites intermingled with binding sites for other transcription factors which act synergistically to yield the observed expression and inducibility pattern.

Possible mechanisms for synergism

Transcriptional synergism between a CACCC-box and a GRE/PRE, as well as cooperative binding to adjacent GREs, shows a clear cyclic distance dependence, suggesting a requirement for stereospecific alignment and a direct or indirect protein-protein interaction of the two factors (Schüle et al, 1988b; Schmid et al, 1989). The simplest explanation for the synergism between two transcription factors would be cooperative binding to two adjacent binding sites. This possibility has been investigated by analyzing DNA binding in vitro. Cooperative binding to a double GRE by purified glucocorticoid receptor has been demonstrated (Schmid et al, 1989). Using DNA binding studies and transfection experiments, it has been shown that the DNA-binding domain of the GR is sufficient to mediate cooperative binding to and synergism on two adjacent binding sites (Baniahmad et al, 1991). Thus it appears that the functional synergism of two HREs is due, at least in part, to the increase of binding affinity of the receptor for the regulatory element.

Transfection experiments using fusion proteins between GR-domains outside of the DBD and the DBD of the yeast transcription factor GAL-4 have revealed synergism on two adjacent GAL-4 binding sites. These fusion proteins do not bind this DNA element cooperatively, indicating in this case a different mechanism of synergism, which probably also contributes to the synergism of wild-type GR on neighboring GREs in addition to cooperative binding. The requirement of GR domains for synergism with other transcription factors also has been examined. Analysis of a HRU composed of an ERE and a GRE/PRE reveals that the N-terminal domains of both GR or PR are required, but deletion of the ER C-terminal half only affects synergism with the GR (Cato and Ponta, 1989). These results indicate that the GR/ER synergism is mediated by a different mechanism than is the ER/PR synergism.

The GR domains required for synergism with the CACCC-box have been determined by measuring the activation mediated by receptor mutants on a reporter gene carrying a CACCC-box adjacent to the GRE. This synergism has been compared to the intrinsic transactivation capacity of the mutants measured on a reporter construct containing one isolated GRE (Muller et al, 1991). Deletion of N-terminal parts of the GR similarly affect transactivation and synergism, suggesting a common pathway for these functions. In contrast, deletion of the HBD strongly reduces synergism without affecting transactivation. These results suggest that synergism and transactivation constitute, at least in the latter case, two separable functions of the same molecule.

The characterization of a receptor accessory factor (RAF) that enhances specific DNA binding of the androgen receptor (AR) and GR (Kupfer et al, 1993) has given rise to the idea that interaction partners of GR may not only synergize for increasing transactivation potency by mediating contact with the transcription machinery or binding to another sequence element, but may also stabilize DNA binding of the receptor (and therefore increase the receptor mediated transcription). The characterized RAF has a molecular mass of 130 kDa and enhances GR binding to DNA about sixfold. Desoxycholate treatment reveals that the RAF is directly interacting with the AR because the detergent inhibits RAF-AR-DNA-complexes but not AR-DNA-complexes. This is evidence for direct protein-protein interaction of AR and RAF and possibly for the related GR.

Factors mediating the GR response to the transcription machinery

Studies in yeast have revealed that rGR cooperates with a complex of multiple SWI-proteins (Yoshinaga et al, 1992). Wild-type yeast cells stably transfected with rGR exhibit a glucocorticoid response. Nevertheless, the GR fails to activate transcription in strains with mutations in the SWI 1, SWI 2 or SWI 3 genes. Immunoprecipitation of GR from wt-yeast extracts leads to SWI 3 coprecipitation. This may be a hint that GR interacts with SWI 1 and SWI 2 which are complexed with SWI 3 (Figure 5.2).

A homologue of SWI 2 has been found in *Drosophila* and named brm (Tamkun et al, 1992). This protein is similar to the yeast factor in both size and structure, and exhibits 57 percent amino acid identity within a putative helicase domain. The brm gene product also seems to be involved in transcriptional activation and appears to have a positive effect on the expression of several homeotic genes. Starting from this point, several groups began searching for SWI homologues in higher eukaryotes. The screening of a human cDNA bank with brm cDNA probes has led to the identification of two human genes coding for putative transcription factors. One protein with a molecular mass of 180 kDa, named hbrm, has been detected in humans and mice and could function as a transcriptional activator when fused to a heterologous DNA binding domain (Muchard and Yaniv, 1993). The hbrm protein is present in all mouse organs but not in every human cell line. It cooperates with GR in transcriptional activation, but has no effect on several other transcription factors. The cooperation between GR and hbrm requires the DBD of GR. A direct protein-protein interaction has not been shown.

The second human protein identified as a brm/SWI 2 homologue is a 205 kDa nuclear protein named BRG 1 (Khavari et al, 1993). Experiments have revealed that a SWI 2/BRG 1 chimaera, with the yeast DNA dependent ATPase domain replaced by the corresponding human domain, leads to wt-phenotype when introduced into SWI 2^- yeast cells. The same construct can also restore GR driven transcription in SWI 2^- yeast cells. The BRG 1 is exclusively found in the nucleus as part of a high molecular weight complex with a molecular mass of 2×10^6 kDa which may be the human equivalent of the yeast SWI protein complex. The hbrm and BRG gene products, therefore, represent candidates for direct or indirect interaction with GR. They cooperate with GR in glucocorticoid induced transcription and might be the mediating partners bridging between the DNA bound receptor and the transcription-initiation complex. Another possibility for mediating glucocorticoid induction is the direct interaction of GR with the transcription-initiation complex. In yeast it can be shown that the $\tau1$ transactivation domain of hGR is important for transactivation and seems to interact directly with the basal transcriptional machinery (McEwan et al, 1993). The interacting factor is not the yeast TBP, and its identification awaits further analysis.

Modulating factors interacting directly with the GR

Another example for GR interaction with transcription factors, most likely by direct interaction, has been found in mouse erythroleukemia (MEL) cells (Chang et al, 1993). In MEL cells the globin synthesis, which is a marker for cell differentiation, is blocked by glucorcorticoids. The globin genes tested contained a GATA-RE and can be transactivated by the transcription factor GATA-1, which is expressed in these cells. Further experiments have revealed a GRE near the GATA-RE in three globin gene promoters and therefore a GR-GATA-1 interaction has been suggested. In fact, the GR seems to bind GATA-1 and to interfere with its function before any DNA interaction, and the GRE augments this effect. The aminoterminal 106 aa of GR have been found to be essential for this GATA-1 interaction. The GR also cooperates with a putative repressor protein in rat hepatoma cells (Tanaka et al, 1993). In hepatoma cells containing the stably integrated MMTV-DNA, the GR and the putative repressor protein modulate glucocorticoid responsiveness of the MMTV-promoter. It has been determined that the repressor protein binds at the promoter between position-163 and -147. This results in a repression of GR mediated expression of viral mRNA. It appears that the relative level of GR is the main determinant of maximum inducibility of virus expression by hormone. Taken together, this leads to the conclusion that differential expression patterns of GR versus repressor protein control the level of target gene expression. In the last two years, transcriptional control of several eukaryotic genes by both GR and AP-1 has been reported. The activation of the transcription is in some cases induced by one transcription factor and inhibited by the other. For the collagenase 1 gene induced expression by phorbol esters and repression by glucocorticoids can be shown (Jonat et al, 1992). Both types of regulation have been found to be mediated by the major enhancer element of the gene containing an AP-1 binding site. This element is sufficient to transmit positive and negative signals to a reporter gene controlled by a minimal promoter carrying only the TATA-box. Immunoprecipitation of the GR leads to coprecipitation of AP-1 suggesting a direct interaction of both transcription factors. This interaction takes place when GR is activated and might result in the inactivation of AP-1 (Jonat et al, 1992). These results are confirmed by other groups, indicating that GR interacts with jun via direct protein-protein interaction (Kerppola et al, 1993). In this study, it can also be shown that the DBD of GR is needed for inhibition of AP-1 DNA binding, but the inhibition is independent from GR binding to DNA. This provides a mechanism of diverse interactions between the AP-1 family with the GR and probably also with other nuclear hormone receptors at different promoters in different cell types. AP-1-GR interference is not mediated by inhibition of DNA-binding; rather the AP-1 response element remains occupied by AP-1 in vivo (König et al, 1992).

Transcriptional control via the phosphatidylinositol and the glucocorticoid signal transduction pathway also has been suggested by others, who point out that this interaction is cell specific (Maroder et al, 1993). Whereas the phorbol ester TPA inhibits the glucocorticoid induced transcriptional activation in NIH 3T3 cells, it augments the GR-driven transcription in several T-cell lines. The TPA/GR synergism takes place at the GRE and requires an intact GR-DBD; the amino and carboxy terminal domains are dispensable, as shown by deletion mutants. In T-cells, TPA treatment leads to increased levels of c-jun, junB and junD expression whether glucocorticoids are added or not, and consequently results in increased transcriptional activity of CAT vectors containing AP-1/TPA responsive elements. Conversely, c-jun and junB transfection blunts GR-dependent transcription in HeLa cells. C-fos has a negative influence on GR function, as exogeneous c-fos expression results in an inhibitory effect on GR dependent transcription from GRE in T cells. Taken together this leads to the hypothesis that jun plays a bifunctional role in GR dependent transcriptional activation, selecting either synergistic or antagonistic activity depending on the cell specific microenvironment.

Different mechanisms are responsible for the AP-1 effects on a composite GRE containing both AP-1 and GR binding sites (Pearce and Yamamoto, 1993). In this study it has been shown that only GR and not MR repress AP-1 stimulated transcription and that the N-terminal part (aa 105–440) of GR is required for repression, as shown by MR-GR chimeras.

Involvement of GR in HIV-Regulation

Recent reports indicate that the GR is involved in the regulation of HIV gene expression. The identification of GRE's in the long terminal repeat (LTR) and in the open reading frame of the *vif* gene suggest that glucocorticoids might, at least partially, control retroviral expression. The first GRE has been found in the LTR of the HIV type 1 (HIV-1) provirus (Kolesnitchenko and Snart, 1992). This GRE shows similarity to the GRE of the MMTV LTR and is inducible by hormone as shown in transfection studies. GR binding to LTR-DNA of HIV-1 has been determined by gel retardation experiments. The second GRE found in HIV-1 is at position $+5002$ in the open reading frame of the *vif* gene (Soudeyns et al, 1993). Further experiments revealed that this GRE is functional as well. Studies of HIV-1 infected lymphoid and monocytoid cell lines indicate that glucocorticoid treatment results in increased HIV-1 production in culture (Soudeyns et al, 1993). These observations suggest that HIV-1 has evolved an interaction with the GR signal transduction pathway to gain replicative advantage in target cells.

GR and Apoptosis

Glucocorticoids are known to induce programmed cell death in immature T cells through a process called apoptosis. Several studies have revealed that glucocorticoid induced apoptosis of thymocytes is due to interaction with the cAMP dependent signal transduction pathway. The process also seems to correlate with intracellular $Ca++$ concentration which is at least partially receptor mediated (Lam et al, 1993). It has been shown that thymocytes undergo cell death when induced by glucocorticoids and that the cAMP-dependent pathway synergistically interacts with the glucocorticoid signaling pathway (Gruol and Altschmied, 1993). In this case, even the glucocorticoid antagonist RU 486 seems to act as agonist and the cooperation with cAMP seems to allow limited portions of RU 486 bound GR to translocate to the nucleus and contribute to a loss of cell viability. The role of cAMP remains unclear but it seems to play an important role in the induction of apoptosis at a step beyond receptor transformation by promoting an interaction between GR and other gene specific regulatory proteins. These results are in agreement with the finding that cAMP potentiated glucocorticoid induced endogeneous exonuclease activation in thymocytes (McConkey et al, 1993). The cAMP dependent cooperating factor for GR may be AP-1, since it is activated in apoptotic cells (Sikora et al, 1993). This has led to a model in which GR interacts with AP-1 in dexamethasone mediated apoptosis. In contrast to the agonistic activity of RU 486 which cooperates with a cAMP mediated increase in cell death, it has been found that RU 486 blocks the T cell receptor mediated apoptosis in immature thymocytes (Jondal et al, 1993). In this study the negative selection in thymus, which results in apoptosis has been investigated. The data supports a model in which thymic negative selection depends on a defined set of transduction signals which potentiate the GR to become responsive to endogeneous levels of glucocorticoid. Another factor in glucocorticoid induced apoptosis is the intracellular Ca + + concentration. In lymphocytes Ca + + fluxes can be induced by GR (Lam et al, 1993). In addition, glucocorticoid induces a significant decrease in cellular Ca + + followed by a modest increase of cellular and cytosolic Ca + + concentration. It can be shown that the depletion of internal Ca + + stores is receptor mediated. Taken together, the glucocorticoid signal transduction pathway is involved in apoptosis of thymocytes and is probably interacting with other transcription pathways.

Conclusions and Perspectives

In recent years the glucocorticoid receptor has been investigated quite intensively, enhancing dramatically our knowledge of its structure and function. Domains responsible for hormone binding, nuclear translocation, DNA-binding and transactivation have been correlated to precise areas of the receptor. Recent studies indicate that in addition to the association of the inactive cytoplasmic GR with several heat shock proteins, there is interaction of activated nuclear GR with other nuclear proteins. This crosstalk with transcription factors like AP-1 has led to the idea that different signaling pathways control specific gene expression. Since AP-1 activity is influenced by PKC mediated phosphorylation, IP3-metabolism seems to be involved. Other important GR interactions involve the direct or indirect binding to the preinitiation complex and interaction or competition with chromatin components, an important aspect this review could not touch. Protein-Protein interaction provides an important field for future research, and the function of GR phosphorylation remains to be determined.

Acknowledgements

This work was supported by grants from the Deutsche Forschungsgemeinschaft and from the Fonds der Chemischen Industrie.

References

Ankenbauer W, Strähle U, Schütz G (1988): Synergistic action of glucocorticoid and estradiol responsive elements. *Proc Natl Acad Sci USA* 85: 7526–7530

Baniahmad C, Muller M, Altschmied J, Renkawitz R (1991): Co-operative binding of the glucocorticoid receptor DNA binding domain is one of at least two mechanisms for synergism. *J Mol Biol* 222: 155–165

Becker PB, Gloss B, Schmid W, Strähle U, Schütz G (1986): In vivo protein-DNA interactions in a glucocorticoid response element require the presence of the hormone. *Nature* 324: 686–688

Bellingham DL, Sar, M, Cidlowski JA (1992): Ligand-dependent down-regulation of stably transfected human glucocorticoid receptors is associated with the loss of functional glucocorticoid responsiveness. *Mol Endocrinol* 6: 2090–2102

Ben-Dror I, Havezelt N, Vardimon L (1993): Developmental control of glucocorticoid receptor transcriptional activity in embryonic retina. *Proc Natl Acad Sci USA* 90: 1117–1121

Ben-Or S, Okret S (1993): Involvement of a C/EBP-like protein in the acquisition of responsiveness to glucocorticoid hormones during chick neural retina development. *Mol Cell Biol* 13: 331–340

Bocquel MT, Kumar V, Stricker C, Chambon P, Gronemeyer H (1989): The contribution of the N- and C-terminal regions of steroid receptors to activation of transcription is both receptor and cell-specific. *Nucl Acids Res* 7: 2581–2595

Bodwell JE, Orti E, Coull JM, Pappin DJC, Smith LI, Swift F (1991): Identification of phosphorylated sites in the mouse glucocorticoid receptor. *J Biol Chem* 266: 7549–7555

Bresnick EH, Dalman FC, Sanchez ER, Pratt WB (1989). Evidence that the 90-kDa heat shock protein is necessary for the steroid binding conformation of the L-cell glucocorticoid receptor. *J Biol Chem* 264: 4992–4997

Buetti E, Diggelmann H (1983): Glucocorticoid regulation of mouse mammary tumor virus: identification of a short essential region. *EMBO J* 2: 1423–1429

Buetti E, Kühnel B (1986): Distinct sequence elements involved in the glucocorticoid regulation of the mouse mammary tumor virus promoter identified by linker scanning mutagenesis. *J Mol Biol* 190: 379–389

Carlstedt-Duke J, Stromstedt P-E, Persson B, Cederlund E, Gustafsson J-A, Jornvall H (1988): Identification of hormone-interacting amino acid residues within the steroid-binding domain of the glucocorticoid receptor in relation to other steroid hormone receptors. *J Biol Chem* 263: 6842–6846

Cato AC, Ponta H (1989): Different regions of the estrogen receptor are required for synergistic action with the glucocorticoid and progesterone receptors. *Mol Cell Biol* 9: 5324–5330

Cato ACB, Geisse S, Wenz M, Westphal HM, Beato M (1984): The nucleotide sequences recognized by the glucocorticoid receptor in the rabbit uteroglobin gene region are located far upstream from the initiation of transcription. *EMBO J* 3: 2731–2736

Cato ACB, Skroch P, Weinmann J, Butkeraitis P, Ponta H (1988): DNA sequences outside the receptor binding sites differentially modulate the responsiveness of the mouse mammary tumor virus promoter to various steroid hormones. *EMBO J* 7: 1403–1407

Chalepakis G, Arnemann J, Slater E, Brüller H, Gross B, Beato M (1988): Differential gene activation by glucocorticoids and progesterone through the hormone regulatory element of mouse mammary tumor virus. *Cell* 53: 371–382

Chang TJ, Scher BM, Waxman S, Scher W (1993): Inhibition of mouse GATA-1 function by the glucocorticoid receptor: possible mechanism of steroid inhibition of erythroleukemia cell differentiation. *Mol Endocrinol* 7: 528–542

Dahlman-Wright K, Grandien K, Nilsson S, Gustafsson JA, Carlstedt-Duke J (1993): Protein-protein interactions between the DNA-binding domains of nuclear receptors: influence on DNA-binding. *J Steroid Biochem Mol Biol* 45: 239–250

Dahlmann-Wright K, Wright A, Gustafsson J-A, Carlstedt-Duke J (1991): Interaction of the glucocorticoid receptor DNA-binding domain with DNA as a dimer is mediated by a short segment of five amino acids. *J Biol Chem* 266: 3167–3112

Danesch U, Gloss B, Schmid W, Schütz G, Schüle R, Renkawitz R (1987): Glucocorticoid induction of the rat tryptophan oxygenase gene is mediated by two widely separated glucocorticoid-responsive elements. *EMBO J* 6: 625–630

Danielsen M, Hinck L, Ringold GM (1989): Two amino acids within the knuckle of the first zinc finger specify DNA response element activation by the glucocorticoid receptor. *Cell* 57: 1131–1138

Danielsen M, Northrop JP, Ringold GM (1986): The mouse glucocorticoid receptor: mapping of functional domains by cloning, sequencing and expression of wildtype and mutant receptor proteins. *EMBO J* 5: 2513–2522.

Danielsen M, Northrop JP, Jonklaas J, Ringold GM (1987): Domains of the glucocorticoid receptor involved in specific and nonspecific deoxyribonucleic acid binding, hormone activation, and transcriptional enhancement. *Mol Endocrinol* 1: 816–822

Denis M, Poellinger L, Wikström A-C, Gustafsson J-A (1988): Requirement of hormone for thermal conversion of the glucocorticoid receptor to a DNA binding state. *Nature* 333: 686–688

Denner LA, Schrader WT, O'Malley BW, Weigel NL (1990): Hormonal regulation and identification of chicken progesterone receptor phosphorylation sites. *J Biol Chem* 265: 16548–16555

Denton RR, Eisen LP, Elsasser MS, Harmon JM (1993): Differential autoregulation of glucocorticoid receptor expression in human T- and B-cell lines. *Endocrinology* 133: 248–256

DeVack CC, Lupp B, Nichols M, Kowenz-Leutz E, Schmid W, Schütz G (1993): Characterization of the nuclear proteins binding the CACCC element of a glucocorticoid-responsive enhancer in the tyrosine aminotransferase gene. *Eur J Biochem* 211: 459–465

Drouin J, Sun YL, Chamberland M, Gauthier Y, De-Lean-A, Nemer M, Schmidt TJ (1993): Novel glucocorticoid receptor complex with DNA element of the hormone-repressed POMC gene. *Embo J* 12: 145–156

Evans RM (1988): The steroid and thyroid hormone receptor superfamily. *Science* 240: 889–895

Fawell SE, Lees JA, White R, Parker MB (1990): Characterization and colocalization of steroid binding and dimerization activities in the mouse estrogen receptor. *Cell* 60: 953–962

Forman BM, Yang C-R, Au M, Casanova J, Ghysdael J, Samuels HH (1989): A domain containing leucine-zipper-like motifs mediate novel in vivo interactions between the thyroid hormone and retinoic acid receptors. *Mol Endocrinol* 3: 1610–1626

Freedman LP, Luisi BF, Korszun ZR, Basavappa R, Sigler PB, Yamamoto KR (1988): The function and structure of the metal coordination sites within the glucocorticoid receptor DNA binding domain. *Nature* 334: 543–546

Gehring U (1993): The structure of glucocorticoid receptors. *J Steroid Biochem Mol Biol* 45: 183–190

Giguere V, Hollenberg SM, Rosenfeld MG, Evans RM (1986): Functional domains of the human glucocorticoid receptor. *Cell* 46: 645–652

Glass CK, Lipkin SM, Devary OV, Rosenfeld MG (1989): Positive and negative regulation of gene transcription by a retinoic acid-thyroid hormone receptor heterodimer. *Cell* 59: 697–708

Godowski PJ, Didier P, Yamamoto KR (1988): Signal transduction and transcriptional regulation by glucocorticoid receptor-LexA fusion proteins. *Science* 241: 812–816

Gruol DJ, Altschmied J (1993): Synergistic induction of apoptosis with glucocorticoids and 3′, 5′-cyclic adenosine monophosphate reveals agonist activity by RU 486. *Mol Endocrinol* 7: 104–113

Härd T, Kellenbach E, Boelens R, Maler BA, Dahlman K, Freedman LP, Carlstedt-Duke J, Yamamoto KR, Gustafsson J-A, Kaptein R (1990): Solution structure of the glucocorticoid receptor DNA-binding domain. *Science* 249: 157–160

Hoeck W, Groner B (1990): Hormone-dependent phosphorylation of the glucocorticoid receptor occurs mainly in the amino-terminal transactivation domain. *J Biol Chem* 265: 5403–5408

Hoeck W, Rusconi S, Groner B (1989): Down-regulation and phosphorylation of glucocorticoid receptors in cultured cells. *J Biol Chem* 264: 14396–14402

Hollenberg SM, Evans RM (1988): Multiple and cooperative trans-activation domains of the human glucocorticoid receptor. *Cell* 55: 899–906

Hollenberg SM, Giguere V, Segui P, Evans RM (1987): Colocalization of DNA-binding and transcriptional activation functions in the human glucocorticoid receptor. *Cell* 49: 39–46

Hollenberg SM, Weinberger C, Ong ES, Cerelli G, Oro A, Lebo R, Thompson EB, Rosenfeld MG, Evans RM (1985): Primary structure and expression of a functional human glucocorticoid receptor cDNA. *Nature* 318: 635–641

Housley PR, Pratt WB (1983): Direct demonstration of glucocorticoid receptor phosphorylation by intact L-cells. *J Biol Chem* 258: 4630–4635

Howard KJ, Holley SJ, Yamamoto KR, Distelhorst CW (1990): Mapping the Hsp90 binding region of the glucocorticoid receptor. *J Biol Chem* 265: 11928–11935

Hutchison KA, Scherrer LC, Czar MJ, Stancato LF, Chow-YH, Jove R, Pratt WB (1993a): Regulation of glucocorticoid receptor function through assembly of a receptor-heat shock protein complex. *Ann NY Acad Sci* 684: 35–48

Hutchison KA, Scherrer LC, Czar MJ, Ning Y, Sanchez ER, Leach KL, Deibel MR, Pratt WB (1993b): FK506 binding to the 56-kilodalton immunophilin (Hsp56) in the glucocorticoid receptor heterocomplex has no effect on receptor folding or function. *Biochemistry* 32: 3953–3957

Hynes N, van Ooyen AJJ, Kennedy N, Herrlich P, Ponta H, Groner B (1983): Subfragments of the large terminal repeat cause glucocorticoid-responsive expression of the mouse mammary tumor virus and of an adjacent gene. *Proc Natl Acad Sci USA* 80: 3637–3641

Ing NH, Beekman JM, Tsai SY, Tsai M-J, O'Malley BW (1992): Members of the steroid hormone receptor superfamily interact with TFIIB (S300-II). *J Biol Chem* 267: 17617–17623

Jantzen HM, Strähle U, Gloss B, Stewart F, Schmid W, Boshart M, Miksicek R, Schütz G (1987): Cooperativity of glucocorticoid response elements located far upstream of the tyrosine aminotransferase gene. *Cell* 49: 29–38

Jonat C, Stein B, Ponta H, Herrlich P, Rahmsdorf HJ (1992): Positive and negative regulation of collagenase gene expression. *Matrix Suppl* 1: 145–155

Jondal M, Okret S, McConkey D (1993): Killing of immature CD4+ CD8+ thymocytes in vivo by anti CD3 or 5'-(N-ethyl)-carboxamide adenosine is blocked by glucocorticoid receptor antagonist RU-486. *Eur J Immunol* 23: 1246–1250

Kerppola TK, Luk D, Curran T (1993): Fos is a preferential target of glucocorticoid receptor inhibition of AP-1 activity in vitro. *Mol Cell Biol* 13: 3782–3791

Khavari PA, Peterson CL, Tamkun JW, Mendel DB, Crabtree GR (1993): BRG1 contains a conserved domain of the SWI2/SNF2 family necessary for normal mitotic growth and transcription. *Nature* 366: 170–174

Klein-Hitpaβ L, Kaling M, Ryffel GU (1988): Synergism of closely adjacent estrogen-responsive elements increases their regulatory potential. *J Mol Biol* 201: 537–544

Kolesnitchenko V, Snart RS (1992): Regulatory elements in the human immunodeficiency virus type 1 long terminal repeat LTR (HIV-1) responsive to steroid hormone stimulation. *Aids Res Hum Retroviruses* 8: 1977–1980

König H, Ponta H, Rahmsdorf HJ, Herrlich P (1992): Interference between pathway-specific transcription factors – glucocorticoids antagonize phorbol ester-induced AP-1 activity without altering AP-1 site occupation *in vivo*. *EMBO J* 11: 2241–2246

Kühnel B, Buetti E, Diggelmann H (1986): Functional analysis of the glucocorticoid regulatory elements present in the mouse mammary tumor virus long terminal repeat _ a synthetic distal binding site can replace the proximal binding domain. *J Mol Biol* 190: 367–378

Kupfer SR, Marschke KB, Wilson EM, French FS (1993): Receptor Accessory factor enhances specific DNA binding of androgen and glucocorticoid receptors. *J Biol Chem* 268: 17519–17527

LaCasse EC, Lochman HA, Walker P, Lefebvre YA (1993): Identification of binding proteins for nuclear localization signals of the glucocorticoid and thyroid hormone receptors. Endocrinology 132: 1017–1025

Lam M, Dubyak G, Distelhorst CW (1993): Effect of glucocorticosteroid treatment on intracellular calcium homeostasis in mouse lymphoma cells. *Mol Endocrinol* 7: 686–693

Laudet V, Hänni C, Coll J, Catzeflis F, Stehelin D (1992): Evolution of the nuclear receptor gene superfamily. *EMBO J* 11: 1003–1013

Leers J, Steiner Ch, Renkawitz R, Muller M (1994): A thyroid hormone receptor dependent glucocorticoid induction. *Mol Endocrinol:* 8: 440–447

Madan AP, DeFranco DB (1993): Bidirectional transport of glucocorticoid receptors across the nuclear envelope. *Proc Natl Acad Sci USA* 90: 3588–3592

Maroder M, Farina AR, Vacca A, Felli MP, Meco D, Screpanti I, Frati L, Gulino A (1993): Cell-specific bifunctional role of Jun oncogene family members on glucocorticoid receptor-dependent transcription. *Mol Endocrinol* 7: 570–584

Mason SA, Housley PR (1993): Site-directed mutagenesis of the phosphorylation sites in the mouse glucocorticoid receptor. *J Biol Chem* 268: 21501–21504

McConkey DJ, Orrenius S, Okret S, Jondal M (1993): Cyclic AMP potentiators glucocorticoid-induced endogeneous endonuclease activation in thymocytes. *FASEB-J* 7: 580–585

McEwan IJ, Wright AP, Dahlman-Wright K, Carlstedt-Duke J, Gustafsson JA (1993): Direct interaction of the tau 1 transactivation domain of the human glucocorticoid receptor with the basal transcriptional machinery. *Mol Cell Biol* 13: 399–407

Miesfeld R, Godowski PJ, Maler BA, Yamamoto KR (1987): Glucocorticoid receptor mutants that define a small region sufficient for enhancer activation. *Science* 236: 423–427

Miesfeld R, Rusconi S, Godowski PJ, Maler BA, Okret S, Wikström A-C, Gustafsson JA, Yamamoto KR (1986): Genetic complementation of a glucocorticoid receptor deficiency by expression of cloned receptor cDNA. *Cell* 46: 389–399

Miksicek R, Borgmeyer U, Nowock J (1987): Interaction of the TGGCA-binding protein with upstream sequences is required for efficient transcription of mouse mammary tumor virus. *EMBO J* 6: 1355–1360

Moyer ML, Borror KC, Bona BJ, DeFranco DB, Nordeen SK (1993): Modulation of cell signaling pathways can enhance or impair glucocorticoid-induced gene expression without altering the state of receptor phosphorylation. *J Biol Chem* 268: 22933–22940

Muchard C, Yaniv M (1993): A human homologue of Saccharomyces cerevisiae SNF2/SW12 and Drosophila brm genes potentiates transcriptional activation by the glucocorticoid receptor. *EMBO J* 12: 4279–4290

Muller M, Baniahmad C, Kaltschmidt C, Renkawitz R (1991): Multiple domains of the glucocorticoid receptor involved in synergism with the CACCC box factor(s). *Mol Endocrinol* 5: 1498–1503

Ning YM, Sanchez ER (1993): Potentiation of glucocorticoid receptor-mediated gene expression by the immunophilin ligands FK506 and rapamycin. *J Biol Chem* 268: 6073–6076

Nishio Y, Isshiki H, Kishimoto R, Akira S (1993): A nuclear factor for interleukin-6 expression (NF-IL6) and the glucocorticoid receptor synergistically activate transcription of the rat alpha 111-acid glycoprotein gene via direct protein-protein interaction. *Mol Cell Biol* 13: 1854–1862

Oakley RH, Cidlowski JA (1993): Homologous down regulation of the glucocorticoid receptor. The molecular machinery. *Crit Rev Eukaryt Gene Expr* 3: 63–88

Okamoto K, Hirano H, Isohashi F (1993): Molecular cloning of rat liver glucocorticoid-receptor translocation promoter. *Biochem Biophys Res Commun* 193: 848–854

Orti E, Hu LM, Munck A (1993): Kinetics of glucocorticoid receptor phosphorylation in intact cells. Evidence for hormone-induced hyperphosphorylation after activation and recycling of hyperphosphorylated receptors. *J Biol Chem* 268: 7779–7784

Orti E, Mendel DB, Smith LI, Munck A (1989): Agonist-dependent phosphorylation and nuclear dephosphorylation of glucocorticoid receptors in intact cells. *J Biol Chem* 264: 9728–9731

Payvar F, DeFranco D, Firestone GL, Edgar B, Wrange Ö, Okret S, Gustafsson JA, Yamamoto KR (1983): Sequence-specific binding of glucocorticoid receptor to MTV-DNA at sites within and upstream of the transcribed region. *Cell* 35: 381–392

Pearce D, Yamamoto KR (1993): Mineralcorticoid and glucocorticoid receptor activities distinguished by nonreceptor factors at a composite response element. *Science* 259: 1161–1165

Picard D, Yamamoto KR (1987): Two signals mediate hormone dependent nuclear localization of the glucocorticoid receptor. *EMBO J* 6: 3333–3340

Picard D, Salser SJ, Yamamoto KR (1988): A movable and regulable inactivation function within the steroid binding domain of the glucocorticoid receptor. *Cell* 54: 1073–1080

Ponta H, Kennedy N, Skroch P, Hynes NE, Groner, B (1985): Hormonal response region in the mouse mammary tumor virus long terminal repeat can be dissociated from the proviral promoter and has enhancer properties. *Proc Natl Acad Sci USA* 82: 1020–1024

Pratt WB, Jolly DJ, Pratt DV, Hollenberg SM, Giguere V, Cadepond FM, Schweizer-Groyer G, Catelli M-G, Evans RM, Baulieu E-E (1988): A region in the steroid binding domain determines formation of the non-DNA-binding, 9S glucocorticoid receptor complex. *J Biol Chem* 263: 267–273

Renkawitz R, Schütz G, von der Ahe D, Beato M (1984): Sequences in the promoter region of the chicken lysozyme gene required for steroid regulation and receptor binding. *Cell* 37: 503–510

Renoir J-M, Radanyi C, Faber LE, Baulieu E-E (1990): The non-DNA-binding heterooligomeric form of mammalian steroid hormone receptors contains a hsp90-bound 59-Kilodalton protein. *J Biol Chem* 265: 10740–10745

Rexin M, Busch W, Gehring U (1988a): Chemical cross-linking of heteromeric glucocorticoid receptors. *Biochemistry* 27: 5593–5601

Rexin M, Busch W, Segnitz B, Gehring U (1988b): Tetrameric structure of the nonactivated glucocorticoid receptor in cell extracts and intact cells. *FEBS Lett* 241: 234–238

Rusconi S, Yamamoto KR (1987): Functional dissection of the hormone and DNA binding activities of the glucocorticoid receptor. *EMBO J* 6: 1309–1315

Sanchez ER, Meshinchi S, Tienrungroj W, Schlesinger MJ, Toft DO, Pratt WB (1987): Relationship of the 90 kDa murine heat shock protein to the untransformed and transformed states of the L cell glucocorticoid receptor. *J Biol Chem* 262: 6986–6991

Scheidereit C, Geisse S, Westphal HM, Beato M (1983): The glucocorticoid receptor binds to defined nucleotide sequences near the promoter of mouse mammary tumor virus. *Nature* 304: 749–752

Scherrer LC, Pratt WB (1992): Energy-dependent coversion of transformed cytosolic glucocorticoid receptors from soluble to particulate-bound form. *Biochemistry* 31: 10879–10886

Scherrer LC, Picard D, Massa E, Harmon JM, Simons SS, Yamamoto KR, Pratt WB (1993): Evidence that the hormone binding domain of steroid receptors confers hormonal control on chimeric proteins by determining their hormone-regulated binding to heat-shock protein 90. *Biochemistry* 32: 5381–5386

Schmid W, Strähle U, Schütz G, Schmitt J, Stunnenberg H (1989): Glucocorticoid receptor binds cooperatively to adjacent recognition sites. *EMBO J* 8: 2257–2263

Schüle R, Muller M, Kaltschmidt C, Renkawitz R (1988a): Many transcription factors interact synergistically with steroid receptors. *Science* 242: 1418–1420

Schüle R, Muller M, Otsuka-Murakami H, Renkawitz R (1988b): Cooperativity of the glucocorticoid receptor and the CACCC-box binding factor. *Nature* 332: 87–90

Schwartz JA, Mizukami H, Skafar DF (1993): A metal-linked gapped zipper model is proposed for the hsp90-glucocorticoid receptor interaction. *FEBS Lett* 315: 109–113

Severne Y, Wieland S, Schaffner W, Rusconi S (1988): Metal binding 'finger' structures in the glucocorticoid receptor defined by sites-directed mutagenesis. *EMBO J* 7: 2503–2508

Sikora E, Grassili E, Bellesia E, Troisano L, Franceschi C (1993): Studies of the relationship between cell proliferation and cell death. III. AP-1 DNA-binding activity during concanavalin A-induced proliferation or dexamethasone-induced apoptosis of rat thymocytes. *Biochem Biophys Res Commun* 192: 386–391

Simons SS, Pumphrey JG, Rudikoff S, Eisen HI (1987): Identification of cysteine 656 as the amino acid of hepatoma tissue culture cell glucocorticoid receptors that is covalently labeled by dexamethasone 21-mesylate. *J Biol Chem* 262: 9676–9680

Soudeyns H, Geleziunas R, Shyamala G, Hiscott J, Wainberg MA (1993): Identification of a novel glucocorticoid response element within the genome of the human immunodeficiency virus type-1. *Virology* 193: 758–768

Spanjaard RA, Chin WW (1993): Reconstitution of ligand-mediated glucocorticoid receptor activity by trans-acting functional domains. *Mol Endocrinol* 7: 12–16

Strähle U, Schmid W, Schütz G (1988): Synergistic action of the glucocorticoid receptor with transcription factors. *EMBO J* 7: 3389–3395

Tai PK, Albers MW, Chang H, Faber LE, Schreiber SL (1992): Association of a 59-Kilodalton immunophilin with the glucocorticoid receptor complex. *Science* 256: 1315–1318

Tamkun JW, Deuring R, Scott MP, Kissinger M, Pattatucci AM, Kaufman TC, Kennison JA (1992): brahma: A regulator of Drosophila homeotic genes structurally related to the yeast transcriptional activator SNF2/SW12. *Cell* 68: 561–572

Tanaka H, Dong Y, McGuire J, Okret S, Poellinger L, Makino I, Gustafsson JA (1993): The glucocorticoid receptor and a putative repressor protein coordinately modulate glucocorticoid responsiveness of the mouse mammary tumor virus promoter in the rat hepatoma cell line M1.19. *J Biol Chem* 268: 1854–1859

Tashima Y, Terui M, Itoh H, Mizunuma H, Kobayashi R, Marumo F (1990): Phosphorylated and dephosphorylated types of non-activated glucocorticoid receptor. *J Biochem* 108: 271–277

Tbarka N, Richard-Mereau C, Formstecher P, Dautrevaux M (1993): Biochemical and immunological evidence that an acidic domain of hsp 90 *is involved in the stabilization of untransformed glucocorticoid receptor complexes. FEBS Lett* 322: 125–128

Tora L, Gronemeyer H, Turcotte B, Gaub M, Chambon P (1988): The N-terminal region of the chicken progesterone receptor specifies target gene activation. *Nature* 333: 185–188

Umesono K, Evans RM (1989): Determinants of target gene specificity for steroid/ thyroid hormone receptors. *Cell* 57: 1139–1146

Urda LA, Yem PM, Simons SS, Harmon JM (1989): Region-specific antiglucocorticoid receptor antibodies selectively recognize the activated form of the ligand-occupied receptor and inhibit the binding of activated complexes to desoxyribonucleic acid. *Mol Endocrinol* 3: 251–260

van der Weijden Benjamin WS, Hendry WJ, Harrison RW (1990): The mouse glucocorticoid receptor DNA-binding domain is not phosphorylated in vivo. *Biochem Biophys Res Commun* 166: 931–936

Vedeckis WV (1983): Subunit Dissociation as a possible mechanism of glucocorticoid receptor activation. *Biochemistry* 22: 1983–1989

von der Ahe D, Janich S, Scheidereit C, Renkawitz R, Schütz G, Beato M (1985): Glucocorticoid and progesterone receptors bind to the same sites in two hormonally regulated promoters. *Nature* 313: 706–709

Wieland S, Dobbeling U, Rusconi S (1991): Interference and synergism of glucocorticoid receptor and octamer factors. *EMBO J* 10: 2513–2521

Yen PM, Simons SS (1991): Evidence against posttranslational glycosylation of rat glucocorticoid receptors. *Receptor* 1: 191–205

Yoshinaga SK, Peterson CL, Herskowitz I, Yamamoto KR (1992): Roles of SWI1, SWI2, and SWI3 proteins for transcriptional enhancement by steroid receptors. *Science* 258: 1598–1604

Zilliacus J, Wright AP, Norinder U, Gustafsson JA, Carlstedt-Duke J (1992): Determinants for DNA-binding site recognition by the glucocorticoid receptor. *J Biol Chem* 267: 24941–24947

6

Thyroid Hormone Receptors

F. Javier Piedrafita and Magnus Pfahl

Introduction

Thyroid hormones notably Triiodothyronine (T_3) and thyroxine (T_4) play important roles in many biological processes in all vertebrates (De Groot et al, 1984). They affect growth, development and differentiation, metabolism and homeostasis. A particularly striking observation was made by Gudernatsch in the early part of this century that thyroid hormones, when given to tadpoles, induce rapid metamorphosis (Gudernatsch, 1912). The molecular mechanisms on how thyroid hormones exert their pleiotropic functions have, therefore, been of great interest.

A major step towards comprehension of the signaling pathways that underlie the large diversity of responses to thyroid hormones has been accomplished through the cloning of specific intracellular receptors. The existence of these nuclear receptors had been already predicted by experiments conducted in the 1960s in which a direct effect of thyroid hormones on RNA synthesis was shown (Tata, 1963; Tata and Widnell, 1966). Ten years later direct evidence for nuclear receptors was obtained through binding studies (Oppenheimer et al, 1972; Samuels and Tsai, 1973; Oppenheimer et al, 1974). Interestingly a thyroid hormone receptor, although a mutated form, the v-*erbA* oncogene, was the first member of the large nuclear receptor superfamily to be cloned and sequenced (Graf and Beug, 1983; Debuire et al, 1984). However, the identification of this protein as a member of the nuclear receptor family had to await cloning of the glucocorticoid and estrogen receptors (Weinberger et al, 1986; Sap et al, 1986; Weinberger et al, 1985; Green et al, 1986; Greene et al, 1986).

In mammals thyroid hormone receptors (TRs) are encoded by two genes, TRα and TRβ, from which multiple isoforms can be generated by alternative splicing (Sap et al, 1986; Benbrook and Pfahl, 1987; Thompson et al, 1987; Mitsuhashi et al, 1988; Nakai et al, 1988; Koenig et al,

INDUCIBLE GENE EXPRESSION, VOLUME 2
P.A. Baeuerle, Editor
© 1995 Birkhäuser Boston

1988; Hodin et al, 1989). Like the steroid hormone receptors, it had been assumed that thyroid receptors essentially carry out their signaling functions by binding as dimers to specific DNA regions (termed response elements) located usually in the promoter regions of susceptible genes. This was further supported by the observation that the identified T_3 responsive elements (TREs) revealed twofold symmetries (Umesono et al, 1991; Naar et al, 1991; Pfahl, 1994). By binding to these DNA sequences, the receptors can regulate transcription of specific genes, and the activation or repression of those genes determines the cellular response. TRs were therefore thought to activate responsive genes in the presence of thyroid hormone, similar to the steroid hormone receptors (Evans, 1988; Green and Chambon, 1988; Beato, 1989). More recently, however, it has become apparent that TR action is more complex. It is now clear that the receptors have multiple ways of functioning and that different pathways exist where TRs interact with other receptors and transcription factors (Pfahl, 1993; Zhang and Pfahl, 1993). In the following paragraphs we review gene regulation by thyroid hormone receptors and the network of interactions including the retinoid receptors as well as the transcription factor AP-1.

Synthesis and mechanisms of action of thyroid hormones

Although this review focuses on thyroid hormone receptors, we summarize here a few important facts on the hormones themselves. A more complete picture can be found in DeGroot et al (1984). Thyroid hormones (T_3 and T_4) (Figure 6.1) are synthesized in the thyroid gland in a reaction that requires iodide ions, a peroxidase, H_2O_2 and an iodide aceptor protein. Iodide ions are extracted from the plasma and concentrated in the interior of the thyroid cell. There, they are oxidized by the peroxidase in the presence of H_2O_2 (DeGroot and Davis, 1961) and then bound to tyrosine residues of thyroglobulin (TG), the major protein constituent of the thyroid gland, and also to other proteins. First monoiodotyrosine and diiodotyrosine are formed; then an iodinated hydroxyphenyl group from one iodotyrosine is transferred to the phenolic hydroxyl group of another molecule by a coupling reaction, forming T_3 and T_4 (Johnson and Tewksbury, 1942).

In spite of the size of TG (670,000 dalton) and the number of tyrosyl residues (about 110) only one molecule of T_4 or T_3 is carried per molecule of TG (Robbins and Rall, 1960). Iodination of TG and formation of the thyroid hormones occurs in the colloid. The secretory process of T_3 and T_4 starts with the interaction of TG with membrane receptors, reinterna-

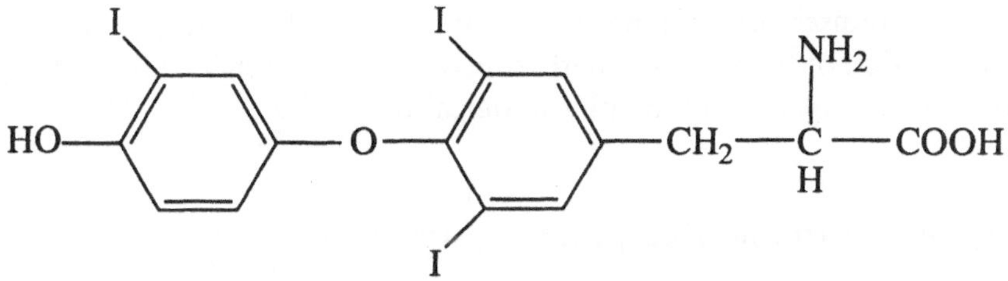

L-3,5,3'-Triiodothyronine (T3)

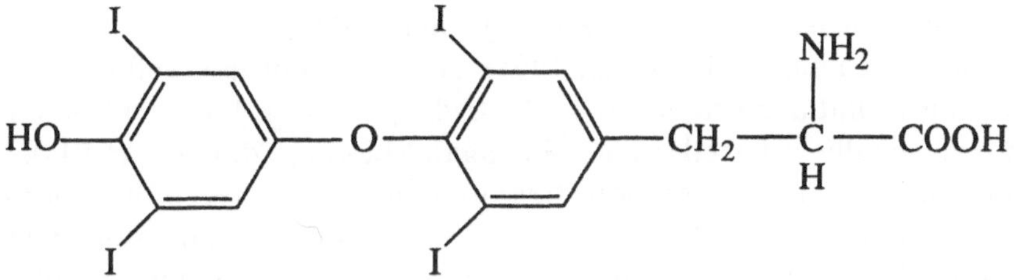

L-Thyroxine (T4)

Figure 6.1 Chemical structure of the thyroid hormones T_3 and T_4.

lization and formation of intracellular colloidal droplets. After fusion with a lysosome, TG is completely degraded and T_4 and T_3 are released and secreted to the blood stream.

In the blood, T_4 is the major iodothyronine found, at concentrations of about four to twelve mg/100 ml, while the concentration of T_3 is only about 100–200 μg/100 ml. The transport of thyroid hormones in the blood towards the target tissues occurs in association with plasma proteins, mainly the thyroxine binding protein, the thyroxine binding prealbumin and albumin. When the thyroid hormones reach their target tissues, cell uptake occurs mostly by passive diffusion although other mechanisms of entry into the cell have been proposed, including an energy-dependent active transport of T_3 and T_4 mediated by the plasma membrane (Eckel et al, 1979; Krenning et al, 1981, 1983; Galton et al, 1986). Also, a membrane receptor-mediated endocytosis mechanism can be important in the T_3 cell uptake (Halpern and Hinkle, 1982; Horiuchi et al, 1982). Once in the cytosol the thyroid hormones are bound in part to soluble binding proteins that include nuclear receptors. Most of the T_4 is deiodinated to T_3, the physiologically most active form. T_3 also diffuses into the nucleus where it binds to nuclear receptors, the main mediators of the T_3 responses. By binding to the TRs, thyroid hormones regulate the transcription of target genes increasing or decreasing their transcriptional rates thereby mediating the physiological responses.

Thyroid Hormone Receptors: Subtypes and Isoforms

Thyroid hormone receptors (TR) belong to a large family of nuclear receptors that also include the steroid hormone and retinoic acid receptors (RAR). Members of this superfamily of nuclear regulatory proteins are characterized by two specific domains, a highly conserved cysteine-rich region, the DNA-binding domain (DBD) that forms two zinc-finger structures which allow protein-DNA as well as protein-protein interaction (Luisi et al, 1991). A second but less well conserved region is found in all receptors in the hydrophobic carboxy-terminal half and is usually referred to as the ligand-binding domain (LBD). In addition to ligand recognition, this domain encodes receptor dimerization (Forman et al, 1989; Zhang et al, 1991a) and transactivation (Zhang et al, 1991b; Zenke et al, 1990), or repressor functions (Zhang et al, 1991b; Disela et al, 1991). TRs are encoded by two genes. TRα and TRβ, from which several isoforms can be expressed (Lazar, 1993). Two TRβ isoforms with different aminoterminal regions have been described (Weinberger et al, 1986; Koenig et al, 1988; Hodin et al, 1989), and at least four isoforms of TRα which all differ in their carboxyterminal sequences, have been isolated (Sap et al, 1986; Benbrook and Pfahl, 1987; Thompson et al, 1987; Mitsuhashi et al, 1988; Nakai et al, 1988). Both TRβ isoforms are ligand-dependent transcriptional activators, but only one of the α-receptor isoforms, TRα1, is a true receptor and functions as a ligand-dependent transcriptional enhancer. The other TRα isoforms are the result of differently spliced mRNAs encoding various forms of an alternative carboxyterminal domain. These carboxyterminal variants are receptor-like proteins that do not bind T_3 (Mitsuhashi et al, 1988; Lazar et al, 1989; Schueler et al, 1990; Lazar, 1993) and little is known about their physiological functions. Their possible roles as negative regulators are discussed below.

A schematic drawing of the TR subtypes and isoforms is shown in Figure 6.2. The viral v-erbA oncoprotein is also shown. v-erbA is a mutated TRα isoform that does not bind T_3 and has lost its transcriptional enhancer function but has maintained its repressor activity (Damm et al, 1989; Sap et al, 1989; Hermann et al, 1993). One major difference between TRs and the steroid hormone receptors (with the exception of ER) is their difference in size, in particular that of the amino terminal

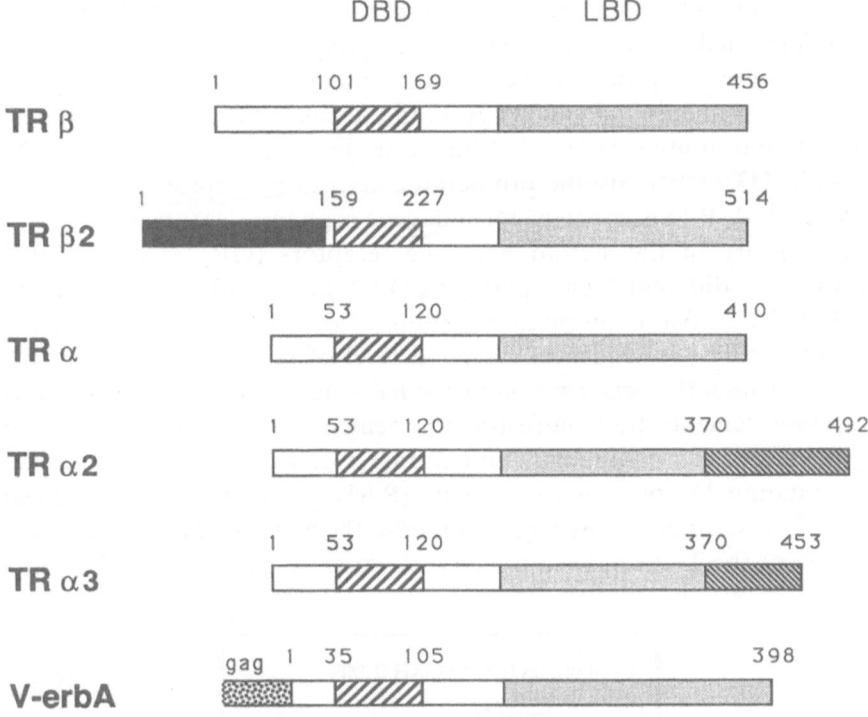

Figure 6.2 Schematic representation of TR proteins and their isoforms, as well as the oncoprotein v-*erb*A. The structures are aligned according to the DNA binding domain (DBD). The ligand binding domain (LBD) is also marked. Different patterns in the N-terminus (TRβ) or in the C-terminus (TRα) indicate differences in the protein sequence due to alternative splicing. A fragment of the viral gag protein fused to v-*erb*A is also shown.

region (Figure 6.2). This region is relatively small in all of the TRs but highly complex in the glucocorticoid, progesterone, mineralocorticoid, and androgen receptor (GR, PR, MR, and AR respectively). In addition carboxyterminal isoforms have so far only been described for TRs. These major differences in building plans are likely to point to different strategies for the signal mediation and specific gene regulation employed for these two distinct subfamilies of nuclear receptors.

Response Elements

One major mechanism by which the thyroid hormone receptors, as well as other members of the superfamily of nuclear hormone receptors achieve regulation of transcription of target genes, is through direct interactions with specific DNA sequences, the hormone response elements (HRE) or, more specifically thyroid hormone response elements (TREs or T$_3$REs).

TREs of a number of genes regulated positively or negatively by T_3 have been identified and a comparison of their sequences (Figure 6.3) shows that they are composed of two or more half-sites, 6 bp in length, with a consensus sequence 5'-PuGGTCA-3'. Interestingly the same half-sites are often found in other HREs that bind estrogen, retinoic acid, retinoid X, vitamin D3 or peroxisome proliferator activator receptors (ER, RAR, RXR, VDR, PPAR, rspectively) and some orphan receptors. In contrast, the majority of the steroid hormone receptors (GR, MR, AR, PR) recognize a different 6 base pair (bp) DNA motif with the sequence 5'-GGTACA-3'. Although many receptors of the TR subfamily recognize the same half-site sequence, a second level of selectivity restricts TRE usage through the organization of the half-sites. In the steroid hormone response elements, both half-sites are oriented as palindromes, separated by 3 nucleotides (Figure 6.2). In contrast TREs as well as retinoic acid and vitamin D3 responsive elements (RAREs and VDREs) are much more diverse, and various arrangements of the half-sites have been found. In case of the TREs at least three types exist: the palindrome (TREpal),

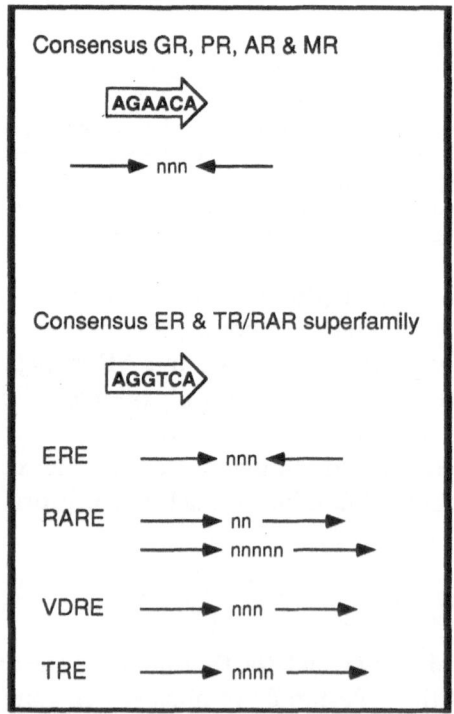

Figure 6.3(a) Hormone response elements. Comparison of the sequences found in different HREs yields consensus sequences for the steroid receptors GR, PR, AR and MR (top) and for ER and the TR/RAR subfamily (bottom).

POSITIVE TREs

PALINDROMES

TREpal	TCAGGTCATGACCTGA
rGH (5')	GAAAGGTAAGATCAGGGACGTGACCGCAG
TK-HSV	AAGGTGACGCGTGTGGCCTCG
rOT	AGGCGGTGACCTTGACCCCAGCCC

DIRECT REPEATS

MHC	CTGGAGGTGACAGGAGGACAGCAGCCCTGA
ME	AGGACGTTGGGGTTAGGGGAGGACAGTGGAC
hCS-B	GGTGGGGTCAAGCAGGGAGAGAG
MMLV	CAGGGTCATTTCAGGTCCTTG
NaK-ATP	AAGGTCACTCCGGGACGCCG

INVERTED PALINDROMES

EM-1	ACTGACCCTAAGGTCCTTT
dTRβ-TRE	GGTGTCCTCCTTAAGGGCAAC
MBP	AGACCTCGGCTGAGGACACGGC
F2	TATTGACCCCAGCTGAGGTCAAGTT
LTR-RSV	TCGTGCCTTATTAGGAAGGCAACA

OTHER

pTRβ-TRE	TGGAGGGAAACCAGGTCACCGGTTGCCAAG
PEPCK	TGACCCCCACCTGACAATTAAGGCAAG

NEGATIVE TREs

rTSH α	GCAGGTGAGGACTTCATT
mTSH β	TGAACGGAGAGTGGGTCATCACAG
rGH (intron)	CTGAGGCTGAGGTAACTTGGGAGTCCCAGGCAGAGGTCACTA
rGH (tata)	AGGTAGGGTATAAAAAGGGCATGC

Figure 6.3(b) Hormone response elements. TREs. The sequences of some representative TREs are shown. In positive TREs, the consensus half-sites can be arranged as palindromes, inverted palindromes or direct repeats; more complex structures also have been found containing usually more than two half-sites. Negative TREs found in different genes show variable arrangements of the AGGTCA consensus motif.

the direct repeat (DR) and the inverted palindrome (IP, also called everted repeat). In combination with different length spacers between the half-sites and some exchanges in the nucleotide sequences, a large number of TREs theoretically can be generated and do appear to exist. Some examples of this are shown in Figure 6.3. Not surprisingly, some TREs also allow activation by other receptors. For instance, a synthetic TREpal, derived from the promoter region of the rat growth hormone gene (rGH) allows activation by TRs in response to T_3 (Glass et al, 1988), as well as activation by RARs (Graupner et al, 1989) and RXRs (Mangelsdorf et al, 1990).

A simple rule, proposed at one time, by which DRs with different spacing confer specificity to TR, RAR and VDR when the half-sites are separated by three, four or five nucleotides respectively (Umesono et al, 1991) was found to apply only to some genes. Examples are TREs with DR-4 like structures found in the rat cardiac myosin heavy chain (MHC) (Flink and Morkin, 1990) and in the rat malic enzyme (ME) (Desvergne et al, 1991). Both genes were shown to be activated selectively by T_3 but not by retinoic acid (RA) or vitamin D3. Other DR-4-TREs have also been identified to bind and be activated by TR, for example in the murine Moloney leukemia virus (MLV) long terminal repeat (Sap et al, 1989), the rat S14 gene (Zilz et al, 1990), the human chorionic somatomammotropin (hCS) gene (Voz et al, 1991), and the human Na,K-ATPase β1-subunit gene (Feng et al, 1993).

A TRE composed out of an inverted palindromic sequence with a 6 bp spacer between the half-sites, was discovered in the F2 element of the chicken lysozyme silencer (Baniahmad et al, 1990) and also in the promoter region of the myelin basic protein (Farsetti et al, 1991). Other elements belonging to this class of TREs are the EM-1, (Ribeiro et al, 1992) an IP with 2 bp spacer found in the embryonic myosine gene, as well as an IP with 5 bp spacer located in the promoter region of the gene coding for TRβ (Sakurai et al, 1992). IPs with 4–6 bp spacers bind TR-homodimers and have major roles as negative elements (see below).

Other TREs do not appear to belong to any of these three classes described above. In addition several DNA sequences found to be necessary for the T_3-dependent regulation of genes, contain three or even more AGGTCA-related motifs arranged in variable forms. The promoter for the TRβ gene also contains one of these more complex TREs composed of three half-sites, two of them arranged as a DR-3 and the third one as a palindrome with 5 bp spacer (Sakurai et al, 1992). Recently, a palindrome with 6 bp spacer has been located in the herpes simplex virus thymidine kinase gene promoter and shown to be activated and bound by TRs (Park et al, 1993). Other complex TREs have been found in the 5' flanking regions of the keratin 10, keratin 14 and keratin 5 promoters with four to five half-site motifs arranged in different arrays (Ohtsuki et al, 1992; Tomic-Canic et al, 1992). Furthermore, many other genes are known to be regulated by T_3, but do not appear to contain typical TRE sequences. The analysis of the DNA sequences responsible for the T_3-dependent regulation of those genes will probably show other, more complex structures and may even contain sequences of half-sites not related to the consensus AGGTCA. The NF-κB and Sp1 motifs of the human immunodeficiency virus LTR also function as TRE (Desai-Yajnik and Samuels, 1993). Another example of a novel class of TREs, is the recently described response element in the Rous sarcoma virus LTR (Saatcioglu et al, 1993b), which contains a different DNA motif (AAGGCA) arranged as an IP with 6 bp spacer. This TRE also responds to TR and T_3 in a different manner than the classical TREs (see below).

Finally, TREs that mediate negative regulation by T_3 have been reported. However at this point it is not clear how these negative TREs function or what type of TR complexes are required for their activity. The thyroid stimulating hormone (TSH) and the thyrotropin-releasing hormone (TRH) genes are negatively regulated by T_3. A palindromic TRE has been identified immediately downstream from the TATA box in the TSHα gene (Chatterjee et al, 1989), while a cis-acting element located in the first exon of TSHβ functions also as a negative element (Wondisford et al, 1989; Wood et al, 1989). The rat growth hormone gene also appears to contain a negative TRE which overlaps with the TATA box (Crone et al, 1990). The glycoprotein hormone α-subunit gene is also negatively regulated by T_3 and different sequence fragments in the promoter have been proposed to be responsible for this T_3-regulation (Burnside et al, 1989; Gurr and Kourides, 1989; Pennathur et al, 1993); however, the proposed sequences are unrelated to other TREs. In summary, in contrast to the steroid hormone receptors, TRs function in combination with a large variety of DNA recognition sequences. Thus different principles of gene regulation through response elements appear to be applied.

Mechanisms of Action

Heterodimers bind and activate most TREs

Unlike the estrogen receptor and other steroid hormone receptors, analysis of the DNA binding properties of TRs indicated that high affinity binding of the receptors to their response elements requires interactions with additional nuclear proteins (also called TRAP for TR auxiliary protein) (Rosen et al, 1991). To identify the nuclear factor(s) or coreceptors that enhanced TR response element interaction, a number of approaches were used by various laboratories (Yu et al, 1991; Zhang et al, 1991a; Zhang et al, 1992b; Kliewer et al, 1992a; Leid et al, 1992; Bugge et al, 1992; Marks et al, 1992). The essential proof for having identified the nuclear factor or an equivalent activity, was to show that the factor could significantly enhance TR DNA binding. The heterodimeric complex bound strongly to several responsive elements containing two half-sites and was unstable on a single TREpal half-site (Zhang et al, 1991b), suggesting that the nuclear factor interacted itself with a TRE half-site. Therefore, members of the nuclear receptor subfamily showing a high degree of similarity in their DNA binding domains to TRs and related receptors (including RARs and RXRs) could be considered as possible candidates for the nuclear auxiliary protein or TRAP. When we mixed a variety of receptors, we observed that one of them, RXRα, dramatically enhanced TR interaction with the TREpal while TR and RXR alone bound poorly or not detectably to the response element (Zhang et al,

1992a). Similar results were obtained by others (Yu et al, 1991; Leid et al, 1992; Kliewer et al, 1992a; Marks et al, 1992; Bugge et al, 1992; Zhang and Pfahl, 1993). Contrary to previous reports (Glass et al, 1989) TRs and RARs could not be observed to form heterodimers that interacted effectively with the response elements investigated (Zhang et al, 1992a). TR/RXR complexes could be shown to form in solution and were stabilized by specific DNA binding sites but not by nonspecific sequences (Hermann et al, 1992; Leid et al, 1992; Kliewer et al, 1992a). However, RXRs were also found to enhance DNA binding and form heterodimers with RARs and the vitamin D3 receptor (Zhang et al, 1992a; Yu et al, 1991; Kliewer et al, 1992a; Leid et al, 1992; Bugge et al, 1992; Marks et al, 1992). Subsequently, it was found that RXRs can also interact with the PPAR (Kliewer et al, 1992b) the v-erbA oncogene (Hermann et al, 1993) and orphan receptors (Apfel et al, 1994) while the *Drosophila* RXR homologue *ultraspiracle* forms heterodimers with the ecdysone receptor (Yao et al, 1992). Thus TRAP turned out to be a general factor with an evolutionary conserved role acting as coreceptor for various receptors that are activated by structurally unrelated ligands.

Since RXR formed heterodimers with TRs as well as RARs and other receptors in solution, an important question was whether the different heterodimers had distinct DNA binding characteristics allowing to distinguish for instance, between TREs and RAREs. When DNA binding and transcriptional activation by TR-RXR and RAR-RXR heterodimers with natural TREs and natural RAREs was analyzed (Hermann et al, 1992), it was observed that TR-RXR heterodimers mostly bound natural TREs, but not natural RAREs. Similarly, RAR-RXR heterodimers only bound RAREs, but not TREs. An exception to this were response elements known to be activated by several hormones like the TREpal. Results from binding studies correlated well with results from transcriptional activation studies showing that heterodimers are not only required for optimal DNA binding, but also for optimal transcriptional activation (Hermann et al, 1992).

The scheme shown in Figure 6.4A is consistent with a large set of recent data from several laboratories. In solution, monomeric TRs, RARs, (or other receptors) and RXRs are in equilibrium with TR/RXR and RAR/RXR heterodimers. Only the heterodimers bind the response elements with high affinity and receptor binding to DNA is independent of ligand binding, such that the receptors can have dual functions, serving as activators in the presence of specific ligands, while they can be repressors in the absence of ligands (Graupner et al, 1989; Damm et al, 1989). As mentioned above, RXRs also form heterodimers with the v-erbA oncogene, a mutated TRα (Hermann et al, 1993), the VDR (Yu et al,

a   b

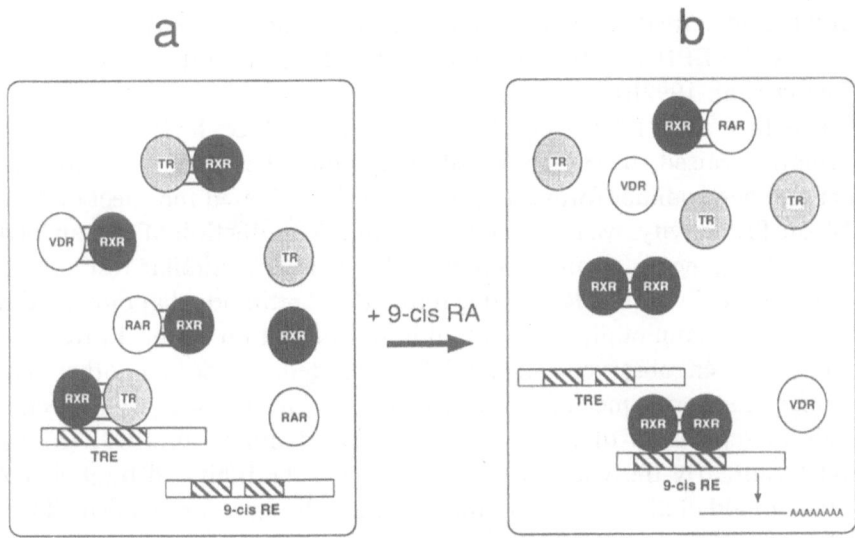

Figure 6.4 Heterodimers and cross-talk via RXR. (A) Heterodimer formation is necessary for efficient binding to most response elements. Heterodimers bind in the absence of ligand. (B) In the presence of 9-cisRA RXR homodimer formation is induced. This reduces the availability of RXR for heterodimer formation with TRs (and also VDRs) and thus leads to a repression of the T_3 response.

1991; Kliewer et al, 1992a; Leid et al, 1992), the PPAR (Kliewer et al, 1992b), and other vertebrate nuclear receptors (Apfel et al, 1994). Since the heterodimers have a much higher affinity for the specific response elements, it is likely that heterodimer formation is the rate limiting step for DNA binding and gene regulation of these receptors. Heterodimers are then the major transcriptional activators; transcriptional activation often observed in transfection experiments using only TRs must therefore be assumed to be due to the ubiquitous presence of endogenous RXRs.

RXR allows hormonal cross talk

From the above described observations, it had become apparent that RXRs have a very central role. They regulate TR-DNA binding and transcriptional activation, as well as the activities of several other hormone receptors that respond to structurally unrelated ligands. In addition it turned out that RXR also has its own ligand, 9-cis retinoic acid (9-cis-RA) (Heyman et al, 1992; Levin et al, 1992). When we investigated RXR DNA interaction, we made the surprising observation that in the presence of 9-cis-RA, RXR no longer requires heterodimerization for DNA binding. 9-cis-RA induces the formation of RXR homodimers that bind with high affinity to DNA (Zhang et al, 1992b). RXR homodimers bind to and activate specific response elements such as the ApoAI-RARE and the CRBPII-RARE, but not the DR-5 or the rat CRBPI-RARE (Zhang et al, 1992b).

The fact that RXR specific ligands can induce RXR homodimer formation raised the possibility of competition between RXR homodimer and heterodimer formation. When we investigated the effect of 9-cis-RA on TR activity, we indeed observed that T_3 induction of certain TRE containing genes is strongly repressed by 9-cis-RA. Similar results have been obtained with RXR selective retinoids — retinoids that only bind to RXRs (Lehmann et al, 1993). In addition, inhibition of the T_3 response only has been observed when RXR is limited. These and other data strongly support a mechanism in which sequestering of RXR molecules leads to repression of the T_3 response (Lehmann et al, 1993) (Figure 6.4B). Similarly the vitamin D3 response can be inhibited by 9-cis-RA (MacDonald et al, 1993) involving probably the same mechanism. Thus, a ligand-induced squelching mechanism exists which can control the availability of RXR molecules for heterodimerization and thereby allow hormonal cross talk (Figure 6.4A, B). However, not all RXR containing heterodimers are negatively affected by RXR specific retinoids. For instance, RAR-RXR heterodimers are not inhibited by 9-cis-RA, but in fact, are (in most cases), superactivated (Lehmann et al, 1992, 1993). The activity of PPAR-RXR heterodimers also appears to be increased when both PPAR and RXR specific ligands are present (Kliewer et al, 1992b). Thus, 9-cis-RA, and other retinoids leading to RXR homodimer formation, inhibit some heterodimers while they enhance others and by doing so, enable cross talk between various hormones and vitamin effectors.

Homodimers

Steroid hormone receptors form ligand-dependent homodimers which bind to the corresponding palindromic HREs to regulate transcription of target genes (Kumar and Chambon, 1988; Guiochon-Mantel et al, 1989; Wrange et al, 1989). However, as mentioned above, TRs and RARs require the presence of RXR to effectively bind and activate TREs (Zhang and Pfahl, 1993). The fact that 9-cis-RA induces RXR homodimer formation (Zhang et al, 1992b) and that some orphan receptors, such as COUP (Tran et al, 1992) and HNF-4 (Sladek et al, 1990), form homodimers with their respective HREs in the absence of ligand, raised the question whether TRs, and other RXR-heterodimer partners, might also regulate gene expression through homodimers. Indeed, activation by

VDR homodimers has now been observed (Carlberg et al, 1993) while an RXR-independent RAR function remains to be demonstrated.

In the case of the TRs, homodimer binding has been controversial. It is now clear that only one class of TRE binds TR homodimers with high affinity. For instance, homodimers have recently been reported to form with the chicken lysozyme F2 element (Yen et al, 1992), and we now have shown that the affinity with which these TR-homodimers to bind the F2-TRE as well as a synthetic IP-6 is equivalent or higher than that observed for the TR-RXR heterodimers. In contrast, although the TREpal and DR-4s are also able to bind TR homodimers under certain conditions, they bind TR-RXR heterodimers with much higher affinity (one or two orders of magnitude higher). Importantly, TR homodimers formed with IP-like sequences are inhibited by T_3 (Piedrafita et al, 1994; Yen et al, 1992), while in the absence of T_3, they bind with high affinity and are one order of magnitude more stable than the TR-RXR heterodimers on the IP-6. The spacing sequence between both AGGTCA motifs in the IP element appears to play an important role in the binding of TR-homodimers, since altering the spacer sequence can reduce dramatically homodimer binding without altering significantly TR-RXR heterodimer binding (Piedrafita et al, 1994). Interestingly, reducing the spacer by one or two nucleotides (IP-5 and IP-4 respectively) does not affect significantly the TR-homodimer binding but abolishes almost completely the TR-RXR heterodimer binding. The opposite effect has been observed when one or more nucleotides are added to the spacer (from IP-7 to IP-9) (Piedrafita et al, 1994).

To explore possible regulatory roles for TR homodimer binding sites, we have recently inserted an IP-6 3' of the TATA box, and observed that in such a situation TR homodimers can function as strong ligand sensitive repressors, pointing to a new role for TR homodimers in gene regulation. Ligand sensitive repression of transcription by TR homodimers through IP-TREs could be a general mechanism for T3 dependent gene regulation, functioning also from sites 5' of the TATA box, since direct interaction of TR with the general transcription factors has been demonstrated (Fondell et al, 1993; Baniahmad et al, 1993).

TRα has also recently been proposed to bind as homodimers or heterodimers to a new class of TRE: two AAGGCA motifs arranged as an inverted palindrome with a 6 bp spacer (Saatcioglu et al, 1993b). However, this new class of element found in the Rous sarcoma virus long terminal repeat functions as a constitutive enhancer. Cotransfection of HeLa cells with a reporter construct containing this TRE together with a TRα expression vector results in an about a 50-fold induction of the reporter gene in the absence of T_3, while the presence of ligand reversed this activation. Interestingly, the RSV-TRE is very specific for the TRα isoform, with no activation observed by TRβ. However the strong constitutive activation observed with this TRE by TRα in transfection assays does not correlate well with in vitro binding data. In vitro translated TRα does not bind to this RSV-TRE, and only a very weak binding can be observed in the presence of RXR, and the magnitude of this binding is not comparable to the binding observed with TREpal, F2 or other TREs (Piedrafita and Pfahl, 1994). In contrast cell extracts made from different cell lines show strong binding, which may suggest that other factor(s) are involved in this novel regulatory mechanism in which unliganded TR functions as a constitutive enhancer and is inhibited by T_3.

TR-AP-1 interaction

A variety of important biological responses to fat soluble hormones, like the steroid and thyroid hormones, appears to involve gene repression. Analogous to the positive elements for transcriptional activation, negative response elements that mediate hormone-induced repression have also been reported (Drouin et al, 1989; Akerblom et al, 1988; Rentoumis et al, 1990). Although the mechanism of receptor action at these negative elements is not well understood at present, it is assumed that receptor binding to such negative response elements in the presence of ligand somehow prevents the receptor from functioning as a transcriptional activator but allows negative interference with adjacent or overlapping binding sites that bind other transcription factors (Drouin et al, 1989; Akerblom et al, 1988; Rentoumis et al, 1990).

However, an additional important regulatory mechanism exists for nuclear receptors which does not require receptor-DNA interaction. This mechanism is based on protein-protein interaction and allows hormone-dependent repression (or in some cases activation of gene transcription) through interaction of the nuclear receptors and the transcription factor AP-1, a protein complex composed of c-*jun* and c-*fos* (Angel and Karin 1991). Its activity is modulated by growth factors, cytokins, oncogenes, and tumor promoters that activate protein kinase C. AP-1 induces transcriptional activation through interaction with a specific DNA recognition sequence, the 12-O–tetra-decanoyl-phorbol-13–acetate (TPA) responsive element (TRE or AP-1 binding site). The AP-1 binding site is recognized by c-*jun* and c-*fos* heterodimers, or c-*jun* homodimers that are formed through the leucine zipper domain of both proteins (Angel et al, 1987). c-*jun* and c-*fos*, when stimulated by extracellular signals, induce expression of genes involved in cell proliferation.

Evidence for nuclear receptor-AP-1 interaction was first obtained in the case of the glucocorticoid receptor (GR) (Jonat et al, 1990) and subsequently also reported for RARs (Yang Yen et al, 1991; Schüle et al, 1991), RXRs (Salbert et al, 1993) and ERs (Tzukerman et al, 1991). The hormone responsive isoforms of TRα and TRβ-1 (Zhang et al, 1991c) also have been found to inhibit AP-1 activity. For instance TRβ-1 effectively inhibits induction of the c-*fos* promoter by TPA in CV-1-cells. In addition, TRα and TRβ have been shown to prevent induction of the collagenase promoter by TPA and c-*jun* or c-*fos*. TR-anti-AP-1 activity is dependent of the receptor and ligand (T_3) concentrations (Zhang et al, 1991c). Interestingly, a deletion of only 17 amino acids from the TRβ-1 carboxyterminal end, a region that can form an amphipatic α-helix (Zhang et al, 1991c), substantially reduces anti-AP-1 activity (Zhang et al, 1991c).

c-*jun* and c-*fos* were also found to inhibit induction of T_3-responsive reporter genes by TRs. TRs did not bind to AP-1 sites, while AP-1 did not bind to a T_3-responsive element (Zhang et al, 1991c). Like GR and RARs, TRs strongly inhibited Jun/Fos binding to the AP-1 site. Inhibition of TRβ binding to a T_3-responsive element by c-*jun*, however, was relatively weak. Desbois and coworkers, in addition, examined the activities of the v-*erb*A oncogene (Desbois et al, 1991). V-*erb*A failed to repress AP-1 activity and acts as a dominant negative factor, inhibiting AP-1 repression by TRα and RARs (Hermann et al, 1993). Similar data also have been reported by Saatcioglu et al (1993a). These investigators also have identified a nine amino acid sequence that when deleted, causes the loss of interference with AP-1. The sequence is conserved in many receptors but deleted in v-*erb*A and is responsible for normal biological activities in TRα as well as RARα (Zenke et al, 1990; Saatcioglu et al, 1993a). Attempts to coprecipitate TRs and c-*jun* and c-*fos* with specific antibodies against either of the proteins, however, have not been successful (Zhang et al, 1991c; Desbois et al, 1991), suggesting a relatively weak interaction between TRs and AP-1 (although it cannot be excluded that the nature of the particular antibodies used precluded coprecipitation). However, as seen also with GR and RAR, TRs and AP-1 clearly can antagonize the activity of each other in vivo and in vitro by a mechanism that does not involve competition for the same (or overlapping) binding site(s) or interference from adjacent binding sites.

The observation that a number of nuclear receptors interact with the AP-1 transcription factor suggests that this is a general feature of the ligand-controlled nuclear receptors. However the action of the receptors as well as c-*jun* and c-*fos* appear to be cell type-, receptor-, and promoter specific. This is also supported by a report (Shemshedinini et al, 1991) investigating the effect of c-*jun* and/or c-*fos* coexpression on transcriptional repression by several steroid hormone receptors, using different reporter gene constructs and different cell lines. In case of TRs, in addition, synergism as well as antagonisms with AP-1 has been observed (Zhang et al, 1991c; Lopez et al, 1993). These observations point to a complex mechanism on how the two classes of proteins, the nuclear receptors and the AP-1 components, interact. One possibility is that both proteins can function as bridging factors that depending on the cell type can act positively or negatively or not at all on the other factor as discussed elsewhere in detail (Pfahl, 1993). However, a simpler mechanism in which the proteins can inhibit each other by forming mixed heterodimers is also possible and is supported by recent in vitro data for the retinoid receptors.

Recently it has been observed that in vitro, inhibition of AP-1 DNA binding by retinoid receptors is ligand dependent when all proteins used are obtained by in vitro translation (Salbert et al, 1993; Fanjul et al, 1994). This is in concert with our observation that all nuclear receptors produced in this manner have optimal activities. It is, therefore, quite possible that direct interactions between TR and AP-1 components so far have not been demonstrated in vitro, simply because the proteins used in these assays have not been fully active. The data from Salbert et al and Fanjul et al suggest a very simple model for nuclear receptor AP-1 interaction as shown in Figure 6.5 (Salbert et al, 1993; Fanjul et al, 1994). In the absence of T_3, TRs bind as heterodimers or homodimers to response elements, while the AP-1 components c-*jun* and c-*fos* bind the AP-1

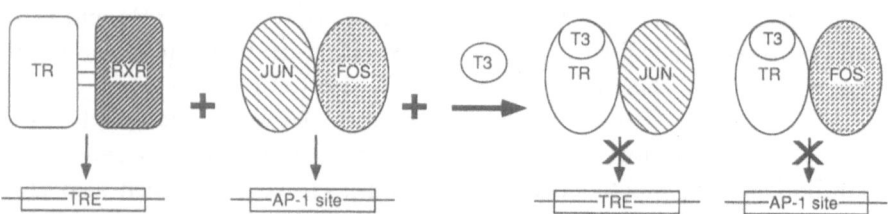

Figure 6.5 TR-AP-1 interaction. Based on recent results for retinoid receptors (Salbert et al, 1993; Fanjul et al, 1994), a model for cross-talk between thyroid hormone receptors and the transcription factor AP-1 (Jun/Fos) is proposed. TRs and RXRs form heterodimers in solution that bind with high affinity to TREs. Similarly, the components of AP-1, Jun and Fos, form heterodimers that bind with high affinity to AP-1 sites. In the presence of T_3, TRs undergo a conformational change that allows them to form complexes with Jun and/or Fos. These receptor/Jun or receptor/Fos complexes do not bind to DNA. Thus, when the TRs are in excess, AP-1 activity is inhibited, while an excess of Jun or Fos can lead to the inhibition of the T_3 receptors. (For other models see Pfahl, 1993).

site. In the absence of T_3, TRs no longer bind as homodimers but in addition can form mixed heterodimer complexes with c-*jun* or c-*fos*. These mixed heterodimers no longer bind to either T_3REs or to AP-1 sites. Thus when TRs are in excess, c-*jun* and c-*fos* activity will be repressed by T_3, whereas when c-*jun* and c-*fos* are in excess, TR activity is inhibited. Other more complex interactions between the two families of transcription factors, allowing synergism between the receptors and AP-1, also can be imagined and have been discussed elsewhere (Pfahl, 1993). Whatever the precise mechanism of interaction between the two families of transcription factors, it is based mainly on protein-protein interaction and not on competition for overlapping DNA binding sites.

Negative regulatory roles for TR isoforms

Several TRα isoforms have been described that do not appear to function as transcriptional activators. The carboxyterminal variants TRα-2 and TRα-2v are identical to the T_3 responsive TRα-1 for the first 370 amino acids, but then diverge as a result of alternative splicing (Figure 6.2) (Benbrook and Pfahl, 1987; Mitsuhashi et al, 1988a, b; Lazar et al, 1988; Izumo and Mahdavi, 1988). The TRα-2 isoforms do not bind T_3 but have been reported to have some repressor activity (Koenig et al, 1989; Rentoumis et al, 1990; Hermann et al, 1991; Nakai et al, 1990). However, DNA binding of the TRα-2 isoforms was found to be weak (Hermann et al, 1991) and was not enhanced by RXR on the TREpal (Zhang et al, 1992a) questioning the biological significance of the weak repressor activities observed with the TRα carboxyterminal variants. On the other hand, these isoforms are major TR forms expressed during postnatal brain development (Wills et al, 1991; Mellstrom et al, 1991) and in the testis (Benbrook and Pfahl, 1987) and are therefore likely to have specific biological roles. A recent observation may explain how TRα-2 isoforms can function as limited repressors (Nagaya and Jameson, 1993). These investigators reported that TRα-2 can bind efficiently as a heterodimer with RXR to certain DR-4 TREs but not to TREs of other structures, such as the TREpal and IP-TREs. In cotransfection assays TRα-2 inhibits only those TREs to which it could bind in the presence of RXR. These data suggest that the TRα-2 isoforms can selectively inhibit certain DR-4 type TREs and thus when coexpressed with TRα-1 and/or TRβ-1, restrict the T_3 response to a subset of responsive genes. Thus expression of TRα isoforms in tissues like brain, testis and the immune system (Strait et al,

1990) may allow a limited T_3 response even at ambiant hormone concentrations. Similarly inactive variants have also been observed for other transcription factors (Benezra et al, 1990; Foulkes et al, 1991) although in these cases the inhibitory factors are encoded by different genes. The unique ability of TRs to bind to a diversity of response elements provides a special situation in which repression of a subset of TREs can be accomplished by specific splice variants.

Thyroid Hormone Receptors and Disease

Many clinical disorders are associated with a dysfunction of the thyroid gland, and abnormal levels of circulating thyroid hormones have been detected in some of these syndromes. A detailed description of these disorders is beyond the scope of this chapter, and the reader is referred to standard textbooks (DeGroot et al, 1984; Wilson and Foster, 1981). However, in one disease TRs are clearly affected. Here we describe briefly the syndrome of generalized resistance to thyroid hormones (GRTH) which segregates as an autosomal dominant disorder (Magner et al, 1986) and is tightly linked to mutations found in the gene coding for TRβ (Usala et al, 1990). Patients with GRTH characteristically have a reduced response to the thyroid hormones and elevated circulating levels of free T_3 and T_4 as well as unusual levels of TSH. In GRTH responsiveness to T_3 is observed in all tissues, but selective pituitary and a peripheral tissue resistance to thyroid hormones have also been found (Refetoff et al, 1993).

The disorder was first described in 1967 (Refetoff et al, 1967), and more than 300 subjects have been identified since then to suffer from this syndrome. An important advance in the molecular basis of understanding the disorder has been the discovery of its association with the TRβ gene (Usala et al, 1990). At least 28 mutations have so far been identified; except for one that is located in the hinge region (Behr and Loos, 1992), all mutations map in two defined regions of the ligand binding domain of TRβ (Parilla et al, 1992; Refetoff et al, 1993) (Figure 6.6). The mutant receptors associated with GRTH have a reduced affinity for T_3 and some of them, as kindreds S and PV, are unable to bind the hormone. Most of the mutant receptors do not function as transcriptional activators in transfection experiments while some can regain their normal function at high concentrations of T_3. Others show only a partial recovery of their function with saturating concentrations of the hormone. All mutants studied in vitro so far, with the exception of the kindred GH (Geffner et al, 1993), show some

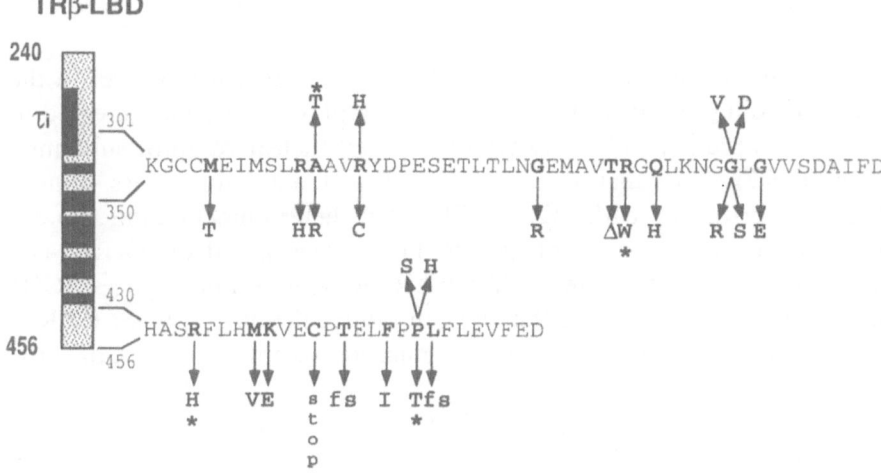

Figure 6.6 Mutations in the TRβ gene associated with GRTH. The structure of the TRβ-LBD is shown (aminoacids 240 to 456). The τi domain and the nine heptad repeats are boxed. The nucleotide sequences of regions 301–350 and 430–456 are shown, corresponding to the two hot spot regions where most of the point mutations have been found. The residues mutated are indicated in bold letters. In some cases different substitutions have been described for the same aminoacid, and the same mutation has been found in different families (indicated with asterisk); fs indicates a frame shift.

dominant negative activity and are able to mask gene activation by wild-type receptor (TR-RXR), although very high concentrations of the mutant receptor were required to see this effect. In our laboratory we have recently compared the binding properties of several mutant receptors and found important differences in their ability to form homodimers with specific DNA sequences (mainly IP-TREs); the mutant homodimers respond to different concentrations of T_3 in gel retardation and in transcriptional activation experiments. We have observed for instance that mutant GH that has been shown not to have a dominant negative effect binds DNA as homodimer very weakly; it, therefore, may have lost its repressor activity. However, none of the mutants has lost its ability to heterodimerize with RXR on any of the TREs assayed. Overall our results suggest that the mutant receptors have altered homodimer activities and that for the majority of them homodimer binding to IP TREs is no longer or is less sensitive to T_3. This observation could also explain their dominant phenotype (in the patients). Importantly, the finding that a defect in the homo-dimer activity of TRβ can be associated with a specific disease (GRTH), strongly suggests a prominant physiological role for TR homodimers in regulating transcription by a T_3 sensitive repressor.

Concluding Remarks

The cloning of nuclear T_3 receptors has allowed recent advances in the understanding of TR and T_3 action. A complex picture has emerged in which TRs are part of a network of a larger nuclear receptor subfamily (Figure 6.7). To function as transcriptional activators, TRs require heterodimerization with RXRs. TR-RXR heterodimers bind to and function from a large set of diverse TREs. The ligand of RXR, 9-cis-RA, does not contribute to TR-RXR activation but induces RXR homodimer formation and thereby can inhibit the T_3 response. RXR is also a partner for several other nuclear receptors and can thus

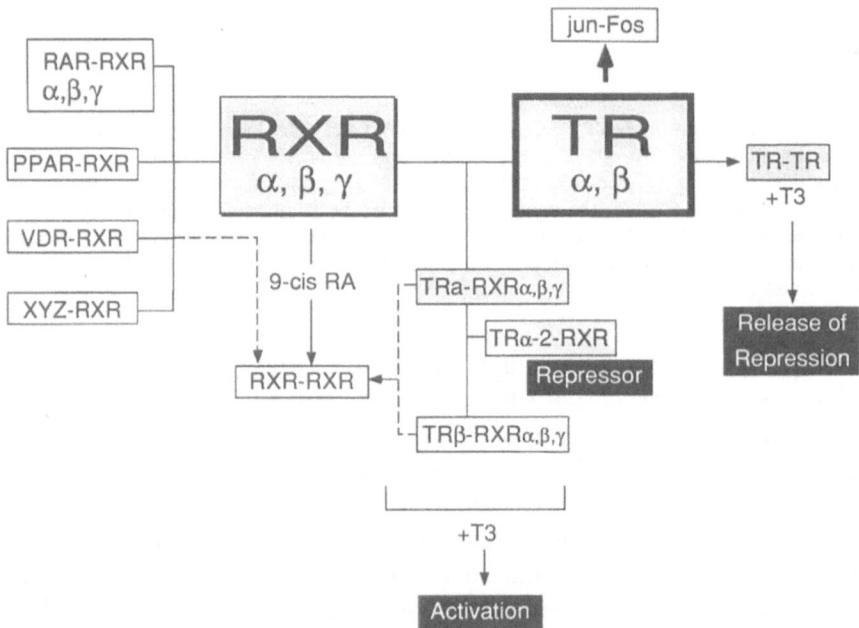

Figure 6.7 Mechanisms of TR action. To function as transcriptional activators TRs require heterodimerization with RXRs. TR-RXR heterodimers bind to and function from a large set of diverse TREs. The ligand of RXR, 9-cisRA, does not contribute to TR-RXR activation but induces RXR homodimer formation and thereby can inhibit the T_3 response. RXR is also a partner for several other nuclear receptors and can thus potentially mediate cross talk between different hormones and vitamin derivatives. TRα isoforms like TRα-2 and TRα-2v are not T_3 responsive activators but (as heterodimers with RXR) appear to bind to a subset of TREs and function on those as repressors, allowing for a restriction of the T_3 induction to a subset of responsive genes. TR homodimers are also not transcriptional activators but bind with high affinity to IP-TREs and can function on those as T_3 sensitive silencers or repressors. TRs, like other nuclear receptors also interact with the transcription factor AP-1 or its components *jun* and *fos*. This allows for an important cross talk between two major signaling systems, one that generally promotes cell proliferation (AP-1) and one (mediated by nuclear receptors) that often promotes cell differentiation.

potentially mediate cross talk between different hormones and vitamin derivatives. TRα isoforms like TRα-2 and TRα-2v are not T_3 responsive activators but (as heterodimers with RXR) appear to bind to a subset of TREs and function on those as repressors, allowing for a restriction of the T_3 induction to a subset of responsive genes. TR homodimers also are not transcriptional activators but bind with high affinity to IP-TREs and can function on those as T_3 sensitive silencers or repressors in a similar fashion like the prokaryotic *lac* repressor (B. Müller-Hill).

Finally TRs, like other nuclear receptors, interact with the transcription factor AP-1. This allows for an important cross talk between two major signaling systems, one that generally promotes cell proliferation (AP-1) and one (mediated by nuclear receptors) that often promotes cell differentiation. The scheme presented here represents a reasonable model system to allow for the complex biological roles of thyroid hormones; however further additions and refinements of this model are certain to be necessary in the future.

References

Akerblom IE, Slater EP, Beato M, Baxter JD, Mellon PL (1988): Negative regulation by glucocorticoids through interference with a cAMP responsive enhancer. *Science* 241: 350–353

Angel P, Karin M (1991): The role of Jun, Fos and the Ap-1 complex in cell-proliferation and transformation. *Biochim Biophys Acta* 1072: 129–157

Angel P, Imagawa M, Chiu R, Stein B, Imbra RJ, Rahmsdorf HF, Jonat C, Herrlich P, Karin M (1987): Phorbol ester-inducible genes contain a common cis element recognized by a TPA modulated trans-acting factor. *Cell* 49: 729–739

Apfel R, Benbrook D, Lernhardt E, Ortiz MA, Pfahl M (1994): A novel orphan receptor with a unique ligand binding domain and its interaction with the retinoid/thyroid hormone receptor subfamily. *Mol Cell Biol* 14

Baniahmad A, Steiner C, Kohn AC, Renkawitz R (1990): Modular structure of a chicken lysozyme silencer: Involvement of an unusual thyroid hormone receptor binding site. *Cell* 61: 505–514

Baniahmad A, Tsai SY, O'Malley BW, Tsai MJ (1992): Kindred S thyroid receptor is an active and constitutive silencer and a repressor for thyroid hormone and retinoic acid responses. *Proc Natl Acad Sci USA* 89: 10633–10763

Beato M (1989): Gene regulation by steroid hormones. *Cell* 56: 335–344

Behr M, Loos U (1992): A point mutation (A1a[229] to Thr) in the hinge domain of the c-*erbAβ* thyroid hormone receptor gene in a family with generalized thyroid hormone resistance. *Mol Endocrinol* 6: 1119–1126

Benbrook D, Pfahl M (1987): A novel thyroid hormone receptor encoded by a cDNA clone from a human testis library. *Science* 238:788–791

Benezra R, Davis RL, Lockshon D, Turner DL, Weintraub H (1990): The protein Id: a negative regulator of helix-loop-helix DNA binding proteins. *Cell* 61: 49–59

Bugge TH, Pohl J, Lonnoy O, Stunnenberg HG (1992): RXRα, a promiscuous partner of retinoic acid and thyroid hormone receptors. *EMBO J* 11: 1409–1418

Burnside J, Darling DS, Carr FE, Chin WW (1989): Thyroid hormone regulation of the rat glycoprotein hormone α-subunit gene promoter activity. *J Biol Chem* 264: 6886–6891

Carlberg C, Bendik I, Wyss A, Meier E, Sturzenbecker LJ, Grippo JF, Hunziker W (1993): Two nuclear signaling pathways for vitamin D. *Nature* 361: 657–660

Chatterjee VKK, Lee J-K, Rentoumis A, Jameson JL (1989): Negative regulation of the thyroid-stimulating hormone α gene by thyroid hormone: Receptor interaction adjacent to the TATA box. *Proc Natl Acad Sci USA* 86: 9114–9118

Crone DE, Kim HS, Spindler SR (1990): α and β thyroid hormone receptors bind immediately adjacent to the rat growth hormone gene TATA box in a negatively hormone-responsive promoter region. *J Biol Chem* 265: 10851–10856

Damm K, Thompson CC, Evans RM (1989): Protein encoded by v-*erb*A functions as a thyroid-hormone receptor antagonist. *Nature* 339: 593–597

Debuire B, Henry C, Bernissa M, Biserte G, Claverie JM, Saule S, Martin P, Stehelin D (1984): Sequencing the erbA gene of avian erythroblastosis virus reveals a new type of oncogene. *Science* 224: 1456–1459

DeGroot LJ, Davis AM (1961): Studies on the biosynthesis of iodotyrosines. *J Biol Chem* 236: 2009

DeGroot LJ, Larsen PR, Refetoff S, Stanbury JB (1984): *The Thyroid and Its Diseases*. New York: John Wiley & Sons

Desai-Yajnik V, Samuels HH (1993): The NF-κB and Sp1 motifs of the human immunodeficiency virus type 1 long terminal repeat function as novel thyroid hormone response elements. *Mol Cell Biol* 13: 5057–5069

Desbois C, Aubert D, Legrand C, Pain B, Samarut J (1991): A novel mechanism of action for v-*erb*A: Abrogation of the inactivation of transcription factor AP-1 by retinoic acid and thyroid hormone receptors. *Cell* 67: 731–740

Desvergne B, Petty KJ, Nikodem VM (1991): Functional characterization and receptor binding studies of the malic enzyme thyroid hormone response element. *J Biol Chem* 266: 1008–1013

Disela C, Clineur C, Bugge T, Sap J, Stengl G, Dodgson J, Stunnenberg H, Beug H, Zenke M (1991): v-*erb*A overexpression is required to extinguish c-*erb*A function in erythroid cell differentiation and regulation of the *erb*A target gene CAII. *Genes Dev* 5: 2033–2047

Drouin J, Trifiro M, Plante R, Nemer M, Eriksson P, Wrange O (1989): Glucocorticoid receptor binding to a specific DNA sequence is required for hormone-dependent repression of pro-opiomelanocortin gene transcription. *Mol Cell Biol* 9: 5305–5314

Eckel J, Rao GS, Rao ML, Breuer H (1979): Uptake of L-tri-iodothyronine by isolated rat liver cells. *Biochem J* 182: 473

Evans RM (1988): The steroid and thyroid hormone receptor family. *Science* 240: 889–895

Fanjul A, Hobbs P, Graupner G, Zhang X-K, Dawson MI, Pfahl M (1994): A novel class of retinoids with selective anti-oncogene activity and their mechanism of action. *Nature* (in press)

Farsetti A, Mitsuhashi T, Desvergne B, Robbins J, Nikodem VM (1991): Molecular basis of thyroid hormone regulation of a myelin basic protein gene expression in rodent brain. *J Biol Chem* 266: 23226–23232

Feng J, Orlowski J, Lingrel JB (1993): Identification of a functional thyroid hormone response element in the upstream flanking region of the human Na,K-ATPase $\beta 1$ gene. *Nucl Acids Res* 21: 2619–2626

Flink IL, Morkin E (1990): Interaction of thyroid hormone receptors with strong and weak cis-acting elements in the human α-myosin heavy chain gene promoter. *J Biol Chem* 265: 11233–11237

Fondell JD, Roy AL, Roeder RG (1993): Unliganded thyroid hormone receptor inhibits formation of a functional preinitiation complex: Implications for action repression. *Genes Dev* 7: 1400–1410

Forman BM, Yang C-R, Au M, Casanove J, Ghysdael J, Samuels HH (1989): A domain containing leucine-zipper-like motifs between the thyroid hormone and retinoic acid receptors. *Mol Endocrinol* 3: 1610–1626

Foulkes NS, Borrelli E, Sassone-Corsi P (1991): CREM gene: Use of alternative DNA-binding domains generates multiple antagonists of cAMP-induced transcription. *Cell* 64: 739–749

Galton VA, St Germain DL, Whittemore S (1986): Cellular uptake of 3,5,3'-triiodothyronine and thyroxine by red blood and thymus cells. *Endocrinol* 118: 1918–1923

Geffner ME, Su F, Ross NS, Hershman JM, Van Dop C, Menke JB, Hao E, Stanzak RK, Eaton T, Sammuels HH, Usala J (1993): An arginine to histidine mutation in codon 311 of the c-*erbAβ* gene results in a mutant thyroid hormone receptor that does not mediate a dominant negative phenotype. *J Clin Invest* 91: 538–546

Glass CK, Holloway JM, Devary OV, Rosenfeld MG (1988): The thyroid hormone receptor binds with opposite transcriptional effects to a common sequence motif in thyroid hormone and estrogen response elements. *Cell* 54: 313–323

Glass CK, Lipkin SM, Devary OV, Rosenfeld MG (1989): Positive and negative regulation of gene transcription by a retinoic acid-thyroid hormone receptor heterodimer. *Cell* 59: 697–708

Graf T, Beug H (1983): Role of the v-*erbA* and v-*erbB* oncogenes of avian erythroblastosis virus in erythroid cell transformation. *Cell* 34: 7–9

Graupner G, Wills KN, Tzukerman M, Zhang X-K, Pfahl M (1989): Dual regulatory role for thyroid-hormone receptors allows control of retinoic-acid receptor activity. *Nature* 340: 653–656

Green S, Chambon P (1988): Nuclear receptors enhance our understanding of transcription regulation. *Trends Genet* 4: 309–314

Green SP, Walter P, Kumar B, Krust A, Bornert J-M, Argos PK, Chambon P (1986): Human estrogen receptor cDNA: Sequence, expression, and homology to v-*erbA*. *Nature* 320: 134–139

Greene GL, Gilna P, Waterfield M, Baker A, Hort Y, Shine J (1986): Sequence and expression of human estrogen receptor complementary DNA. *Science* 231: 1150–1154

Gudernatsch JF (1912): Feeding experiments on tadpoles. I. The influence of specific organs given as food on growth and differentiation. A contribution to the knowledge of organs with internal secretion. *Arch Entwicklungsmech Organ* 35: 457

Guichon-Mantel A, Loosfelt H, Lescop P, Sar S, Atger M, Perrot-Applanat M, Milgrom E (1989): Mechanisms of nuclear localization of the progesterone receptor: Evidence for interaction between monomers. *Cell* 57: 1147–1154

Gurr JA, Kourides IA (1989): Regulation of the transfected human glycoprotein hormone α-subunit gene by dexamethasone and thyroid hormone. *DNA* 8: 473–480

Halpern J, Hinkle PM (1982): Evidence for an active step in thyroid hormone transport to nuclei: Drug inhibition of L-^{125}-I-triiodothyronine binding to nuclear receptors in rat pituitary tumor cells. *Endocrinol* 110: 1070

Hermann T, Hoffmann B, Zhang X-K, Tran P, Pfahl M (1992): Heterodimeric receptor complexes determine 3,5,3′-triiodothyronine and retinoid signaling specificities. *Mol Endocrinol* 6:1153–1162

Hermann T, Hoffmann B, Piedrafita FJ, Zhang X-K, Pfahl M (1993): V-*erb*A requires auxiliary proteins for dominant negative activity. *Oncogene* 8: 55–65

Hermann T, Zhang X-K, Tzuckerman M, Wills KN, Graupner G, Pfahl M (1991): Regulatory functions of a non-ligand binding thyroid hormone receptor isoform. *Cell Regul* 2: 565–574

Heyman RA, Mangelsdorf DJ, Cyck JA, Stein RB, Eichele G, Evans RM, Thaller C (1992): 9-cis retinoic acid is a high affinity ligand for the retinoid X receptor. *Cell* 68: 397–406

Hodin RA, Lazar MA, Wintman BI, Darling DS, Koenig RJ, Larsen PR, Moore DD, Chin WW (1989): Identification of a thyroid hormone receptor that is pituitary-specific. *Science* 244: 76–78

Horiuchi R, Cheng S-Y, Willingham M, Pastan I (1982): Inhibition of the nuclear entry of 3,3′,5-triiodo-L-thyronine by monodansylcadaverine in GH$_3$ cells. *J Biol Chem* 257: 3139

Izumo S, Mahdavi V (1988): Thyroid hormone receptor α isoforms generated by alternative splicing differentially activate myosin HC gene transcription. *Nature* 334: 539–542

Johnson TB, Tewksbury LB (1942): The oxidation of 3,5-diiodotyrosine to thyroxine. *Proc Natl Acad Sci USA* 28: 73

Jonat C, Rahmsdorf HJ, Park K-K, Cato ACB, Gebel S, Ponta H, Herrlich P (1990): Antitumor promotion and anti-inflammation: Down-modulation of AP-1 (Fos/Jun) activity by glucocorticoid hormone. *Cell* 62: 1189–1204

Kliewer SA, Umesono K, Mangelsdorf DJ, Evans RM (1992a): Retinoid X receptor interacts with nuclear receptors in retinoic acid, thyroid hormone and vitamin D3 signaling. *Nature* 355: 446–449

Kliewer SA, Umesono K, Noonan DJ, Heyman RA, Evans RM (1992b): Convergence of 9-cis retinoic acid and peroxisome proliferator signaling pathways through heterodimer formation of their receptors. *Nature* 358: 771–774

Kliewer SA, Umesono K, Heyman RA, Mangelsdorf DJ, Dyck JA, Evans RM (1992c): Retinoid X receptor-COUP-TF interactions modulate retinoic acid signaling. *Proc Natl Acad Sci USA* 89: 1448–1452

Koenig RJ, Lazar MA, Hodin RA, Brent GA, Larsen GA, Chin WW, Moore DD (1989): Inhibition of thyroid hormone action by a non-hormone binding c-*erb*A protein generated by alternative mRNA splicing. *Nature* 337: 659–661

Koenig RJ, Warne RL, Brent GA, Harney JW, Larsen PR, Moore DD (1988): Isolation of a cDNA clone encoding a biologically active thyroid hormone receptor. *Proc Natl Acad Sci USA* 85: 5031–5035

Krenning EP, Doctor R, Bernard B, Visser T, Hennemann G (1981): Characteristics of active transport of thyroid hormone into rat hepatocytes. *Biochim Biophys Acta* 676: 314–320

Krenning EP, Doctor R, Visser TJ, Henneman G (1983): Plasma membrane transport of thyroid hormone: Its possible pathophysiologic significance. *J Endocrinol Invest* 6: 59

Kumar V, Chambon P (1988): The estrogen receptor binds tightly to its responsive element as a ligand-induced homodimer. *Cell* 55: 145–156

Lazar MA (1993): Thyroid hormone receptors: Multiple forms, multiple possibilities. *Endocr Rev* 14: 184–193

Lazar MA, Hodin RA, Darling DS, Chin WW (1988): Identification of a rat c-erbAα related protein which binds DNA but does not bind thyroid hormone. *Mol Endocrinol* 2: 893–901

Lazar MA, Hodin RA, Chin WW (1989): Human carboxyl-terminal variant of α-type c-erbA inhibits transactivation by thyroid hormone receptors without binding thyroid hormone. *Proc Natl Acad Sci USA* 86: 7771–7774

Lehmann JM, Jong L, Fanjul A, Cameron JF, Lu X-P, Haefner P, Dawson MI, Pfahl M (1992): Retinoids selective for retinoid X receptor response pathways. *Science* 258: 1944–1946

Lehmann JM, Zhang X-K, Graupner G, Lee M-O, Hermann T, Hoffmann B, Pfahl M (1993): Formation of RXR homodimers leads to repression of T_3 response: hormonal cross-talk by ligand induced squelching. *Mol Cell Biol* 13: 7698–7707

Leid M, Kastner P, Lyons R, Nakshatri H, Saunders M, Zacharewski T, Chen J-Y, Staub A, Garnier JM, Mader S, Chambon P (1992): Purification, cloning, and RXR identity of the HeLa cell factor with which RAR or TR heterodimerizes to bind target sequences efficiently. *Cell* 68: 377–398

Levin AA, Sturzenbecker LJ, Kazmer S, Bosakowski T, Huselton C, Allenby G, Speck J, Kratzeisen C, Rosenberger M, Lovey A, Grippo JF (1992): 9-cis retinoic acid steroisomer binds and activates the nuclear receptor RXRα. *Nature* 355: 359–361

Lopez G, Schaufele F, Webb P, Holloway JM, Baxter JD, Kushner PJ (1993): Positive and negative modulation of jun action by thyroid hormone receptor at a unique AP-1 site. *Mol Cell Biol* 13: 3042–3049

Luisi BF, Xu WX, Otwinowski Z, Freedman LP, Yamamoto KR, Sigler PB (1991): Crystallographic analysis of the interaction of the glucocorticoid receptor with DNA. *Nature* 352: 497–505

MacDonald PN, Dowd DR, Nakajima S, Galligan MA, Reeder MC, Haussler CA, Ozato K, Hausler MR (1993): Retinoid X receptors stimulate and 9-cis retinoic acid inhibits 1,25-dihydroxyvitamin D3-activated expression of the rat osteocalcin gene. *Mol Cell Biol* 13: 5907–5917

Magner JA, Petrick P, Menezes-Ferreira MM, Stelling M, Weintraub BD (1986): Familial generalized resistance to thyroid hormones: Report of three kindreds and correlation of patterns of affected tissues with the binding of [125I] triiodothyronine to fibroblast nuclei. *J Endocrinol Invest* 9: 459–470

Mangelsdorf DJ, Ong ES, Dyck JA, Evans RM (1990): Nuclear receptor that identifies a novel retinoic acid response pathway. *Nature* 345: 224–229

Marks MS, Hallenbeck PL, Nagata T, Segars JH, Appella E, Nikodem VM, Ozato K (1992): H-2RIIBP (RXRβ) heterodimerization provides a mechanism for combinatorial diversity in the regulation of retinoic acid and thyroid hormone responsive genes. *EMBO J* 11: 1419–1435

Mellstrom B, Naranjo JR, Santos A, Antonio MG, Bernal J (1991): Independent expression of the α and β c-erbA genes in developing rat brain. *Mol Endocrinol* 56: 1339–1350

Mitsuhashi T, Tennyson GE, Nikodem VM (1988a): Nucleotide sequence of novel cDNAs generated by alternative splicing of a rat thyroid hormone receptor gene transcript. *Nucl Acids Res* 16: 5697

Mitsuhashi T, Tennyson GE, Nikodem VM (1988b): Alternative splicing generates messages encoding rat c-*erb*A proteins that do not bind thyroid hormone. *Proc Natl Acad Sci USA* 85: 5804–5808

Naar AM, Boutin J-M, Lipkin SM, Yu VC, Holloway JM, Glass CK, Rosenfeld MG (1991): The orientation and spacing of core DNA-binding motifs dictate selective transcriptional responses to three nuclear receptors. *Cell* 65: 1267–1279

Nagaya T, Jameson JL (1993): Distinct dimerization domains provide antagonist pathways for thyroid hormone receptor action. *J Biol Chem* 268: 24278–24282

Nakai A, Sakurai A, Bell GI, DeGroot LJ (1988): Characterization of a third human thyroid hormone receptor co-expressed with other thyroid hormone receptors in several tissues. *Mol Endocrinol* 2: 1087–1092

Nakai A, Sakurai A, Macchia E, Fang V, DeGroot LJ (1990): The roles of three forms of human thyroid hormone receptor in gene regulation. *Mol Cell Endocrinol* 72: 143–148

Ohtsuki M, Tomic-Canic M, Freedberg IM, Blumenberg M (1992): Nuclear proteins involved in transcription of the human K5 keratin gene. *J Invest Dermatol* 99: 206–215

Oppenheimer JH, Koerner K, Schwartz HL, Surks MI (1972): Specific nuclear triiodothyronine binding sites in rat liver and kidney. *J Clin Endocrinol Metab* 35: 330–333

Oppenheimer JH, Schwartz HL, Surks MI (1974): Tissue differences in the concentration of triiodothyronine nuclear binding sites in the rat: liver, kidney, pituitary, heart, brain, spleen, and testis. *Endocrinology* 95: 897–903

Parilla R, Mixson AJ, McPherson JA, McClaskey JH, Weintraub BD (1991): Characterization of seven novel mutations of the c-*erb*Aβ gene in unrelated kindreds with generalized thyroid hormone resistance. *J Clin Invest* 88: 2123–2130

Park H-Y, Davidson D, Raaka BM, Samuels HH (1993): The herpes simplex virus thymidine kinase gene promoter contains a novel thyroid hormone response element. *Mol Endocrinol* 7: 319–330

Pennathur S, Madison LD, Kay TWH (1993): Localization of promoter sequences required for thyrotropin-releasing hormone and thyroid hormone responsiveness of the glycoprotein hormone a-gene in primary cultures of rat pituitary cell. *Mol Endocrinol* 7: 797–805

Pfahl M (1993): Nuclear receptor/AP-1 interaction. *Endocr Rev* 14: 651–658

Pfahl M (1994): Vertebrate Receptors: Molecular Biology, dimerization and response elements. In: *Seminars in Cell Biology*, 5: 95–103

Piedrafita FJ, Pfahl M (1994): Unpublished data

Piedrafita FJ, Bendik I, Ortiz MA, Pfahl M (1994): Thyroid hormone receptor homodimers can function as ligand-sensitive repressors on specific TREs. (submitted)

Refetoff S, Weiss RE, Usala SJ (1993): The syndromes of resistance to thyroid hormone. *Endocrine Rev* 14: 348–399

Refetoff S (1982): Syndromes of thyroid hormone resistance. *Am J Physiol* 243: 88–98

Refetoff S, DeWind LT, DeGroot LJ (1967): Familial syndrome combining deaf-mutism, stippled epiphyses, goiter, and abnormally high PBI: Possible target organ refractoriness to thyroid hormone. *J Clin Endocrinol Metab* 27: 279–294

Rentoumis A, Krishna V, Chatterjee K, Madison LD, Datta S, Gallagher GD, DeGroot LJ, Jameson JL (1990): Negative and positive transcriptional regulation by thyroid hormone receptor isoforms. *Mol Endocrinol* 90: 1522–1531

Ribeiro RCJ, Kushner PJ, Apriletti JW, West BL, Baxter JD (1992): Thyroid hormone alters in vitro DNA binding of monomers and dimers of thyroid hormone receptors. *Mol Endocrinol* 6: 1142–1152

Robbins J, Rall JE (1960): Proteins associated with the thyroid hormones. *Physiol Rev* 40: 415

Rosen ED, O'Donnell AL, Koenig RJ (1991): Protein-protein interactions involving *erb*A superfamily receptors, through the TRAPdoor. *Mol Cell Endocrinol* 78: C83–C88

Saatcioglu F, Bartunek P, Deng T, Zenke M, Karin M (1993a): A conserved c-terminal sequence that is deleted in v-*erb*A, is essential for the biological activities of the c-*erb*A/thyroid hormone receptor. *Mol Cell Biol* 13: 3675–3685

Saatcioglu F, Deng T, Karin M (1993b): A novel cis element mediating ligand-independent activation by c-*erb*A: Implications for hormonal regulation. *Cell* 75: 1095–1105

Sakurai A, Miyamoto T, DeGroot LJ (1992): Cloning and characterization of the human thyroid hormone receptor $\beta1$ gene promoter. *Biochem Biophys Res Commun* 185: 78–84

Sakurai A, Takeda K, Ain K, Ceccarelli P, Nakai A, Seino S, Bell GI, Refetoff S, DeGroot LJ (1989): Generalized resistance to thyroid hormone associated with a mutation in the ligand-binding domain of the human thyroid hormone receptor β. *Proc Natl Acad Sci USA* 86: 8977–8981

Salbert G, Fanjul A, Piedrafita FJ, Lu X-P, Kim S-J, Tran P, Pfahl M (1993): RARs and RXRα down-regulate the TGF-$\beta1$ promoter by antagonizing AP-1 activity. *Mol Endocrinol* 7: 1347–1356

Samuels HH, Tsai JS (1973): Thyroid hormone action in cell culture: Demonstration of nuclear receptors in intact cells and isolated nuclei. *Proc Natl Acad Sci USA* 70: 3488–3492

Sap J, Munoz A, Damm K, Goldberg Y, Ghysdael J, Leutz A, Beug H, Vennstrom B (1986): The c-*erb*A protein is a high-affinity receptor for thyroid hormone. *Nature* 324: 635–640

Sap J, Munoz A, Schmitt J, Stunnenberg H, Vennstrom B (1989): Repression of transcription mediated at a thyroid hormone response element by the v-*erb*A oncogene product. *Nature* 340: 242–243

Schueler PA, Schwartz HL, Strait KA, Mariash CN, Oppenheimer JH (1990): Binding of 3,5,3'-triiodothryonine (T$_3$) and its analogs to the in vitro translation products of c-*erb*A protooncogenes: differences in the affinity of the α- and β-forms for the acetic acid analog and failure of the human testis and kidney α-2 products to bind T$_3$. *Mol Endocrinol* 4: 227–234

Schüle R, Rangarajan P, Yang N, Kliewer S, Ransone LJ, Bolado J, Verma IM, Evans RM (1991): Retinoic acid is a negative regulator of AP-1 responsive genes. *Proc Natl Acad Sci USA* 88: 6092–6096

Shemshedini L, Knauthe R, Sassone-Corsi P, Pornon A, Gronemeyer H (1991): Cell-specific inhibitory and stimulatory effects of Fos and Jun on transcription activation by nuclear receptors. *EMBO J* 10: 3839–3849

Sladek FM, Zhong WM, Lai E, Darnell JE (1990): Liver-enriched transcription factor HNF-4 is a novel member of the steroid hormone receptor superfamily. *Genes Dev* 4: 2353–2365

Strait KA, Schwartz HL, Perez CA, Oppenheimer JH (1990): *J Biol Chem* 265: 10514–10521

Tata JR (1963): Inhibition of the biological action of thyroid hormones by actinomyocin D and puromycin. *Nature* 197: 1167–1168

Tata JR, Widnell CC (1966): Ribonucleic acid synthesis during the early action of thyroid hormones. *Biochem J* 98: 604–620

Thompson CC, Weinberger C, Lebo R, Evans RM (1987): Identification of a novel thyroid hormone receptor expressed in the mammalian central nervous system. *Science* 237: 1610–1613

Tomic-Canic M, Sunjevaric I, Freedberg IM, Blumenberg M (1992): Identification of the retinoic acid and thyroid hormone receptor-responsive element in the human K14 keratin gene. *J Invest Dermatol* 99: 842–847

Tran P, Zhang X-K, Salbert G, Hermann T, Lehmann JM, Pfahl M (1992): COUP orphan receptors are negative regulators of retinoic acid response pathways. *Mol Cell Biol* 12: 4666–4676

Tzukerman M, Zhang X-K, Pfahl M (1991): Inhibition of estrogen receptor activity by the tumor promoter TPA: a molecular analysis. *Mol Endocrinol* 5: 1983–1992

Umesono K, Murakami KK, Thompson CC, Evans RM (1991): Direct repeats as selective response elements for the thyroid hormone, retinoic acid and vitamin D3 receptors. *Cell* 65: 1255–1266

Usala SJ, Bale AE, Gesundheit N, Weinberger C, Lash RW, Sondisford FE, McBride OW, Weintraub BD (1988): Tight linkage between the syndrome of generalized thyroid hormone resistance and the human c-erbAβ gene. *Mol Endocrinol* 2: 1217–1220

Usala SJ, Tennyson GE, Bale AE, Lash RW, Gesundheit N, Wondisford FE, Accili D, Hauser P, Weintraub BD (1990): A base mutation of the C-erbAβ thyroid hormone receptor in a kindred with generalized thyroid hormone resistance. *J Clin Invest* 85: 93–100

Voz ML, Peers B, Belayew A, Martial JA (1991): Characterization of an unusual thyroid response unit in the promoter of the human placental lactogen gene. *J Biol Chem* 266: 13397–13408

Weinberger C, Hollenberg SM, Rosenfeld MG, Evans RM (1985): Domain structure of human glucocorticoid receptor and its relationship to the v-erbA oncogene product. *Nature* 318: 670–672

Weinberger C, Thompson CC, Ong ES, Lebo R, Gruol DJ, Evans RM (1986): The c-erbA gene encodes a thyroid hormone receptor. *Nature* 324: 641–646

Wills KN, Zhang X-K, Pfhal M (1991): Coordinate expression of functionally distinct TRα isoforms during neonatal brain development. *Mol Endocrinol* 5: 1109–1119

Wilson JD, Foster DW (1981): *Textbook of Endocrinology*. Philadelphia: W.B. Saunders

Wondisford FE, Farr EA, Radovick S, Steinfelder HJ, Moates JM, McClaskey JH, Weintraub BD (1989): Thyroid hormone inhibition of human thyrotropin β-subunit gene expression is mediated by a cis-acting element located in the first exon. *J Biol Chem* 264: 14601–14604

Wood WM, Kao MY, Gordon DF, Ridgway EC (1989): Thyroid hormone regulates

the mouse thyrotropin β-subunit gene promoter in transfected primary thyro-
tropes. *J Biol Chem* 264: 14840–14847

Wrange O, Eriksson P, Pearlmann T (1989): The purified activated glucocorticoid
receptor is a homodimer. *J Biol Chem* 264: 5253–5259

Yang Yen H-F, Zhang X-K, Graupner G, Tzukerman M, Sakamoto B, Karin M,
Pfahl M (1991): Antagonism between retinoic acid receptors and AP-1: Implication for tumor promotion and inflammation. *New Biol* 3: 1206–1219

Yao TP, Segraves WA, Oro AE, McKeown M, Evans RM (1992): Drosophila
ultraspiracle modulates ecodysone receptor function via heterodimer formation. *Cell* 71: 63–72

Yen PM, Darling DS, Carter RL, Forgione M, Umeda PK, Chin WW (1992):
Triiodothyronine (T₃) decreases binding to DNA by T₃-receptor homodimers
but not receptor-auxiliary protein heterodimers. *J Biol Chem* 267: 3565–3568

Yu VC, Delsert C, Andersen B, Holloway JM, Devary OV, Naar AM, Kim SY,
Boutin J-M, Glass CK, Rosenfeld MG (1991): RXRβ: a coregulator that
enhances binding of retinoic acid, thyroid hormone, and vitamin D receptors
to their cognate response elements. *Cell* 67: 1251–1266

Zenke M, Munoz A, Sap J, Vennstrom B, Beug H (1990): V-*erb*A oncogene activation
entails the loss of hormone-dependent regulator activity of c-*erb*A. *Cell* 61: 1035–
1049

Zhang X-K, Pfahl M (1993): Regulation of retinoid and thyroid action through
homo- and heterodimeric receptors. *Trends in Endocrinology and Metabolism* 4:
156–162

Zhang X-K, Hoffmann B, Tran P, Graupner G, Pfahl M (1992a): Retinoid X receptor
is an auxiliary protein for thyroid hormone and retinoic acid receptors. *Nature*
355: 441–446

Zhang X-K, Lehmann J, Hoffmann B, Dawson MI, Cameron G, Graupner G,
Hermann T, Pfahl M (1992b): Homodimer formation of retinoid X receptor
induced by 9-cis retinoic acid. *Nature* 358: 587–591

Zhang X-K, Tran P, Pfahl M (1991a): DNA binding and dimerization determinants
for TRα and its interaction with a nuclear protein. *Mol Endocrinol* 5: 1909–1920

Zhang X-K, Wills KN, Hermann T, Graupner G, Tzukerman M, Pfahl M (1991b):
Ligand-binding domain of thyroid hormone receptors modulates DNA binding
and determines their bifunctional roles. *New Biol* 3: 1–14

Zhang X-K, Wills KN, Husmann M, Hermann T, Pfahl M (1991c): Novel pathways
for thyroid hormone receptor action through interaction with jun and fos
oncogene activities. *Mol Cell Biol* 11: 6016–6025

Zilz ND, Murray MB, Towles HC (1990): Identification of multiple thyroid hormone
response elements located far upstream from the rat S₁₄ promoter. *J Biol Chem*
265: 8136–8143.

7

Retinoic Acid Receptors

MARIE KEAVENEY AND HENDRIK G. STUNNENBERG

Introduction

Retinoids, a class of hydrophobic compounds including retinol (vitamin A), retinoic acid (RA) and a series of natural and synthetic derivatives, exhibit a vast array of profound and diverse effects on vertebrate development from early embryogenesis to maturity. Several families of serum, cytoplasmic and nuclear proteins are involved in the metabolism and biological actions of retinoids. Some of these proteins mediate direct effects of retinoids on gene expression while others are involved in their transport, storage and metabolism. The effects of retinoids on transcription are mediated by a number of nuclear binding proteins of two types: retinoic acid receptors (RARs) and retinoid X receptors (RXRs). RXR serves as an auxiliary factor required by RAR and other nuclear receptors for target gene regulation. These ligand-inducible transcription factors belong to the nuclear receptor superfamily, which also includes receptors for thyroid hormone, vitamin D_3 and steroid hormones. Detailed studies of RAR and RXR function have revealed the existence of a vast elaborate web of gene regulation. This chapter will aim to analyse the diverse and complex pathways of retinoic acid responses.

Biology of Vitamin A

Vitamin A or retinol cannot be synthesized de novo by animals, and its main natural dietary sources are provitamin A carotenoids from vegetables and the long-chain retinyl esters from animal tissues. It is stored in the liver and released into the vascular system as a protein complex of retinol bound to retinol-binding protein (RBP) (Blomhoff et al, 1990; Ragsdale and Brockes, 1991). In many target tissues retinol can be converted in a reversible manner by oxidative metabolism to retinal, which can in turn be further oxidised irreversibly to all-trans retinoic acid

INDUCIBLE GENE EXPRESSION, VOLUME 2
P.A. Baeuerle, Editor
© 1995 Birkhäuser Boston

(RA). Two other metabolites synthesised directly from retinol are 3,4-didehydroretinoic acid (ddRA) and 14-hydroxy-4,14-retro-retinol; however their functional role is less well understood. Intracellularly, RA can be further metabolised via a series of oxidative and isomerization reactions to a number of active retinoids, including 9-cis retinoic acid (9-cis RA), 11-cis RA, 13-cis RA (Warrell et al, 1993) and 4-oxo-RA (Pijnappel et al, 1993). At least four of these endogenous retinoids are known to activate gene transcription, namely all-trans RA, ddRA (Thaller and Eichele, 1990), 9-cis RA and 4-oxo-RA (Figure 7.1) (Pijnappel et al, 1993). However, from the point of view of activation of the nuclear retinoid receptors, all-trans RA and 9-cis RA are probably the most important.

Role of Retinoids in Mammalian Development

Retinoids are a group of signalling molecules that affect a wide array of biological processes (Goodman, 1984; Brockes, 1990; De Luca, 1991; Ragsdale and Brockes, 1991; Tabin, 1991; Mendelsohn et al, 1992; Morriss-Kay, 1993). These include vertebrate development, cellular differentiation and homeostasis. Vitamin A is required for proper epithelial cell differentiation, maintenance of reproductive capacity and functioning of the visual cycle. All-trans retinoic acid (RA), the most biologically active vitamin A analogue, modulates cellular differentiation, pattern formation and embryonic development. More specifically, RA is believed to play a fundamental role in the development of the vertebrate nervous system, limbs and craniofacial features as well as in the regeneration of amphibian limbs.

Early studies on the role of RA in limb development indicated that retinoids may be natural morphogens (Tickle et al, 1982; Chytil, 1984; Slack, 1987; Thaller and Eichele, 1987; Brockes, 1989; Eichele, 1989a, 1989b; Smith et al, 1989; Brockes, 1990; Summerbell and Maden, 1990). However, more recent data challenge this point of view (Brockes, 1991; Noji et al, 1991; Wanek et al, 1991), the results of which would suggest that the ectopic administration of RA may act during development to regulate the expression of other factors, which themselves may be true morphogens. Thus, whilst RA itself may not constitute a morphogenetic agent, it plays an important role in limb development (Tabin, 1991; Mendelsohn et al, 1992), inducing profound effects when it is ectopically administered either during development or in the adult, or on cells in culture, where it has been associated with alterations in patterns of gene expression and/or levels of expression of specific genes (Nicholson et al,

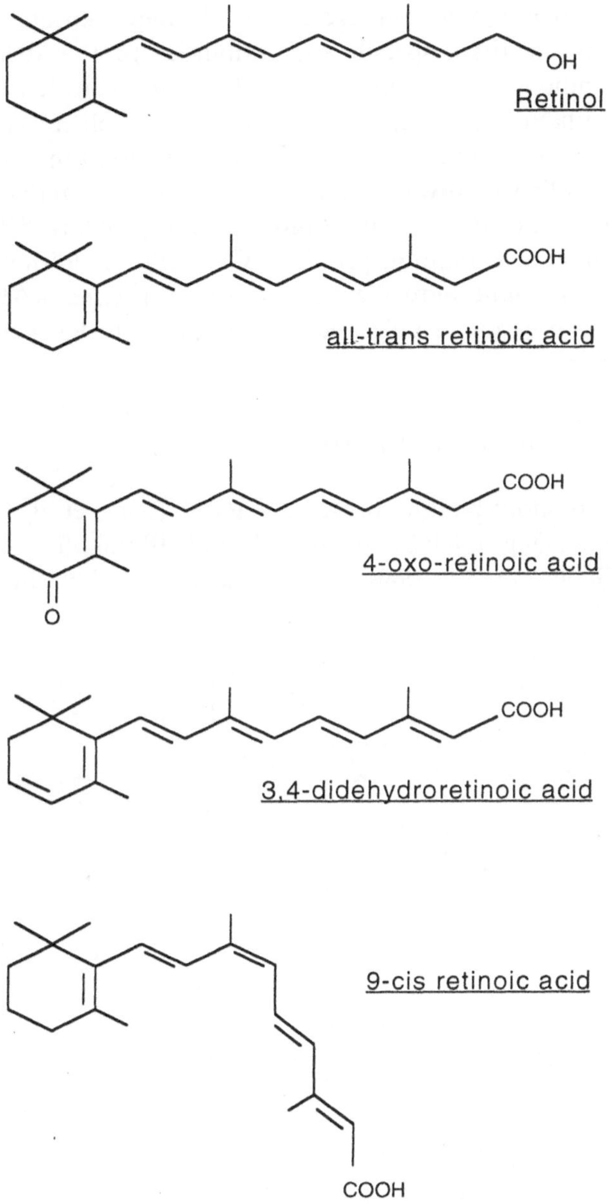

Figure 7.1 Chemical structures of retinol and some naturally occurring retinoids known to be involved in the retinoid signalling process.

1990; Simeone et al, 1990; Wang et al, 1990; Izpisua-Belmonte et al, 1991; Kessel and Gruss, 1991).

Therefore, in contrast to the physiological requirements of vitamin A for many biological processes, retinoids may be toxic or severely teratogenic when applied in excess or at inappropriate times. The differences in the tissue-specific susceptibilities to RA deficiency and RA excess indicate that in normal development there is a mechanism whereby available RA levels are regionally controlled. To date, two general classes of retinoid binding proteins have been implicated in the retinoid signalling pathway: first, the cytoplasmic retinoid binding proteins which include cellular retinol binding proteins (CRBPs), and cellular retinoic acid binding proteins (CRABPs). Additionally, at the nuclear level the ligand-inducible transactivator proteins, RAR and RXR regulate the expression of target genes in response to RA (Figure 7.2).

Cellular Retinoid Binding Proteins

The cellular retinoid binding proteins, CRBPs (Saari et al, 1982; Chytil and Ong, 1984; Ong, 1984; Sundelin et al, 1985; Blomhoff et al, 1990) and CRABPs (Bailey and Siu, 1988; Stoner and Gudas, 1989; Blomhoff et al,

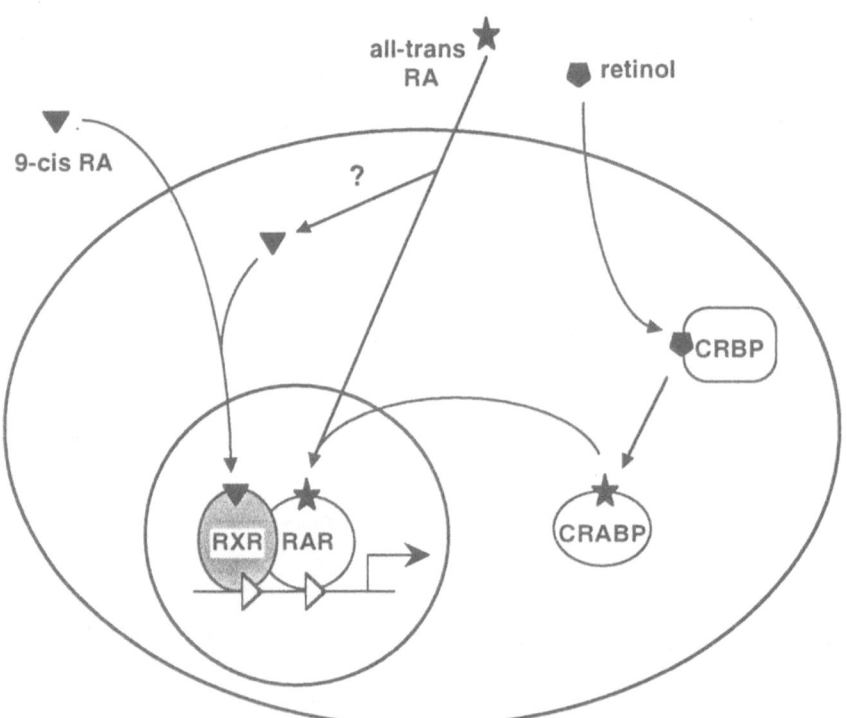

Figure 7.2 Schematic representation of the components of the retinoid signalling pathway. CRBP, cellular retinol-binding protein: CRABP, cellular retinoic acid-binding protein; RAR, retinoic acid receptor; RXR, retinoid X receptor.

1990, 1991; Giguere et al, 1990a) bind retinol and all-trans retinoic acid (RA) respectively (Figure 7.2), with high affinity and specificity. The initial finding that these cytoplasmic proteins shared structural similarities with a number of intracellular fatty-acid binding proteins implied that they may have a possible role in transport and storage of retinoid molecules.

There are at least two genes for each of these binding proteins: CRBP I and II (Ong, 1987) and CRABP I and II (Chytil and Ong, 1984; Sundelin et al, 1985; Demmer et al, 1987; Shubeita et al, 1987; Wei et al, 1987; Bailey and Siu, 1988; Stoner and Gudas, 1989; Giguere et al, 1990a), which are expressed in a spatial and temporal restricted manner during embryogenesis and in the adult (Shubeita et al, 1987; Wei et al, 1987; Maden et al, 1988, 1989, 1990; Dollé et al, 1989, 1990; Perez-Castro et al, 1989; Dencker et al, 1990; Giguere et al, 1990a; Vaessen et al, 1990; Ruberte et al, 1992). The expression patterns of CRBP and CRABP are linked with tissues susceptible to vitamin A deficiency and retinoid excess respectively. Tissues expressing high levels of CRBP include those that are most vulnerable to vitamin A deficiency, which suggests that CRBP functions to store retinol in sites where RA is eventually required at high concentrations for developmental processes. CRABP, on the other hand, has been proposed to control the delivery of physiological amounts of RA to the nucleus and in conjunction with this, its expression pattern correlates well with known target tissues of excess RA-induced teratogenesis (Morriss-Kay, 1993). CRABP is theorised to be instrumental in protecting cells from the developmentally important action of RA by prohibiting aberrant activation of RA-responsive genes during critical stages of development. In support of this notion is the observation that in the developing chick limb bud, there is a CRABP gradient across the anteroposterior axis and a similar gradient has been detected along the proximodistal axis of the murine limb. This gradient could serve to modulate the levels of free RA available to nuclear receptors (Tabin, 1991).

Furthermore, transcripts encoding CRBP I (Smith et al, 1991) and CRABP II (Giguere et al, 1990a; Durand et al, 1992) are RA inducible, perhaps providing an in vivo feedback mechanism whereby excess ligand could be sequestered, thus preventing inappropriate expression of RA-responsive genes (De Luca, 1991; Morriss-Kay, 1993). The RA inducible CRABP II has a lower affinity for RA than CRABP I, and there is some evidence to suggest that CRABP II does not bind RA until the binding capacity of CRABP I is saturated. It is thus interesting in this respect that during development CRABP II expression is much more restricted than is CRABP I. The specific expression of CRABP II in tissues known to be targets for RA action during embryogenesis such as the limb bud, suggests that this regulatory control feedback mechanism functions to protect the developing limb from RA induced abnormalities. In this way CRABP would provide a fine regulation of the amount of free RA.

The intracellular concentration of available RA is critical in determining the extent of activation of RA nuclear receptors. Although CRBP and CRABP have been proposed to function in regulating these levels, the precise biological roles of CRBP and CRABP are unclear at present. In addition to this labyrinth of regulation, a number of cell types respond to retinoids even though they do not express the corresponding cellular binding proteins. Hence, although the cellular retinoid binding proteins appear to be necessary for some cells to exhibit retinoid responses, other cells seem to function without the binding proteins. On the other hand, unidentified cellular binding proteins may exist that are tissue-specific. Furthermore, it is also probable that particular cellular binding proteins exist that are specific for the many retinoid metabolites. It is interesting in this respect that none of the identified cellular retinoid binding proteins can bind 9-cis RA which suggests the possible existence of additional ligand-specific cellular binding proteins. Alternatively, this may also suggest that 9-cis RA levels are indirectly controlled in the cell by CRABP. As this protein sequesters RA, it may also regulate its conversion to other retinoid metabolites. The regulation of RA isomerization is probably a key step in retinoid physiology and could provide a novel means for differential cell-specific regulation of the activity of retinoid pathways.

Nuclear Retinoid Binding Proteins

The effects of retinoids on gene regulation are mediated by two classes of nuclear receptors, namely the retinoic acid receptors (RARs) and the retinoid X receptors (RXRs). The identification of these proteins as members of the nuclear receptor gene superfamily has led to important insights into the molecular mechanisms of action of retinoids (De Luca, 1991; Glass et al, 1991; Leid et al, 1992a; Lohnes et al, 1992; Mangelsdorf and Evans, 1992; Yu et al, 1992; Parker, 1993; Stunnenberg, 1993). Both classes of receptors are ligand-inducible transcription factors and appear to control RA-dependent gene expression by interacting with cis-acting DNA elements known as RA responsive elements (RAREs) present in the promoter regions of target genes. RARs can be efficiently activated by either RA or 9-cis RA at 5×10^{-8}M concentrations (Allenby et al, 1993). Members of the RXR subfamily respond to much higher concentrations of RA ($> 1 \times 10^{-6}$M), but do not bind this ligand. As their natural ligand has been determined to be 9-cis RA (Heyman et al, 1992; Levin et al, 1992), activation at high levels of RA probably reflects the cellular conversion of RA to 9-cis RA. Both RARs and RXRs consist of three receptor subtypes, α, β and γ. Each RAR subtype can be further divided into several different isoforms generated through alternative splicing mechanisms and dual promoter usage. A number of distinct isoforms of RXRs have recently been shown to exist but their characterisation is as yet incomplete.

Retinoic Acid Receptors

The three human RAR types, α, β and γ, are encoded by separate genes with chromosomal locations 17q21.1, 3p24 and 12q13 respectively (Mattei et al, 1991). The three murine RAR genes are situated on homologous chromosomes and homologues of all three RARs appear to exist across all vertebrate species (Blumberg et al, 1992). Comparison of the amino acid sequences of the three human RARs with the murine RARs shows that the interspecies conservation of a member of the RAR subfamily is much higher than the conservation of all three receptors within a given species, suggesting that each subtype has a specific function (Figure 7.3) (Leid et al, 1992a). The fact that each subtype encodes variable isoforms arising by differential splicing of two primary RNA transcripts adds further complexity to possible regulatory mechanisms. The structures of the isoforms share a common motif: that is, for any RAR type, each isoform diverges only in the 5′ untranslated and A regions. The sequence from the B region to the 3′ untranslated region remains identical (Figure 7.3). A total of seven different isoforms have been isolated and characterised for both the RAR α (α1 to 7) (Giguere et al, 1987; Petkovich et al, 1987; Brand et al, 1988, 1990; Zelent et al, 1989; Leroy et al, 1991a, 1991b) and RARγ genes (γ1 to 7) (Krust et al, 1989; Giguere et al, 1990b; Kastner et al, 1990; Lehmann et al, 1991b). However, there are two predominant isoforms of each subtype, namely α 1 and 2, and γ 1 and 2. Four major isoforms of the RARβ gene (β1 to 4) are known to exist (Benbrook et al, 1988; Brand et al, 1988; De Thé et al, 1989; Zelent et al, 1989, 1991; Nagpal et al, 1992b).

The transcripts of all three RAR genes display specific spatio-temporal patterns of distribution during embryogenesis and in the adult (Dollé et al, 1989, 1990; Ruberte et al, 1991; Smith and Eichele, 1991). The RARα1 expression has been found to be almost ubiquitous, in agreement with the housekeeping type promoter structure of this gene. RARα2 has a more

(A)

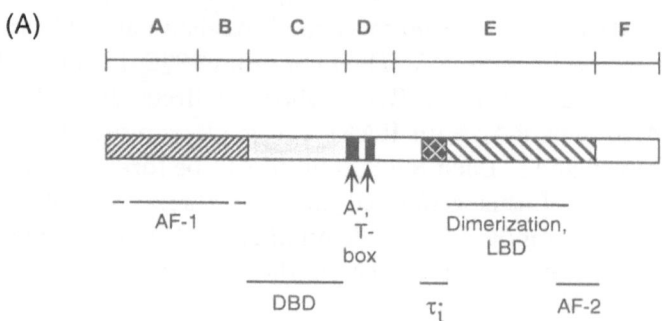

(B)

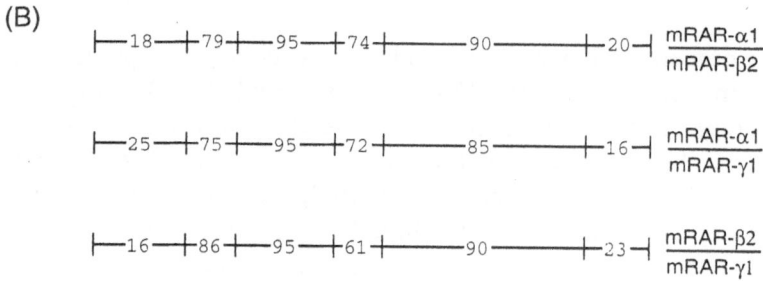

(C)

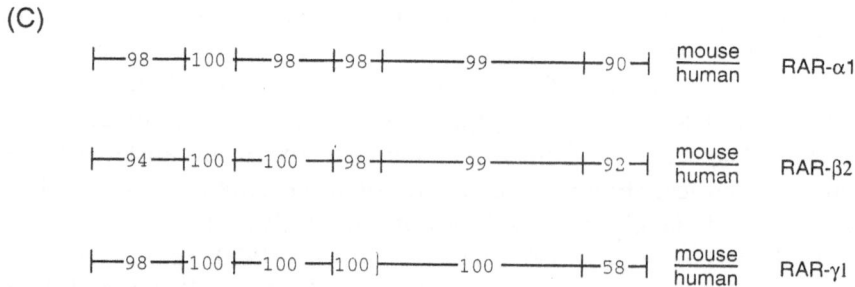

Figure 7.3 (A) Schematic representation of the domain structure (A to F) of the nuclear receptor superfamily. The functional roles of each domain are depicted underneath; AF-1, activation function 1; AF-2, activation function 2; DBD, DNA binding domain; LBD, ligand binding domain. **(B)** Comparison of the intraspecies homology of the murine RAR subtypes. **(C)** Comparison of the amino acid sequence homology between the human and mouse RARα, β and γ subtypes.

restricted pattern of expression, and this transcript is inducible in embryonal carcinoma (EC) cells, consistent with the presence of a RARE in the promoter of this gene (Leroy et al, 1991b; Davis and Lazar, 1993). In contrast to the distribution of RARα, RARβ and RARγ exhibit stage and tissue-specific developmental patterns of expression (Dollé et al, 1989, 1990; Ruberte et al, 1990, 1991). The restricted expression pattern of RARβ transcripts indicates that they may be involved in the ontogenesis of the nervous system, as well as in the differentiation of certain epithelia. They are abundant in embryonic and adult brain, consistent with a requirement for these isoforms in the development of the CNS. They are also highly expressed in the inter-digital mesenchyme during the period of digit separation, and it has been proposed that they may function in regulating programmed cell death (Dollé et al, 1989, 1990; Mendelsohn et al, 1991; Smith and Eichele, 1991). Transcripts for RARβ2 and 4 are readily inducible in EC cells in response to RA, but in contrast to RARα2, RARβ2 transcripts are induced from undetectable to very high expression levels (De Thé et al, 1989, 1990; Sucov et al, 1990). This transcription profile from the downstream RA responsive promoter (P2) of the RARβ gene suggests an involvement in the early stages of development. The expression pattern of RARγ suggests a role in morphogenesis, chondrogenesis and in the differentiation of squamous epithelia. RARγ is found almost exclusively in the skin and is mainly restricted to precartilage and cartilage, and to keratinizing squamous epithelia (Dollé et al, 1989; Ruberte et al, 1990). Once ossification begins, transcripts are no longer detectable, thereby indicating a possible role for RARγ in controlling the synthesis of molecules important in chondrogenesis or alternatively, repressing the expression of factors which inhibit cartilage formation. The RARγ2 is the predominant form in the early embryo whereas RARγ1 predominates in late embryogenesis as well as in newborn and adult skin.

Retinoid-X-Receptors

The classification of the nuclear retinoid receptors into the RAR and RXR subfamilies is mainly based on the differences in primary structure and sensitivity to different retinoid molecules. Three subtypes for RXR exist: α, β and γ, and as is the case with RARs, there is a higher conservation for a given RXR type across species that among all three RXRs within a species (Mangelsdorf et al, 1990, 1992).

Although no apparent insect homologue of the RAR has been identified, the *Drosophila* homologue of the RXR gene is known and maps to the *ultraspiracle (usp)* locus (Oro et al, 1990). The discovery in *Drosophila* of an RXR gene homologue indicates the ancient evolution-ary origin of this gene family and indeed may indicate that RXR preceded RAR as the original retinoid signalling system. This idea is supported by the central unique role played by RXR in interacting with nuclear receptors responsive to a variety of ligands. RXR serves as an auxiliary factor not only for RAR but also for thyroid hormone receptor (T_3R), vitamin D_3 receptor (VD_3R) and peroxisome proliferator activated receptor (PPAR) (Bugge et al, 1992; Kliewer et al, 1992b; Leid et al, 1992b; Zhang et al, 1992). Analogous to RXR, *Drosophila* USP hetero-dimerizes with the ecdysone receptor (EcR) (Yao et al, 1992; Thomas et al, 1993).

In contrast to the RARs, RXRα and β mRNAs show diffuse expression patterns in early embryogenesis, whereas RXRγ expression appears more restricted (Mangelsdorf et al, 1992). However, in late organogenesis and in the adult, RXRα expression becomes more restricted and is mainly found in the skin, liver and digestive tract epithelia. RXRβ remains ubiquitously expressed at low levels even in the adult (Mangelsdorf et al, 1992). The major sites of expression of RXRγ correspond to the myogenic cell lineages and discrete areas of the developing CNS (Dollé et al, 1994).

As RARs and RXRs regulate common target genes, it is interesting to compare the expression patterns of the different family subtypes. Both RARα and RXRβ are expressed almost ubiquitously, whereas RARβ and γ and RXRα and γ show a much more restricted expression pattern. RXRγ expression in developing brain tissue specifically overlaps with that of RARβ, suggesting a role for these receptors in CNS differentiation. However, in the myogenic lineage there is apparently no coexpression of either RARβ or γ with RXRγ (Dollé et al, 1994). RXRα shows abundant expression in liver, kidney, spleen and a variety of visceral tissues. This is in marked distinction to the RAR subtypes that are detected at very low levels at these corresponding sites. Such an expression pattern may indicate a role for RXRα in retinoid and lipid metabolism. This theory is further supported by the finding that a number of genes, which function in the absorption and transport of retinol, the dietary precursor of RA, are RXR responsive (Kliewer et al, 1992b; Mietus et al, 1992). RXRα and RARγ are both expressed abundantly in the skin indicating that these two receptors may be responsible for the dermatological effects of retinoids.

Gene Targeting of Nuclear Retinoid Receptors

The distinct spatio-temporal distribution of the various retinoid binding molecules in the embryo and adult organism, and the correlation of their presence with important biological processes in development, highlights important functional roles. This, taken together with the strict evolutionary conservation, especially in the N-terminal region, is consistent with the idea that each RAR and RXR type (and isoform) may perform unique functions. To establish their role, mutant mice have been created by gene targeting in which RAR genes and/or isoforms have been functionally inactivated. Analyses of the phenotypic changes brought about by these null mutations provides valuable insights in to the processes that are regulated by each of these proteins (Lohnes et al, 1992). To date, mutant mice have been generated in which all isoforms of the RARα or γ have been disrupted. Null mutant mice have also been created in which single predominant isoforms have been targeted, namely RARα1, RARβ2/β4 or RARγ2. Elimination of the RARα and RARγ types results in mutants that exhibit distinct phenotypes including abnormalities similar to those observed in vitamin A deficiency syndrome (Lohnes et al, 1993; Lufkin et al, 1993). RARα gene disruption results in early postnatal lethality and testis degeneration. RARγ mutants exhibit growth deficiency, early lethality and male sterility.

One very important group of genes whose expression pattern is altered by exposure to RA are the homeobox (Hox) containing genes, which function to define the identity of body segments. Mutations in these lead to homeotic alterations in which body parts are transformed to structures normally found in heterologous segments (Tabin, 1991). Exposure of mouse embryos to teratogenic doses of RA has been shown to generate homeotic transformations of vertebrae along the entire body axis accompanied by alterations in Hox gene expression (Kessel and Gruss, 1991). Interestingly, RARγ null mutant mice also have homeotic transformations, mainly anteriorizations, which occur with variable frequency along the anteroposterior axis, demonstrating that its presence and also presumably RA is required for proper specification of some cervical and thoracic vertebrae (Lohnes et al, 1993).

Extending this analysis, it is interesting that targeted disruption of both RARγ alleles in F9 embryonal carcinoma cells has identified several differentiation-specific genes, including two Hox family members, which are regulated either directly or indirectly by RARγ in these cells. RARγ gene disruption specifically results in a drastically reduced induction of Hoxa-1 RNA and in the lack of induction of the laminin B1 and collagen type IV (α1) RNAs. In contrast, Hoxb-1 and CRABP II show normal RA induction in the RARγ minus cell line but are incorrectly down-regulated (Boylan et al, 1993). However, even though these F9-derived cells fail to exhibit a normal differentiation morphology upon treatment with RA, not all differentiation-specific genes are affected.

In contrast to the complete gene disruptions, mice lacking either RARα1 (Li et al, 1993; Lufkin et al, 1993), RARβ2 (Mendelsohn et al, 1994) or RARγ2 (Lohnes et al, 1993) isoforms are apparently normal, exhibiting no detectable malformations. These findings suggest a high degree of functional redundancy, both among the three RAR types and among the isoforms of a particular RAR type. On the one hand, this redundancy of the RAR isoforms is surprising, especially with regard to the high degree of interspecies conservation of the divergent N-terminal sequences. Furthermore, although RARα1 is the most abundant RAR isoform in the majority of tissue types, it shows complete redundancy. A possible explanation is that any specific function encoded by the amino terminus is phenotypically very subtle and in the absence of other genetic lesions is not essential. On the other hand, this extensive functional redundancy is not so astonishing when one considers the vast number of RAR molecules in the cell, which, with the exception of the A/B region, are all very similar in sequence and thus may be capable of performing a number of common regulatory functions, albeit with different efficiencies (Lohnes et al, 1993). The idea of functional redundancy is further supported by preliminary observations made on mutant mice with double gene disruptions: that is, RARγ/RARα1 null animals are more severely affected than RARγ null mutants (Chambon, 1994).

As RXR acts as a coregulator in many diverse pathways, one can envisage that RXR gene disruption would result in a much more severe phenotype. As a first step to address the role of RXR in retinoid-mediated gene regulation, dominant negative RXRβ embryonal carcinoma (EC) P19 cells have been generated which express a RXRβ lacking a DNA binding domain. These receptors are still capable of dimerizing with partner receptors resulting in the formation of nonfunctional dimers. Results show that such mutant RXRs fail to activate target genes in response to RA and affect the growth and differentiation of EC cells (Minucci et al, 1994). Analyses from the in vivo disruption of the RXR gene are currently in progress, and preliminary data verify the fundamental role of this protein, as RXRα gene targeting results in an embryonic lethal phenotype (Chambon, 1994). This is consistent with its unique role as a heterodimeric partner for RAR, T_3R, VD_3R and PPAR thus allowing it to function in various signalling pathways. The results to date from this extensive gene targeting analysis have been very informative if somewhat unexpected, but further light will be shed on this complex system from the generation of mutant mice with multiple gene disruptions. It is highly probable that the compensatory effects observed with single gene disruptions mask the subtle functional role of each individual RAR type.

Evolution of the Nuclear Receptor Family

The nuclear receptor gene family appears to be of ancient evolutionary origin, and this is reflected in its representation in all major species of metazoans, insects, amphibians and higher eucaryotes. Phylogenetic trees derived from this family of genes reveal that three subfamilies of nuclear receptor genes emerged at an early stage, evolving mainly through a simple duplication model. However, in addition to canonical gene duplication of the ancestral nuclear receptor gene, other events such as recombination, translocation or exon shuffling may also have arisen in order to yield the complex gene family known at present. The divergence of the nuclear receptor subfamilies appears to have occurred more than 500 million years ago, preceding the division of the arthropod and vertebrate lineages (Laudet et al, 1992).

Phylogenetic trees were constructed by comparing the most highly conserved domains of the nuclear receptor superfamily: namely, the DNA binding domain and the τi domain which is implicated in transcriptional regulation (Figure 7.3) (Forman and Samuels, 1990b; O'Donnell and Koenig, 1990). Comparisons of the highly conserved DNA binding domain have led to the conclusion that this multigene family can be broadly categorised into three subfamilies: (1) retinoic acid and thyroid hormone receptors and the *erb-a* related subgroup of genes (EAR) (2) the majority of the orphan receptors and (3) steroid hormone receptors (Amero et al, 1992; Laudet et al, 1992). The phylogenetic tree derived from the τi domain shows the same distribution in the three subfamilies but places two groups of receptors in a different position. The exceptions, which apparently cannot be classified in one of the definite subfamilies, are *knirps, knirps-related* and *egon* in *Drosophila*, which encode proteins whose carboxyl terminal domain is unrelated to that of other nuclear receptors (Laudet et al, 1992). These nuclear receptor-like proteins, therefore, appear to be devoid of a dimerization function and possibly a ligand binding domain. The subgroup of genes consisting of VD$_3$R (Baker et al, 1988), EcR (Koelle et al, 1991), the *fushi tarazu* protein (FTZ-F1) (Lavorgna et al, 1991), nerve growth factor I-B (NGFI-B) (Wilson et al, 1992) (also referred to as *nur77* and *N10*) (Hazel et al, 1988; Milbrandt, 1988; Ryseck et al, 1989), *tailless* (Pignoni et al, 1990) and hepatocyte nuclear factor 4 (HNF-4) (Sladek et al, 1990) represent the second receptor branch that appear to contain DNA binding and τi domains that evolved independently. It can be hypothesised that an incomplete gene duplication event or a homologous recombination-like event has given rise to such chimeric receptors (Laudet et al, 1992).

Based on functional analysis, the receptor superfamily can also be divided into three subtypes that largely agree with phylogenetic studies (Stunnenberg, 1993). These are classified as type I which consists mainly of the steroid receptor family members, type II which includes receptors for thyroid hormone, vitamin D_3 and RA, and type III consisting of a number of orphan receptors such as NGFI-B (Wilson et al, 1992), the embryonal long terminal repeat binding protein (ELP) (Ueda et al, 1992) and the steroidogenic factor (SF-1) (Ikeda et al, 1993). Receptors within each of the three delineated subfamilies share many related properties including DNA binding requirements (Figure 7.4). Type I receptors bind DNA as homo-dimeric species whereas type II receptors form heterodimers, with RXR acting as a common binding partner (Figure 7.5). In contrast, most of the characterised type III receptors bind DNA as monomeric species (Figure 7.4). The remainder of the nuclear receptor members can be collectively grouped into a fourth subfamily defined as type IV, which includes the majority of the orphan receptors whose function is unknown at present. Orphan receptors, which are so called because no known ligand has yet been identified, comprise the vast majority of the nuclear receptor multi-gene family. Within the superfamily the monomer binding receptors appear to be the progenitors, and the most highly evolved members are those regulated by steroid and thyroid hormones and vitamins (Amero et al, 1992; Laudet et al, 1992; O'Malley and Conneely, 1992).

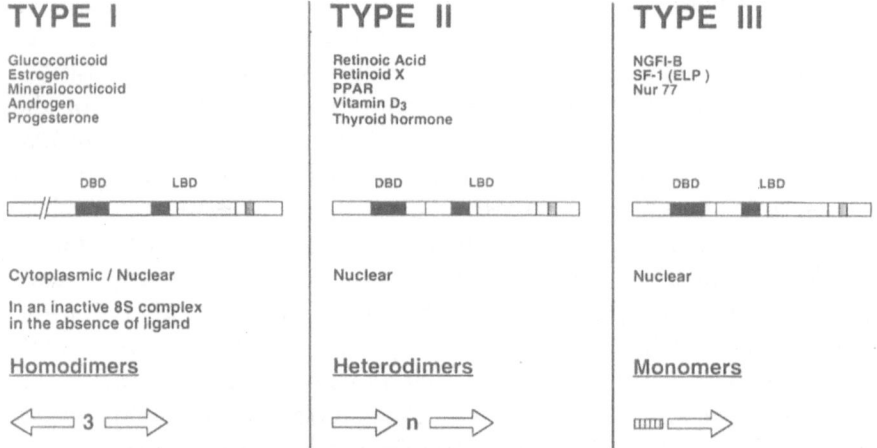

Figure 7.4 Functional classification of the nuclear receptor gene family into three subtypes. Type I consists of the steroid receptor members. Type II includes RAR, RXR, PPAR (peroxisome proliferator activated receptor), VD_3R (vitamin D_3 receptor) and T_3R (thyroid hormone receptor) and type III consists of the monomer binding receptors, NGFI-B (nerve growth factor I-B), ELP (embryonal long terminal repeat binding protein) and SF-1 (steroidogenic factor).

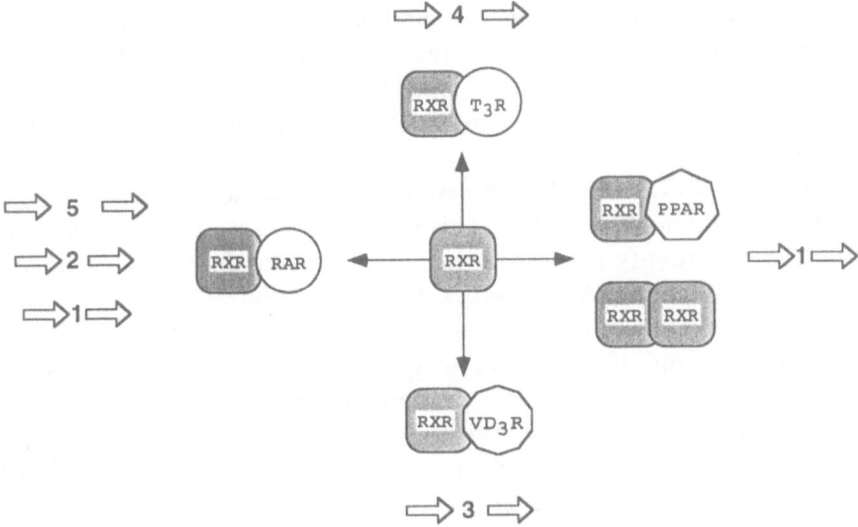

Figure 7.5 RXR plays a central unique role in modulating several ligand-activation pathways by acting as an auxiliary factor for type II nuclear receptors.

The remarkable structural unity among the different members of the nuclear receptor superfamily stands in striking contrast to the diversity of the chemical structures of their ligands. However, although steroids and retinoids have no obvious structural similarity, they are synthesised by remarkably similar pathways. Both are terpenes, derived by the assembly of isoprene units. This similarity, combined with the presence of terpenoid hormones in plant and insect species has led to the hypothesis that such molecules may act as ligands for the orphan receptor family members (Moore, 1990). However, whether these proteins bind ligand or not is an open question. The possibility remains that they may have no specific ligand(s) and may thus act as hormone-independent transcriptional regulators.

Domain Structure of the Nuclear Receptor Family

The primary amino acid sequences of all nuclear receptors can be divided into six regions, labelled A to F, based on homology (Figure 7.3) Glass et al, 1991; Leid et al, 1992a; Gronemeyer, 1993). These regions may harbour one or more functional domains. Domain C, which is the most highly conserved region, corresponds to the DNA binding domain (DBD) (Green and Chambon, 1988). Between the RAR subtypes this region is 93% to 95% conserved. This is a sixty-eight amino acid region which contains two zinc fingers modules (CI and CII), each finger structure containing four cysteine residues forming a tetrahedral coordination complex with a zinc metal ion. The DBD is largely responsible for specific response element recognition (Schwabe et al, 1993; Zechel et al, 1994a; 1994b).

The second most highly conserved region, domain E, is termed the ligand binding domain (LBD). This region, which is approximately two hundred and twenty residues in length, is greater than 85% conserved among the different RAR subtypes, but interestingly, shows only a 27% identity with RXR in this domain. Although this domain is not highly conserved among different receptor types, it contains a short stretch of twenty amino acids termed the τi domain that exhibits 20% to 45% conservation among all receptors (Wang et al, 1989; Forman and Samuels, 1990b; O'Donnell and Koenig, 1990). Furthermore, its structural architecture is similar in that this domain possesses nine heptad repeats of hydrophobic residues which constitute a dimerization motif (Forman and Samuels, 1990b). Region E is functionally complex since in addition to the ligand binding function and a dimerization interface, it also contains a ligand-dependent transcriptional activation function (Figure 7.3). It has been suggested for most of the type II receptors that the two extremities of the E domain bind the ligand whereas the central part of this region is devoted to dimerization. The precise functional role of the conserved τi domain is unclear at present. Studies on the glucocorticoid receptor (GR), however, show that this region appears to contain at least part of the binding site for heat shock protein 90 (hsp 90), the dissociation of which has been implicated in activation at least of the steroid hormone receptor members (Denis et al, 1988; Pratt et al, 1988).

Region D exhibits much less similarity among receptor types, and until recently no definite functional role had been assigned to this domain although many activities have been associated with it. Recent data however shows that a sequence at the amino extremity of the D domain is involved in determining the specificity and polarity of receptor heterodimers on their cognate DNA response elements (Perlmann et al, 1993; Zechel et al, 1994a, 1994b). Due to its location between the DBD and the LBD, the D region is also referred to as the hinge domain. Its location and proposed flexibility allows one to speculate that it may function to accommodate the receptor dimers when bound to DNA and ligand. There is some evidence to suggest that conformational changes which take place upon ligand binding occur mainly within the D and E domains and that these are essential to enable the receptor complex to transactivate through its LBD activation function (Keidel et al, 1994). It is also thought that region D may contain a nuclear localization signal

(NLS) (Ylikomi et al, 1992). This is based on the fact that this region contains a high number of basic residues, and similar sequences in other members of the nuclear receptor superfamily have been shown to function as nuclear localization signals (Picard and Yamamoto, 1987; Guiochon-Mantel et al, 1989; Dingwell and Laskey, 1991).

The B domain shows a high degree of homology among the three RARs (75% to 86%). In contrast, the two terminal regions A and F show the greatest diversity among the α, β and γ subtypes; however, within a given subtype there is a high degree of conservation across species boundaries (Krust et al, 1989). For example, the A domains of RAR α and β share approximately 15% identity, whereas this region of human RARα exhibits 90% identity with its murine counterpart (Glass et al, 1991). Interestingly, RAR is the only member of the type II receptor family that contains an extra C-terminal F domain, and although no specific function(s) have yet been assigned to this region, its conservation points to a functional role (Figure 7.3). The N-terminal domain has been the focus of intense studies as investigations on the steroid members of the nuclear receptor hormone family have revealed two transactivation functions: a hormone-dependent transactivation domain as previously described in the E region, and a ligand-independent function located in the A/B domain. The latter constitutive transactivation function operates in both a cell-type and a promoter specific manner, and a recent study has confirmed that the N-terminal domains of RARs and RXRs also contain an autonomous, ligand-independent activation function (Folkers et al, 1993; Nagpal et al, 1993). In view of this, it has been speculated that the many isoforms of each RAR subtype, which diverge in the A domain, could, therefore, each elicit a distinct transactivation function depending on promoter and cell type via their distinct A region sequence.

Homodimers versus Heterodimers versus Monomers

The extensive analyses of the steroid receptor subfamily provided a basis for analysis of type II receptors, although, many of the observations made on the type I receptors could not simply be extended to the type II subfamily. This was initially demonstrated with regard to the location and association status of receptors prior to ligand binding. Type I receptors are located either in the cytoplasm or nucleus in a complex with hsp 90 and other factors. Ligand binding causes an allosteric change resulting in the release of the receptor from the protein complex, which in turn, allows nuclear translocation and homodimerization to take place, facilitating receptor binding to its hormone response element (HRE). In contrast, type II receptors are nuclear-localized in an apparently uncomplexed form, and can bind to their cognate response elements in the absence of ligand.

Previous studies on the steroid hormone members of the nuclear receptor family also demonstrated that they bind as homodimer species to their cognate response elements. These HREs are arranged as palindromes with an invariable spacing of three nucleotides (Figure 7.4). In contrast, type II receptors all share the same consensus binding motif and specificity is largely determined by the half-site spacings (Parker, 1993; Stunnenberg, 1993). Furthermore, members of this subfamily bind predominantly to response elements which are arranged as direct repeats (Figure 7.4). However, perhaps the most distinguishing difference between these two receptor types is their dimer requirements. Steroid hormones all bind their cognate HRE as homodimer complexes. However, type II receptors bind as heterodimers with RXR acting as a common binding partner for RAR, T_3R, VD_3R and PPAR (Figure 7.5) (Yu et al, 1991; Bugge et al, 1992; Kliewer et al, 1992b; Leid et al, 1992b; Marks et al, 1992; Zhang et al, 1992). RXR may also form a weak heterodimeric interaction with the orphan receptor, chicken ovalbumin upstream promoter transcription factor (COUP-TF) (Kliewer et al, 1992b). Additionally, RXRs are functionally unique in that they can form homodimeric as well as heterodimeric complexes. The ability of RXR to interact with receptors responsive to a variety of ligands establishes a central role for this protein in modulating multiple response pathways.

The ability of RXR to form heterodimers with individual receptors clearly offers the potential to generate many different combinations of receptor dimers, especially given that RAR, RXR and PPAR are each encoded by three genes and T_3R and COUP-TF by two genes. The complexity of the retinoid receptors is further increased because each gene gives rise to several isoforms. Furthermore, as the expression of different isoforms depends on cell type and varies during growth and development, it follows that heterodimerization results in the formation of an extremely diverse group of receptor dimers with an enormous number of potential activation regions (Parker, 1993). Therefore, a number of mechanisms have evolved to facilitate both cell-type and developmental-stage-specific regulation of hormone responses.

Response Elements

HREs in target genes have evolved to impart high specificity of hormonal response as well as cross-talk between hormonal pathways. There are a number of guidelines for specific recognition in response elements: namely, the sequence and directional arrangement of the half-sites (palindrome versus inverted palindrome or direct repeat (DR)) and also the spacing between these motifs all help to establish the binding specificity.

Type I receptors all contain a related binding site consisting of a palindromic repeat with an invariant three nucleotide spacing. For the estrogen receptor (ER) the estrogen response element (ERE) half-site motif consists of the consensus sequence AGGTCA, whereas the other steroid receptors, that is, glucocorticoid receptor (GR), progesterone receptor (PR), androgen receptor (AR) or mineralocorticoid receptor (McR), all bind to a common HRE with a consensus motif of AGAACA (Yu et al, 1992; Gronemeyer, 1993). Target specificity is determined by the amino acid sequence in the proximal box (P box) of the first zinc finger module, **EGCKA** for the ER and **GSCKV** for the other steroid members, whereas the five amino acids at the base of the second zinc finger module, the so-called distal box (D box) confer the specificity for half-site spacing and function as a DNA dependent dimerization domain in type I receptors.

Type II response elements are direct repeats composed of two core motifs identical to the estrogen response element (ERE) or closely related degenerate motifs of the form $PuG^G/_TTCA\,(x)_n\,PuG^G/_TTCA$ where the spacing n is variable. Spacers of 3, 4 and 5 nucleotides initially were proposed to serve as the optimal response elements for the VD_3R/RXR, T_3R/RXR and RAR/RXR respectively, a pattern denoted as the 3 - 4 - 5 rule (Näär et al, 1991; Umesono et al, 1991). Subsequently, a significant amount of degeneracy of the above rule has been observed especially with regard to RAR/RXR binding. RAR/RXR heterodimers bind to a variety of RAREs (DR + 1 to DR + 5) in which the repeated motifs vary in sequence and orientation; however, DR + 5 and DR + 2 are the most efficient RAREs (Figure 7.5). Examples of natural response elements include those in CRABP I which contains two RAREs with internal half-site spacings of one and two nucleotides respectively (DR + 1 and DR + 2), and which have been shown to mediate RAR/RXR transactivation (Durand et al, 1992). However, DR + 1 appears to be a promiscuous binding site, since it is also recognized by other members of the nuclear receptor family, for example RXR homodimers, PPAR (Tugwood et al, 1992), COUP-TF (Kliewer et al, 1992a) and the related apolipoprotein A1 regulatory protein 1 (Arp-1) (Widom et al, 1992) and EAR 2 (Kliewer et al, 1992a). Consistent with the similarity between the ERE and type II response element sequences, type II receptors also contain a related P box composed of the residues **EGCKG**.

In addition to the inverted and direct repeats, there is a third type of nuclear receptor binding site, which does not contain a repeat element, but rather consists of an extended half-site as exemplified by the binding of rat NGFI-B (Wilson et al, 1992), the *Drosophila* protein FTZ-F1 (Lavorgna et al, 1991) and its murine developmentally regulated homologues, ELP (Ueda et al, 1992) and SF-1 (Ikeda et al, 1993). These orphan receptors seem to function by binding to a single AGGTCA motif preceded by three adenine nucleotides in the case of NGFI-B, or PyCA for FTZ-F1, ELP and SF-1, and as such, represent a third subclass of the nuclear receptor multigene family (type III) in that they bind DNA as monomers (Figure 7.4). Non-zinc finger residues in the DBD participate in binding to the sequence upstream of the conserved half-site (Wilson et al, 1992). The majority of the orphan receptors so far described have not been classified among the types I, II or III subfamilies and although their functional role is unknown at present, it is interesting to note that all orphan receptors exhibit closely related P boxes (**EGCKS** or **EGCKG**) with the exception of HNF4/*tailless* and FTZ-F1 (Laudet et al, 1992).

Discrimination of Target Response Elements

The subclass of nonsteroid receptors, including RAR, RXR T_3R, VD_3R and certain orphan receptors, which recognise the core half-site (AGGTCA) may distinguish targets by recognising the relative orientation and spacing of two such sites (Stunnenberg, 1993). However, unlike type I receptors, members of this subclass do not form symmetrical dimers on the DNA target and are instead arranged in a head-to-tail orientation. From X-ray crystallography of type I DBDs, it is known that the subunits of a type I receptor dimer lie on the same surface of the DNA making extensive protein-protein contacts (Schwabe et al, 1990, 1993; Luisi et al, 1991). The individual amino-terminal zinc finger module regions expose recognition α-helices into adjacent major grooves and direct contacts with the bases of the target site. The carboxy-terminal finger module forms a dimerization surface (D box). Cooperativity results from favourable protein-protein contacts made through an interface which is aligned by DNA binding. If the spacing between the half-sites is increased or decreased by a single nucleotide, the contacts are disrupted, resulting in a loss of cooperativity and abolition of transactivation. In addition to the weak dimerization function present in the DBD of type I receptors, studies on the ER and GR show that while dimerization occurs through the D box, the main dimerization interface lies in the LBD (Kumar and Chambon, 1988; Tsai et al, 1988; Fawell et al,

1990). This region, when aligned with other nuclear receptors, is postulated to contain a conserved heptad repeat of hydrophobic residues (Forman and Samuels, 1990a) that can form a dimerization interface analogous to the coiled-coil interface of leucine zippers. This region has been proposed to direct both homo- and heterodimerization.

However, the models explaining how steroid receptors bind to DNA cannot simply be extended to the type II or the less well defined type III receptors. This was demonstrated initially by showing that the dimerization interface in the RXR receptor differs from that of the steroid family members (Wilson et al, 1992). A region encompassing twelve amino acids, termed the T box, which is located carboxy-terminal to the second zinc finger (CII) of RXR, is required for DNA binding (Figure 7.3). Moreover, the precise binding mechanism of type III receptors was presupposed to be distinct, since they bind as monomers to a response element consisting of a half-site motif preceded by three nonrandom nucleotides. The NFGI-B domain critical for nerve growth factor I-B response element (NBRE) recognition, the A box, a seven residue region, is also located carboxy-terminal to CII but does not co-localize with the RXR T box (Figure 7.3) (Wilson et al, 1992). Interestingly, the critical downstream domains of both RXR and NGFI-B show evolutionary conservation, highlighting their biological significance. These residues downstream of the second zinc finger structural domain, which are not necessary for DNA recognition by many nuclear receptors, are not only critical for sequence recognition by RXR and NGFI-B, but also appear to function by different mechanisms. Whereas the residues of the A box of NGFI-B are necessary for 5' adjacent sequence recognition, the T box of RXR is postulated to form part of a dimerization interface between protein monomers. Subsequently, the resolution of the 3-dimensional structure of the RXR DBD has shown that in contrast to the DBDs of GR (Luisi et al, 1991) and ER (Schwabe et al, 1990, 1993), this region of RXR contains an additional helix immediately after the second zinc finger (Lee et al, 1993). This third helix mediates both protein-protein and protein-DNA interactions required for cooperative dimeric binding of the RXR DBD to DNA. Consistent with the data of Wilson et al, the third helical domain includes the region encompassing the T box of RXR or the A box of NGFI-B (Wilson et al, 1992).

The identification of an extra helix thus defined a potential structural feature required for selective dimerization of the RXR on response elements composed of half-sites arranged as direct repeats. Studies using chimeric DBDs of GR and RXR on composite response elements showed that the location of each partner on natural direct repeat elements is fixed (Mader et al, 1993; Perlmann et al, 1993), that is, RXR is always positioned on the 5′ half-site. However, it is only recently that the dimerization interfaces utilized in each individual type of heterodimer or homodimer complex on its natural response element has been elucidated (Zechel et al, 1994a, 1994b). Two types of dimerization interfaces have been described, which always involve the C-terminal finger region (or parts of it) of a RXR monomer as one of the two dimerization surfaces. In one type of dimerization interface, the second surface is provided by the N-terminal CI fingertip region of RAR (RXR/RAR heterodimer on a DR + 5 element), or T_3R CI finger region (RXR/T_3R heterodimer on a DR + 4 element), whereas in class II, the second surface encompasses the T box region of either a second RXR (RXR/RXR homodimer on a DR + 1 element), or RAR (RXR/RAR on a DR + 2 element). Furthermore, the authors have elaborately distinguished two discrete portions of the RXR C-terminal finger region that specifically contribute to the dimerization interfaces which dictate cooperative binding depending on whether the response element consists of widely spaced (DR + 4 and + 5) or closely spaced (DR + 1 and + 2) half-sites. For DR + 4 and + 5 elements, the D box of RXR is specifically required for formation of the interface, whereas for DR + 1 and + 2 elements the region of RXR required is also the CII finger region but excluding the D box (Figure 7.6) (Zechel et al, 1994a, 1994b). Analogous to type I receptors, it can be concluded that although the major dimerization function present in the LBD has little effect on the selectivity of DR recognition, the free energy released as a consequence of region E dimerization may influence the binding stability of monomeric partners (Mader et al, 1993).

Thus, the model presents a means of discriminating the binding specificities of the multiple heterodimers that may be preformed in the nucleus. The two types of asymmetric interfaces existing in the DBD result in a binding polarity of the asymmetric dimers to specific DR elements, providing a mechanism whereby target response elements are specifically recognised by their particular hetero- or homodimer leading to the regulation of distinct sets of genes. It should be noted that both dimerization interfaces are possible only when the RXR partner is positioned 5′ to the second monomer (Figure 7.6) (Zechel et al, 1994a; 1994b). In addition, although not investigated in detail, this model predicts that a heterodimer consisting of RAR/RXR on the low affinity DR + 1 element necessitates that the RAR partner occupy the 5′ position. It therefore appears that both the specificity of the DR response element repertoire and the polarity of the receptors on DR elements are determined by the asymmetry of their DBD dimerization interfaces

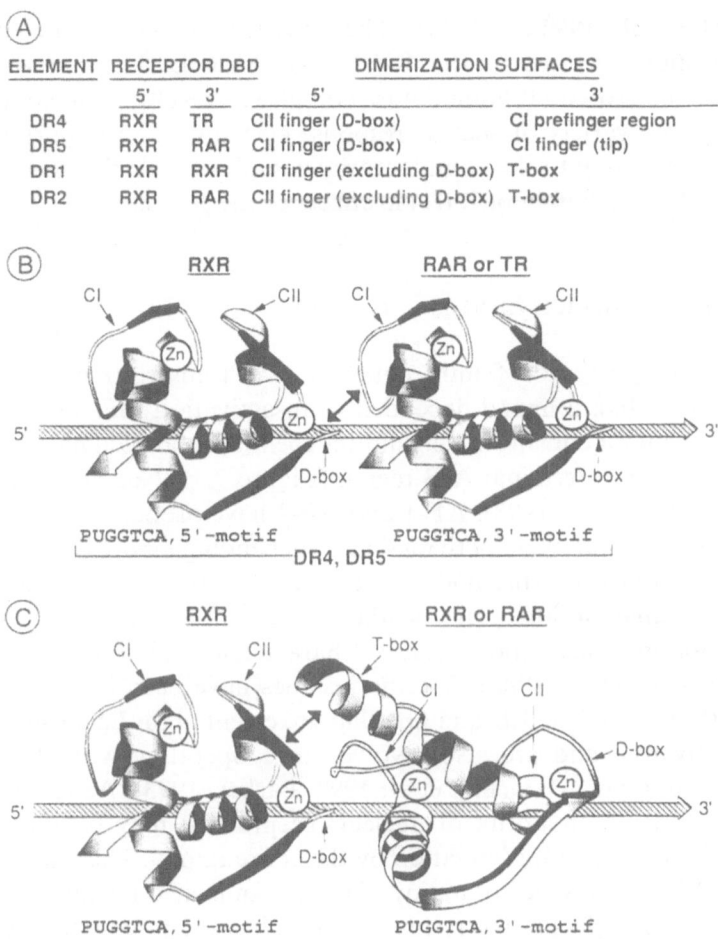

Figure 7.6 Modelling of the two types of dimerization interfaces which result in homo- and heterocooperative interactions among the DBDs of RXR, RAR and T$_3$R. (A) Summary of the protein-protein interactions that are responsible for cooperative DNA binding to DR + 1 (RXR), DR + 2 and DR + 5 (RXR and RAR), and DR + 4 elements (RXR and T$_3$R) and of the orientations of the monomers on the DNA. (B) A ribbon model of two DBD monomers is shown with the monomers facing in the same orientation to reflect the binding of RAR, RXR and T$_3$R DBDs to DR + 5 and DR + 4. The DNA representation has been omitted for the sake of clarity, and only its 5' to 3' polarity is indicated by a hatched arrow. The monomers are represented facing the same side of the DNA helix, assuming identical protein-DNA interactions for each monomer and a centre to centre distance of the PuGGTCA motifs of roughly one helical turn, as would be expected for the binding of DR + 4 (centre to centre distance of 10 bp) and DR + 5 (centre to centre distance of 11 bp) elements. (C) Similar modelling to that in (B) is shown, except that the 3' located monomer was rotated by ~90° against the 5' positioned monomer to reflect their relative orientation on DR + 1 and DR + 2 elements. In addition, the 3' located monomer is schematically represented with a helical T-box region according to a three-dimensional NMR analysis of the structure of the RXR DBD, even though the structure of the corresponding sequence in RAR is unknown. (Reprinted with permission of Oxford University Press from Zechel et al, 1994a, 1994b.)

(Zechel et al, 1994a; 1994b). The presence of several independent dimerization surfaces in the RXR and RAR DBDs that can form multiple hetero- and homodimer complexes resulting in cooperative binding to a variety of specific response elements is an important extra parameter adding to the combinatorial complexity which is necessary to account for the highly pleiotropic effects of the RA signal.

Multiple Transactivations Functions

The nuclear receptor family contains two transcriptional activation functions (AFs): a ligand-dependent transactivation function (AF-2) in the LBD, and a constitutive ligand-independent transactivation domain (AF-1) in the N-terminal A/B region (Figure 7.3) (Nagpal et al, 1992a, 1993; Folkers et al, 1993). AF-1 and AF-2 have properties distinct from one another, and their activities vary depending upon the responsive promoter and cell type, and in some cases both are required for full transcriptional stimulation. Neither of the AFs appear to contain transcriptional activation motifs characteristic of those previously identified (Ptashne, 1988). Specific residues have been identified within the LBD AF-2 of the ER and GR that are essential for ligand-dependent transactivation and are also involved in cooperating with the amino-terminal domain (Danielian et al, 1992; Parker, 1993). As this region is conserved in the majority of nuclear receptors, it may be essential for ligand-dependent transactivation by the entire family. The major feature of AF-2 is the presence of an invariant glutamic acid residue flanked by two pairs of hydrophobic residues, which has the potential to form an amphipathic α-helical structure. Mutations within this element that abolish transcriptional activation do not significantly affect other receptor functions such as ligand or DNA binding (Danielian et al, 1992).

The consensus AF-2 motif is modified or absent in a minority of family members; for example, COUP-TF (Cooney et al, 1992; Tran et al, 1992) and its near relative, Arp-A1 (Widom et al, 1992), contain a conservative substitution of the central glutamic acid residue for aspartic acid. The observation that both these proteins function primarily as transcriptional repressors (Berrodin et al, 1992; Cooney et al, 1992; Kliewer et al, 1992a; Tran et al, 1992; Widom et al, 1992) is interesting in this regard. Other family members including NGFI-B and its homologues *Nur77* and *Nak1* (Hazel et al, 1988; Milbrandt, 1988; Ryseck et al, 1989; Nakai et al, 1990) contain a degenerate version of the AF-2 motif. Its complete absence has been noted in several orphan receptor proteins, including ELP (Tsukiyama et al, 1992), EAR1/*rev-erbA* (Lazar et al, 1989; Miyajima et al, 1989) and a number of *Drosophila* receptors (Barettino et al, 1994).

The AF-2 motif, although it has a net negative charge, does not appear to be related to the VP16 acidic activation domain and may thus represent a novel activation surface (Barettino et al, 1994). The AF-2 domain, which can function autonomously as documented by Gal-fusion experiments, is, however, controlled by ligand in its natural context, which may reflect a ligand induced conformational change necessary for activation to occur. Studies on the chicken $T_3R\alpha$ receptor show that while AF-2 mediates transactivation, it is also involved in transcriptional interference (squelching), not only between T_3R and other type II family members, but also between type II and type I receptors. Furthermore, mutations that interfere with one effect also have an equivalent influence on the other. Therefore, it is likely that AF-2 provides a surface for interaction either with a general transcription factor or a putative coactivator protein utilized by both type I and II receptors (Barettino et al, 1994).

The various RARs and RXRs exhibit differential transcriptional activation activity on several naturally occurring and synthetic RA-response promoters. Interestingly, the synergistic action of AF-1 and AF-2 of a given receptor appears to be dependent on both the promoter context and the nature of the RARE. Thus the same promoter or identical RAREs placed in different promoter contexts may be differently activated by different receptors (Nagpal et al, 1992a). The different functional properties probably reflect the different distribution of specific RAR and RXR isoforms. In addition, AF-1 and AF-2 of a given receptor appear to be significantly different from those of the other receptors since AF-1 of one RAR type may synergise with AF-2 of the same receptor, but not with AF-2 of another RAR type. Since AF-1 is partially contained in the RAR isoform-specific amino-terminal region, the global activation potential of a given receptor type may ultimately be determined by a common ligand-dependent AF-2 and an isoform-specific, cell-type and developmental stage dependent AF-1 (Nagpal et al, 1992a, 1993). The promoter context dependence is also likely to reflect a specificity in the cooperation between the AFs of receptors and other transactivators specificially bound to a given promoter type (Nagpal et al, 1992a). Furthermore, although all three RAR types are activated by RA and 9-cis RA they are differently bound and activated by some synthetic retinoids (Åstrom et al, 1990; Graupner et al, 1991; Lehmann et al, 1991a) which suggests that their LBDs and/or AF-2 may be distinct (Leid et al, 1992a) and supports the idea that the α, β and γ subtypes of RAR and RXR may preferentially control the transcription of different subsets of RA-responsive genes. It is interesting in this respect to note that targeted disruption of the RARγ gene in F9 EC cells results in an altered expression pattern of several differentiation-specific genes (Boylan et al, 1993).

Agonist/Antagonist Action

A vast array of synthetic ligand derivatives have been developed, and their significance with respect to binding and agonist/antagonist properties is the focus of much attention. Agonists can be defined as natural or synthetic ligands that compete with endogenous hormone for receptor binding and upon binding, exert equivalent effects. Their major advantage is that they can be subtype-specific and thus have restricted biological activity, thereby reducing undesired effects. Antagonists or antihormones, on the other hand, also compete with natural ligand for receptor binding, but upon binding are functionally opposing. They can be divided into two categories: those that inactivate receptor function completely, designated pure or type II antagonists, and those that have partial agonistic activity and are thereby referred to as type I antagonists (Gronemeyer et al, 1992). Antagonists may function at three levels. First, they may inhibit or slow the dissociation of receptor from its cytoplasmic complex, a level of action presumably only relevant for the steroid members of the nuclear receptor family. Second, they may prevent receptor binding to response elements, which is the mechanism by which type II antagonists function, or, finally, they may allow DNA binding but interfere with the processes by which the receptor complex activates transcription, namely type I antagonists. The partial agonistic activity seen with the type I antagonists may be due to the activation function (AF-1) present in the N-terminal region of the receptor (Gronemeyer et al, 1992). Therefore, the variable degree of activity observed with such compounds may be explained by differences in the activity of AF-1 in different species and tissues, as well as by the nature of the response element. The complete quenching of the activation function (AF-2) in the LBD appears to result from an altered structural configuration of receptor on binding the antihormone, which completely masks this region (Allan et al, 1992a, 1992b).

The profound effects of retinoids on cell differentiation and proliferation has brought them under intense investigation due to the potential for clinical application as therapeutic agents in a number of areas. This is of considerable interest not only for pharmaceutical drug design, but also as a tool for our understanding of ligand-receptor interaction. The identification of selective retinoids that activate preferentially or exclusively one receptor type, or antagonists that specifically interfere with one receptor is of primary importance in dissecting the complexity of the retinoid signalling system. The synthetic retinoids, Ro41-5253, Ro46-5471 and Ro46-8515 are selective RARα antagonists and inhibit transactivation of this receptor by RA and other retinoids (Apfel et al, 1992; Keidel et al, 1994; Valcárcel et al, 1994). Ro41-5253 has been shown to inhibit RA-induced differentiation of the promyelocytic cell line, HL-60 and to revert retinoid-induced inhibition of mouse B-cell proliferation (Apfel et al, 1992). Analysis of the proteolytic fragmentation pattern of antagonist- and agonist-bound RARα has revealed that, irrespective of the protease used, an altered structural conformation is observed on antagonist binding. The proposed agonist- and antagonist-induced conformational changes are reversible in a concentration dependent manner. More detailed investigations have shown that the major conformational changes, occurring mainly in regions D and E, play an important role both in the activation of RARα by agonists and in its neutralization by antagonists. Agonist-induced conformational changes appear to implicate structural alterations in region D and a more compact folding of the LBD, thus allowing AF-2 to function. Conversely, the structural changes induced by antagonists do not appear to involve region D. Furthermore, it appears that this antagonist-induced conformational change may be a common mechanism of hormone inaction within the complete nuclear receptor family (Figure 7.7) (Allan et al, 1992a; Keidel et al, 1994).

As the three RAR subtypes exhibit different ligand activation specificities, it is possible to design receptor subtype selective retinoids. In this respect it is interesting to note that the LBDs of RARβ and γ are more closely related to each other than to RARα. Therefore, it is not surprising that a number of synthetic retinoids tested show more closely related activation profiles for RARβ and γ than for α. Although there is no evidence to prove this, it is tempting to speculate that the ubiquitous expression of RARα and more restricted distribution of RARβ and γ may reflect availability and tissue specificity of RARβ/γ-specific retinoids (Lehmann et al, 1991a). A further study has identified a series of retinoids (SR11217 and SR11237) that apparently activate RXR homodimers but do not affect RAR/RXR heterodimers, thereby suggesting that both pathways can be independently activated (Lehmann et al, 1992).

Another factor controlling the pleiotropic effects of retinoids may be the existence of multiple ligand activation pathways for the retinoid receptors. At least four endogenous retinoids are known that regulate gene expression: RA, 9-cis RA, ddRA and 4-oxo-RA (Figure 7.1). The role of ddRA, however, is not very well understood. It is biologically active and is found in the chick limb bud (Thaller and Eichele, 1990) and

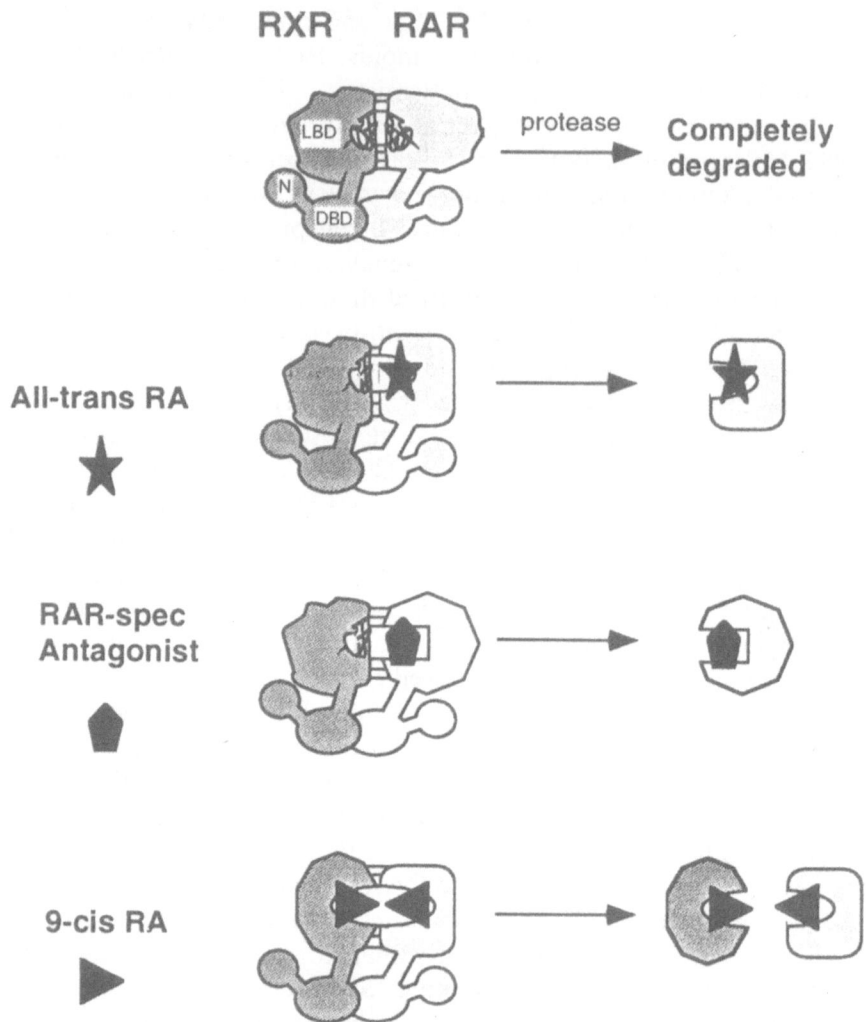

Figure 7.7 Schematic representation of the conformational changes induced in an RXR/RAR heterodimer upon ligand binding. The ligand binding properties of the heterodimer are shown with regard to all-trans RA binding, RAR-specific antagonist binding and 9-cis RA binding. The subsequent proteolytic fragmentation pattern upon protease digestion is depicted on the right of the diagram.

some mammalian tissues. Whether it has a unique receptor type or whether its effects are mediated by interactions with one of the RARs or RXRs, is still not known (Allenby et al, 1993). 4-oxo-RA is a highly active metabolite which can modulate positional specification in the early embryo, having a strong effect on anteroposterior axis specification. It is known to bind at least one RAR type with high affinity, namely RARβ. However, in contrast to its high activity during anteroposterior axis formation, this retinoid is less active than RA in a number of growth and differentiation assays (Pijnappel et al, 1993), supporting the idea that specific retinoid ligands regulate different physiological processes in vivo. It should be noted that 9-cis RA through its ability to bind RXR can activate several distinct pathways. One can therefore envisage that there exists in the cell a mechanism(s) to fine tune the action of retinoids, and in particular, with 9-cis RA, a means to differentially regulate which pathway is activated. The synthesis and natural distribution of retinoid metabolites in different tissues at distinct developmental stages has not yet been fully elucidated, rendering it impossible to predict in which specific areas RARs and RXRs can be activated. Such discoveries will play a pivotal role in understanding the retinoid-dependent developmental processes.

Retinoid Receptor Interaction with the Basal Transcription Machinery

The mechanism by which the DNA bound receptor complex transmits its transactivating signal to the basal transcription machinery is largely unknown. The two general transcription factors predominantly implicated as targets for transactivation are TFIID and TFIIB. TFIID is a protein complex consisting of the TATA-binding protein (TBP) and a distinct set of TBP associated factors (TAFs). Upon binding to the TATA box, TFIID recruits the other general transcription factors resulting in the ordered formation of the basal transcription apparatus. While recombinant TBP can mediate basal level transcription in vitro, the TFIID complex is required for activated transcription, indicating that TFIID provides a target for transcriptional regulation exerted by upstream factors (Hernandez, 1993).

RARβ2 is one of a number of known autoregulated retinoid binding proteins (De Thé et al, 1990; Sucov et al, 1990). It has been most intensively studied as a number of features make this an ideal system for analysing the mechanism of nuclear hormone receptor transcriptional activation. RA-dependent induction of the RARβ2 promoter is mediated by an RARE (DR + 5) located immediately upstream of the TATA element (De Thé et al, 1990; Sucov et al, 1990; Vivanco Ruiz et al, 1991). In contrast to the location of most RAREs, the close apposition between the RARE and the TATA box in the RARβ2 promoter is exceptional (Figure 7.8). As previously mentioned, RA induces the differentiation of EC cells such as F9 or P19, and an early event in this process is the expression of the RARβ2 gene. Indeed,

SUPERACTIVATION PATHWAY

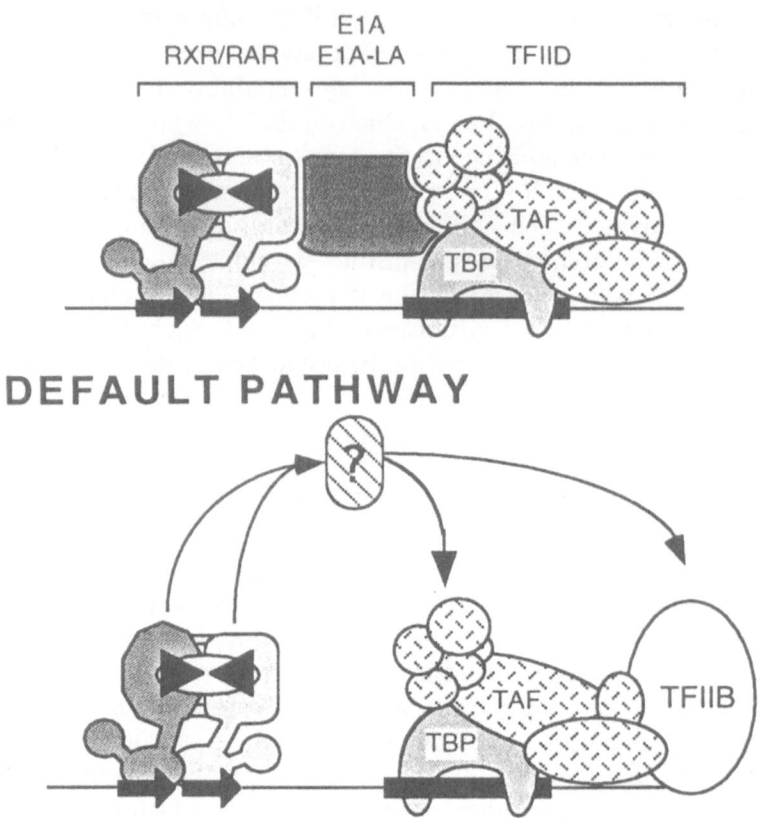

DEFAULT PATHWAY

Figure 7.8 Schematic representation of the two distinct pathways of activation of the RARβ2 promoter. The superactivation pathway is utilised in EC cells in which ligand-activated RXR/ RAR signals via the cell-specific E1A-LA to the basal transcription factor, TFIID. This pathway can be mimicked in non-EC cells by the cotransfection of the adenovirus E1A. TFIID consists of the TATA-binding protein (TBP) and TBP associated factors (TAFs). In differentiated cells RARβ2 is activated via a default pathway. Transactivation from RXR/RAR may be mediated either directly to a component of the basal transcription machinery or may require a bridging factor other than E1A-LA.

RARβ2 appears to be the predominant RAR isoform that is induced in EC cells in response to RA (Zelent et al, 1991).

RA signalling from the RAR/RXR complex on the RARβ2 promoter can occur via two distinct pathways, namely the default pathway used in differentiated cells and the superactivation pathway first observed in EC cells. RA signalling occurs in EC cells via a cell specific E1A-like activity (E1A-LA) giving rise to superactivation of the RARβ2 promoter, and this effect can also be mediated in differentiated cells via the adenoviral E1A (13S) protein (Berkenstam et al, 1992; Keaveney et al, 1993). The expression of an E1A-LA appears to be down-regulated during differentiation of EC cells (La Thangue, 1987), concomitant with a marked decrease in the inducibility of the RARβ2 promoter. Therefore, in this system E1A or E1A-LA act as bridging factors between the activation complex and the basal transcription apparatus (Figure 7.8). In vitro studies have revealed that TBP directly binds E1A via its basic domain (Horikoshi et al, 1991; Lee et al, 1991). Further genetic evidence has shown that a distinct sequence in the core domain of TBP which lies in close proximity to the basic region is essential to provide a surface for E1A-LA interaction (Keaveney et al, 1993). Whether the coactivators E1A-LA or E1A directly contact the RAR/RXR complex in this system remains to be established.

The second general transcription factor known to mediate nuclear hormone transactivation signals is TFIIB. Several members of the receptor superfamily are reported to contact TFIIB directly, including ER, PR, T$_3$Rβ and COUP-TF (Ing et al, 1992; Baniahmad et al, 1993; Fondell et al, 1993). These interactions are thought to facilitate the process of transcriptional activation of responsive genes. Although significant advances have been made, questions concerning how interaction of the receptor complex with the basal transcription machinery promotes or inhibits gene transcription and the exact role(s) of putative coactivators in this process remain unsolved.

Nuclear Receptor Cross-talk

In addition to their role as transcriptional activators, many nuclear receptors have been shown to repress transcription. A number of alternative mechanisms for transcriptional repression have been proposed including competition for common DNA binding sites, formation of inactive complexes, competition within a cell for the common auxiliary protein RXR and binding to negative response elements.

A feature referred to as cross-talk exists between a number of receptor types, which has been most extensively studied in receptors that bind to the DR+1 element. A number of receptors can bind to the DR+1 element including Arp-1 (Ladias and Karathanasis, 1991), HNF-4 (Sladek et al, 1990) COUP-TF, EAR 2 and RXR. All of these are expressed in the liver and small intestine to varying degrees. A number of hepatic or intestinal specific genes have been identified which have DR+1 elements, and, therefore, their expression is likely controlled by the relative abundance and interplay of these receptors (Kliewer et al, 1992b). At least two members of the apolipoprotein (apo) gene family, namely apo A1 and apo CIII, are examples of such genes. Both are lipid binding proteins involved in the transport of cholesterol and other lipids in the plasma. The expression of these genes depends on the intracellular balance of positive and negative regulatory factors. Apo A1 is transcriptionally repressed by Arp-1, and in the absence of RA, Arp-1 alone or Arp-1 and RXRα together dramatically repress expression from the apo A1 promoter. However, in the presence of RA, repression by Arp-1 and RXRα, but not Arp-1 alone, is almost completely alleviated. RXRα alone has little effect on levels of expression regardless of the presence or absence of RA. Therefore, the magnitude of response appears to be regulated by the intracellular ratio of Arp-1 to RXRα (Widom et al, 1992). Three members of the nuclear receptor family bind to the apo CIII promoter region, namely HNF-4, Arp-1 and COUP-TF, and while the former activates transcription, the latter two repress transcription of this gene (Mietus et al, 1992).

The intracellular ratio of positively and negatively acting nuclear receptors may thus represent a general mechanism for switching gene transcription between alternative regulatory pathways. Another example, CRBP II, has a complex promoter structure; however, the major RA response element binds a number of receptor types. CRBP II gene stimulation, mediated by the nuclear receptors HNF-4, RAR and RXR, is repressed by Arp-1 and COUP-TF. This suggests that in intestinal cells CRBP II expression may not be controlled by RA, but rather may depend primarily on the intracellular levels of HNF-4 and Arp-1 acting as positive and negative regulatory factors respectively (Nakshatri and Chambon, 1994). Consequently, these data suggest that transactivation by RXR homodimers can occur only in cells in which RXR levels are much higher than those of competitor receptor binding proteins, or on response elements with higher affinity for RXR homodimers, or alternatively, on response elements in which the binding of RXR homodimers is enhanced by other promoter bound factors.

As COUP-TF, Arp-1 and EAR 2 can specifically modulate the cellular retinoid response by interfering with RXR mediated transactivation they have been assigned negative regulatory functions. However, the data cannot exclude that under appropriate conditions they may also function as positive modulators of transcription, perhaps in combination with other transcription factors or in the presence of as yet unidentified ligands. It is interesting in this respect to note that COUP-TF was originally identified as a positive activator of chicken ovalbumin gene in vitro (Sagami et al, 1986). Further evidence in support of a positive role for these receptors has been provided by the observation that the *Drosophila* homologue of the mammalian COUP-TF, *seven-up*, is required in photoreceptor cell precursors during eye development (Mlodzik et al, 1990). Overall, however, the studies to date suggest that a major role of the orphan receptors is to regulate the activities of ligand-activated receptors and their response elements (Tran et al, 1992). They may achieve this not only by competitive binding to common response elements, but also by forming inactive complexes in the nucleus.

Competition between type II receptors for their common dimerization partner RXR represents another example of cross-talk between receptor family members. As RXR is a central common partner of several type II receptors, it follows that a hierarchy of dimerization preference may exist, and thus competition for RXR would have important implications for the regulation of transcription by type II receptors (Figure 7.5). Furthermore, a number of type II receptors are ubiquitously expressed suggesting that such competition may play a major role in certain cell types or at distinct developmental stages. Studies reveal that $T_3R\alpha$ competes successfully with RAR for RXR, resulting in a T_3-independent suppression of RA-dependent transactivation. This suppression effect can be alleviated in vitro by cotransfection of RXR, implying that the relative abundance of different type II receptors per cell is an important determinant for ligand susceptibility. However, the different affinities of type II receptors for RXR is only relevant to the regulation of RA-responsive genes in cells where RXR is expressed at low levels (Barettino et al, 1993). Thus T_3R and RAR, and probably other type II receptors may play a dual regulatory role by activating transcription through their cognate binding sites in response to their respective ligand and secondly, in the absence of ligand, by modulating RA-dependent regulation by competition for RXR (Barettino et al, 1993; Stunnenberg, 1993).

Negative response elements have been demonstrated in the promoters of several genes and offer yet another mechanism for receptor mediated transcriptional repression. Examples of such elements are the negative GRE (nGRE) in the pro-opiomelanocortin gene (Drouin et al, 1993) or the negative thyroid response elements (nT_3RE) described for the epidermal growth factor (EGF) receptor gene (Thompson et al, 1992) and the α subunit of rat thyrotropin stimulating hormone gene (Burnside et al, 1989; Krishna et al, 1989). Another feature of a number of promoters is the presence of composite response elements which allow integrated responses to a number of hormone or vitamin types. Initial studies had suggested that the rat oxytocin promoter contains a negative RARE (nRARE) (Lipkin et al, 1992). However, more recent data would suggest that this element is a composite response element which can be stimulated by RA, thyroid hormone or estradiol. Such an element would therefore allow selective response to different stimuli at critical periods during development and represents another example of the cross-talk existing between the nuclear receptor family (Adan et al, 1993). Another interesting group of genes directly regulated by RA is the keratin gene family. It is known that there is a retinoid-mediated inhibition of keratin expression in epidermal cells. The RA-induced changes are mediated at the transcriptional level and are postulated to occur via nRAREs (Tomic et al, 1990; Aneskievich and Fuchs, 1992). Such a regulatory mechanism is extremely interesting in view of the role of retinoids in epidermal differentiation and their use in dermatology as therapeutic agents.

Interaction Between Different Signalling Pathways

In recent years much evidence has been presented of a phenomenon referred to as cross-talk between different signalling pathways, especially between the nuclear receptor and the AP-1 signalling pathways. AP-1 is composed of heterodimers either of products of the *Jun* gene subfamily or the *Jun* and *Fos* subfamilies and is responsive to stimulation by a number of agents including phorbol esters, certain inflammatory mediators and growth factors (Kerppola and Curran, 1991; Miner and Yamamoto, 1991). This cross-talk can occur at two different levels, either at the protein-DNA binding or the protein-protein interaction level. This phenomenon was first identified when studies revealed that regulation of proliferin gene expression is dependent on GR and members of the AP-1 family. The levels of expression are dependent on which transcription factors are bound to a composite 25-nucleotide regulatory sequence in the promoter region. The ability of GR to stimulate proliferin gene transcription is enhanced by c-Jun homodimers but repressed by c-Jun/c-Fos heterodimers (Diamond et al, 1990; Miner et al, 1991; Miner and Yamamoto, 1991). Interaction at the composite promoter element is dependent not only on DNA sequence binding, but also on interaction between the bound protein complexes. The protein-protein interaction involves the DBD of GR and the basic zipper (bZip) region of AP-1. Interestingly, combinations of different steroid and AP-1 family members elicit different regulatory effects which presumably reflect interactions with other factors and add further diversity to the system. As the bZip region defines the AP-1 DBD, one can speculate that the evolutionary conservation of the DBDs from different protein families has been exploited for other intermolecular interactions, such as contacts between different signalling pathways that result in combinatorial regulation (Miner and Yamamoto, 1992).

In a mechanism distinct from the above, glucocorticoids have also been shown to regulate collagenase gene transcription. This regulation appears to occur via a direct protein-protein interaction between AP-1 and GR. Although GR does not bind to the AP-1 site in the collagenase promoter, nevertheless, it has been shown that the AP-1 element is essential for GR mediated gene repression. Additionally, RARs can repress transcriptional induction of the collagenase gene. Surprisingly, however, RXRs fail to repress AP-1 activity, demonstrating a distinct difference between the two regulatory systems through which retinoids exert their transcriptional control. Again, the DBDs of both families participate in the protein-protein interactions (Schüle et al, 1991). In agreement with the studies on the collagenase gene promoter, a second member of the metalloproteinase family, stromelysin, is also repressed by RA through an identical AP-1 site (Nicholson et al, 1990). Reciprocally, AP-1 can also mediate repression of hormone-inducible promoters (Jonat et al, 1990; Schüle et al, 1990; Yang-Yen et al, 1990; Desbois et al, 1991) highlighting the existence of mutual antagonistic action between the nuclear receptor and the AP-1 family members in the cell.

Another known system upon which retinoids exert an influence is the transforming growth factor β (TGFβ) family. TGFβ is the prototypical multifunctional growth factor (Masagué, 1990; Sporn and Roberts, 1990), and it occupies a unique position with regard to regulation of both normal and pathologic physiology. Several members of the nuclear receptor family exert significant effects on the expression of TGFβ, both at the transcriptional and posttranscriptional level (Sporn and Roberts, 1990). RA is known to be a potent inducer of TGFβ2: an effect mediated at the posttranscriptional level. It is also known that the TGFβ genes can up-regulate or down-regulate each other and that this regulation is influenced in part, by interactions between TGFβ and the AP-1 complex. Taken together, these data show that there are multiple levels of cross-coupling of transcriptional control by members of the nuclear receptor family and the leucine-zipper classes of transcriptional regulators. These mechanisms directly couple major signal transduction pathways and thus allow them to act in a common integrated pathway to regulate cellular differentiation and proliferation (Sporn and Roberts, 1991).

Therefore, in conjunction with various modes of cross-talk within the nuclear hormone receptor family, there also exists extensive cross-talk between this and other signalling pathways. Such interactions allow retinoids and other nuclear receptors to regulate the function of diverse transcription factors through both DNA-binding-dependent and DNA-binding-independent protein-protein interactions. Nature has clearly selected heterodimerization between transcription factors as a means of generating increased diversity of transcriptional regulation, and when one considers that proteins from both the nuclear receptor and AP-1 families generally form heterodimers within their respective family, further complexity is added to the system. The mechanisms responsible for the functional interaction between distinct signalling pathways may be gene-specific, and it is probable that tissue-specific and perhaps stage-specific differences exist that dictate how these transcription factors interact or whether they interact in all cell types. It is also possible that a mechanism exists to balance the mutual antagonism between transcription factors from different families. Furthermore, the ability to interact with molecules from a different signalling pathway conceivably introduces not only diversity but also specificity in a cell expressing multiple nuclear receptor types. Thus it is plausible that the principle of intracellular cross-talk between different signalling pathways is much more widespread than originally envisaged. The idea that several diverse factors may work in concert, as well as opposition, leading to gene activation or transcriptional inhibition respectively is an attractive one, and would support the involvement of many signalling pathways in numerous complex physiological processes (Miner and Yamamoto, 1991; Burnstein and Cidlowski, 1993; Parker, 1993).

Human Disease Models

The diverse fundamental roles exerted by retinoids in numerous biological functions is clearly without precedent; they function as critical regulators of vision, cell proliferation, differentiation and embryonal morphogenesis. These physiological effects, which are mediated primarily by the nuclear retinoid receptors, are very sensitive to changes in retinoid concentration; thus deficiency or excess results in a number of characteristic abnormalities. Developmental effects resulting from maternal vitamin A deficiency include microphthalmia, cleft palate and/or lip, cardiovascular and urogenital anomalies and malformed limbs. Embryos exposed to excess vitamin A show abnormalities of many organ systems, the most pronounced effects being abnormal limb and craniofacial development (Morriss-Kay, 1993).

The knowledge of the powerful effects mediated by vitamin A deficiency or excess, naturally has led to analyses of retinoids as potential therapeutic agents in the control of several diseases. As carcinogenesis is fundamentally a disorder of cell differentiation and proliferation, the relationship between retinoids and cancer had been ascertained early on. In vitro, the most spectacular effects of RA-induced differentiation has been shown in the mouse F9 and P19 embryonal carcinoma (EC) (Strickland and Mahdavi, 1978; Jones-Villeneuve et al, 1983) and the human promyelocytic leukemia HL60 cell lines (Collins, 1987; Collins et al, 1990), which differentiate upon RA administration into normal parietal endoderm and functional mature granulocytes respectively. It has also been shown that retinoids cause growth inhibition in many hyperproliferating cell lines. Hence, in conjunction with a fundamental role in normal development, retinoids are known to function as antitumour and antipsoriatic agents. They have been shown to have inhibitory effects in tumour development and also to inhibit the growth of neoplastically transformed cells. Their role in the prevention and treatment of cancers of various organs, including skin, bladder, breast and lung has been documented but perhaps their most striking efficacy is in the treatment of patients with acute promyelocytic leukemia (APL).

APL comprises approximately ten per cent of the acute myeloblastic leukemias in adults. Treatment with RA results in complete remission of this disease, which is associated with the differentiation of immature neoplastic cells into mature granulocytes followed by the emergence of normal hematopoietic cells. Paradoxically, a unique protein encoding a fusion gene product between RARα and a myeloid gene product called PML (for promyelocytes) has been identified which participates in the leukemogenesis of APL by blocking normal RA-dependent myeloid differentiation (Chomienne et al, 1989, 1990; Castaigne et al, 1990; Degos et al, 1990; Warrell et al, 1993). This aberrant protein is due to a nonrandom chromosomal abnormality characterised by balanced and reciprocal translocations between the long arms of chromosomes 15 and 17 t(15:17). The t(15:17) (q22, q11-21) breakpoint fusing PML and RARα results in the synthesis of two reciprocal fusion transcripts: PML-RARα found in all APL patients studied to date, and RARα-PML detected in approximaately two-thirds of patients. PML-RARα rather than RARα-PML is known to be the fusion protein contributing to the disease (Grignani et al, 1993a, 1993b). The breakpoint in RARα always occurs in intron two, whereas the PML breakpoints on chromosome 15 are clustered in three regions, termed bcr 1 (intron 6), bcr 2 (exon 6) and bcr 3 (intron 3), leading to distinct fusion products. The differences in the location of the PML breakpoints, however, have not yet been correlated with clinical characteristics such as morphology, response or survival. Molecular studies reveal further heterogeneity of the PML-RARα transcript due to alternative splicing of PML, alternative usage of RARα polyadenylation sites, splice junctions or mini-introns (De Thé et al, 1991; Kakizuka et al, 1991; Kastner et al, 1992; Pandolfi et al, 1992; Matsuoka et al, 1993).

PML belongs to a newly discovered gene family which encodes nuclear proteins with a distinctive cysteine-rich motif, referred to as a cysteine chapel or RING finger motif (Freemont et al, 1991; Hanson et al, 1991). It also contains a coiled-coil domain that shows homology with the leucine zipper of the AP-1 family (Kakizuka et al, 1991). Its role in normal development and more precisely in myeloid differentiation is unknown. PML-RARα contains the amino terminus of PML with its RING finger motif and dimerization domains fused to the DBD and LBD of RARα. This hybrid protein is considerably more abundant than normal RARα in leukemic promyelocytes, suggesting that the aberrant PML-RARα exerts a dominant effect over the wild-type RARα product. The aberrant PML-RARα is RA-responsive although it displays both promoter and cell-specific differences from wild-type RARα (De Thé et al, 1991; Kakizuka et al, 1991; Perez et al, 1993; Rousselot et al, 1994). The exact mechanism of PML-RARα induced leukogenesis remains to be established. A number of alternatives have been proposed, including interference with normal PML function, or interference with the control of genes activated by RARs that commit myeloid precursor cells to granulocytic differentiation. Data from several studies favour the latter mechanism of interference.

In contrast to RARα, which requires heterodimerization with RXR for efficient DNA binding, PML-RARα exists as stable homodimeric complexes and can bind to target sequences in the absence of RXR. PML-RARα also appears to dimerize with normal PML and in the presence of RXR induces the formation of PML-RARα/RXR heterodimers which bind to DNA like the classical RAR/RXR heterodimers. The efficient sequestration of RXR by PML-RARα suggests that dominant inactivation of RXR may be a possible mode of action of this fusion protein. Accordingly, PML-RARα has also been shown to block the differentiation normally induced by vitamin D_3 in the myeloid precursor cell line U937 (Grignani et al, 1993b). It therefore appears that PML-RARα is capable of repressing the normal RA signal transduction system by two mechanisms, one involving competition for DNA binding and another by sequestration of RXR. The DNA binding of PML-RARα homodimers or PML-RARα/RXR heterodimers may lead to inappropriate gene activation or repression, or alternatively, a dominant inactivation of RAR and other nuclear receptors may inhibit promyelocytic differentiation (Perez et al, 1993; Warrell et al, 1993). At high concentrations of RA, it is thus conceivable that normal RARs would simply outcompete the aberrant PML-RARα protein. This hypothesis is supported by the fact that in many cells including promyelocytes, RA increases the expression of RARα (Chomienne et al, 1991; Leroy et al, 1991a).

Further evidence lending support to this theory comes from a recent study showing that the cellular localization of the aberrant fusion protein differs from that of PML and can be reversed by the addition of RA. PML is specifically localized in discrete subnuclear compartments corresponding to nuclear bodies. In APL cells, however, the PML-RARα fusion displays an abnormal localization and furthermore, directs RXR and other nuclear proteins into aberrant structures that are tightly bound to chromatin, clearly suggesting that PML-RARα can exert a dominant negative effect by diverting a subset of proteins from their natural sites of action. Interestingly, treatment of APL cells with RA induces a complete and dramatic relocalization of each of these proteins to their normal positions thereby indicating that the beneficial effect of RA in promoting myeloid differentiation in APL may be related to its ability to restore a normal subnuclear organization. Moreover, RXR no longer colocalizes with PML-RARα but instead reverts to its diffuse nuclear location supporting the argument that sequestration of RXR by the hybrid fusion protein in APL cells is responsible for the observed effects (Dyck et al, 1994; Weis et al, 1994).

Recently a second variant translocation has been observed in APL patients involving RARα and a previously uncharacterised zinc-finger gene termed promyelocytic leukemia zinc finger (PLZF), a putative transcription factor of the *Krüppel* family. In this instance a t(11:17) reciprocal translocation results in the generation of the two fusion proteins PLZF/RARα and RARα/PLZF, again containing identical portions of RARα. Notably, the PLZF and PML sequence in the hybrid proteins are not related to each other except that both contain potential DNA binding regions (zinc fingers) and sequences implicated in transcriptional regulation (proline-rich regions). Similar to PML-RARα, it is highly likely that PLZF-RARα also antagonises the action of normal RARα during differentiation of promyelocytes (Chen et al, 1993). It is extremely intriguing that not only is RARα a common target in both types of translocation associated with APL, but also that the same functional domains of the receptor are present in both, strongly implicating the expression of an abnormal fusion receptor gene in the aetiology of this disease.

It is not surprising that the number of discovered malignancies harbouring an altered retinoid receptor gene is increasing. Recent research suggests a role for RARβ as a tumour suppressor gene in human epidermoid lung cancer cells (Houle et al, 1993). This area of research is currently being expanded to investigate whether other tumour cells also contain anomalous RAR genes. Moreover, considering previous reports about other members of the hormone receptor multigene family, the discovery of a potential proto-oncogenic role for retinoid receptor genes is not astonishing. Probably the best defined example is avian erythroleukemia which is caused by a mutated derivative of T_3R, the v-*erb*A oncogene (Sap et al, 1986; Weinberger et al, 1986). This oncogene blocks the action of the endogenous T_3R, thereby preventing erythroblasts from differentiating in response to triiodothyronine (Zenke et al, 1990; Beug and Vennström, 1991; Barettino et al, 1993).

RA and its analogues have long been used in dermatology to treat such multifactorial skin disorders as acne, psoriasis and photo-induced aging. It has also been shown that retinoids prevent certain cancers of the skin and have efficacy as agents in human malignant and premalignant cutaneous disorders. Retinoids exert profound effects on epidermis cells both in vivo and in vitro. RA deficiency can convert a secretory epithelium to a squamous epithelium, a condition known as squamous metaplasia, whereas retinoid excess converts a stratified squamous epithelium to a secretory epithelium (mucous metaplasia) (Aneskievich and Fuchs, 1992). Therefore, the beneficial roles of retinoids in skin disorders is not surprising considering their striking effects on epithelial cell growth and differentiation and influence on keratin production. The defined role of RA in suppressing terminal differentiation in the epidermis of the skin is remarkable and in many ways paradoxical considering the more typical role of RA in accentuating differentiation. However, a novel dual effect of RA on epidermal differentiation has been observed, which is dependent on ligand concentrations. The mechanism whereby RA both enhances and suppresses differentiation is unclear, but it is possible that retinoid receptors are expressed differently in basal and differentiating keratinocytes and that this difference gives rise to a multifaceted response (Aneskievich and Fuchs, 1992). These recent findings open the possibility that truncated receptors contribute to epidermal cancers or other differentiation suppressing disorders in a fashion perhaps analogous to that of certain leukemias.

Finally, it should be noted that a number of major diseases in addition to cancers are characterized by excessive proliferation of cells, often with excessive accumulation of extracellular matrix material. Such diseases include rheumatoid arthritis, idiopathic pulmonary fibrosis, cirrhosis of the liver as well as artherosclerosis. Thus the possibility exists that retinoids, which can influence cell differentiation and proliferation, may be of therapeutic value in some of these proliferative diseases (Goodman, 1984).

Conclusions and Future Perspectives

The fundamental importance of the retinoid signalling pathway in developmental events, in generating a body plan, in limb development and organogenesis is becoming increasingly clear. Research over the past decade has revolutionized our thinking in this field of research. Much has been learned about the mechanisms employed by retinoids, in combination with their nuclear binding proteins, in achieving the regulation of specific sets of genes thereby triggering a cascade of regulatory events. However, there are still many important issues that need to be answered.

Initially, if we consider the ligand question, there are many topics that need to be addressed. Are there additional endogenous ligands? What are the pathways responsible for these ligand conversions? What is the distribution of endogenous ligand(s), and how is this controlled? Are there cell-specific metabolites that selectively act as agonists or antagonists for individual nuclear receptors? In conjunction with this, we can also ask if there are additional cytoplasmic retinoid binding proteins and what is their precise role? To date, the existence, if any, of a 9-cis RA cellular binding protein has not been established. On a broader scale, the search for orphan receptor ligands is also an area of active research, and their potential roles in indirectly affecting the complex retinoid machinery remain an issue.

Regarding the nuclear receptors, there are several unanswered questions which include: are there other auxiliary factors for nuclear receptors in addition to RXR; are there additional ligand-dependent or -independent nuclear receptors that form functional heterodimers with RXR; what role does the competition between type II receptors for their common partner play in vivo; are their different biological roles for the many isoforms; does the existence of multiple retinoid receptors reflect the need for distinct genetic mechanisms to implement the diverse effects produced by retinoid treatment; do the multiple receptors function interactively, or are they independent, and what is the level of redundancy in the system? Modifications of the receptor proteins remain an issue that is also unclear at present; for example, what role does phosphorylation play in gene regulation?

Analyzing the mechanisms of gene regulation requires answers to the following questions. What are the additional target genes regulated by retinoids? What are the factors that mediate positive and negative interactions with the basal transcription machinery, and how do they function? How complex are the interactions between the retinoid and other signalling pathways? The potential role of chromatin structure in regulating retinoid receptor gene regulation is also an area that has received little attention to date. Finally, we need to address many disease related topics. Are there other disease models in which an aberrant retinoid receptor or an aberrant retinoid fusion product is detected, and what is the role of these in transforming events? Future research should also convey whether all the biological effects of retinoids can be accounted for by the nuclear retinoid receptors.

Notwithstanding that the retinoid signalling pathways are as yet largely undetermined it is conceivable that multiple unique and overlapping mechanisms, which may be cell type and developmental stage specific, exist in the cell allowing regulation of distinct genes to occur. One can therefore envisage that the next decade will generate many more monumental results in this exciting field of research.

Acknowledgements

We would like to thank P. Chambon for communicating unpublished results and H. Gronemeyer for allowing us to reproduce the heterodimer polarity and specificity model. We are also grateful to members of the Stunnenberg laboratory and Francis Stewart for critically reading this manuscript. M. Keaveney is supported by a postdoctoral fellowship from the European Community.

References

Adan RA, Cox JJ, Beischlag TV, Burbach JP (1993): A composite hormone response element mediates the transactivation of the rat oxytocin gene by different classes of nuclear hormone receptors. *Mol Endocrinol* 7: 47–57

Allan GF, Leng X, Tsai SY, Weigel NL, Edwards DP, Tsai MJ, O'Malley BW (1992a): Hormone and antihormone induce distinct conformational changes which are central to steroid receptor activation. *J Biol Chem* 267: 19513–19520

Allan GF, Tsai SY, Tsai MJ, O'Malley BW (1992b): Ligand-dependent conformational changes in the progesterone receptor are necessary for events that follow DNA binding. *Proc Natl Acad Sci USA* 89: 11750–11754

Allenby G, Bocquel MT, Saunders M, Kazmer S, Speck J, Rosenberger M, Lovey A, Kastner P, Grippo JF, Chambon P, Levin AA (1993): Retinoic acid receptors and retinoid X receptors: interactions with endogenous retinoic acids. *Proc Natl Acad Sci USA* 90: 30–34

Amero SA, Kretsinger RH, Moncrief ND, Yamamoto KR, Pearson WR (1992): The origin of nuclear receptor proteins: a single precursor distinct from other transcription factors. *Mol Endocrinol* 6: 3–7

Aneskievich BJ, Fuchs E (1992): Terminal differentiation in keratinocytes involves positive as well as negative regulation by retinoic acid receptors and retinoid X receptors at retinoid response elements. *Mol Cell Biol* 12: 4862–4871

Apfel C, Bauer F, Crettaz M, Forni L, Kamber M, Kaufmann F, LeMotte P, Pirson W, Klaus M (1992): A retinoic acid receptor alpha antagonist selectively counteracts retinoic acid effects. *Proc Natl Acad Sci USA* 89: 7129–7133

Åstrom A, Pettersson U, Krust A, Chambon P, Voorhees JJ (1990): Retinoic acid and synthetic analogs differentially activate retinoic acid receptor dependent transcription. *Biochem Biophy Res Comm* 173: 339–345

Bailey JS, Siu C-H (1988): Purification and partial characterization of a novel binding protein for retinoic acid from neonatal rat. *J Biol Chem* 263: 9326–9332

Baker AR, McDonnell DP, Hughes M, Crisp TM, Mangelsdorf DJ, Haussler MR, Pike JW, Shine J, O'Malley BW (1988): Cloning and expression of full-length cDNA encoding human vitamin D receptor. *Proc Natl Acad Sci USA* 85: 3294–3298

Baniahmad A, Ha I, Reinberg D, Tsai S, Tsai M-J, O'Malley BW (1993): Interaction of human thyroid hormone receptor β with transcription factor TFIIB may mediate target gene derepression and activation by thyroid hormone. *Proc Natl Acad Sci USA* 90: 8832–8836

Barettino D, Bugge TH, Bartunek P, Vivanco RM, Sonntag BV, Beug H, Zenke M, Stunnenberg HG (1993): Unliganded T_3R, but not its oncogenic variant, v-*erb* A, suppresses RAR-dependent transactivation by titrating out RXR. *EMBO J* 12: 1343–1354

Barettino D, Vivanco Ruiz MdM, Stunnenberg HG (1994): Characterization of the ligand-dependent transactivation domain of thyroid hormone receptor. *EMBO J* 13: 3039–3049

Benbrook D, Lenhardt E, Pfahl M (1988): A new retinoic acid receptor identified from a hepatocellular carcinoma. *Nature* 333: 669–672

Berkenstam A, Vivanco Ruiz MdM, Barettino D, Horikoshi M, Stunnenberg HG (1992): Cooperativity in transactivation between retinoic acid receptor and TFIID requires an activity analogous to E1A. *Cell* 69: 401–412

Berrodin TJ, Marks MS, Ozato K, Linney E, Lazar MA (1992): Heterodimerization among thyroid hormone receptor, retinoic acid receptor, retinoid X receptor, chicken ovalbumin upstream promoter transcription factor, and an endogenous liver protein. *Mol Endocrinol* 6: 1468–1478

Beug H, Vennström B (1991): Avian erythroleukaemia: Possible mechanisms involved in v-*erb* A oncogene function. In: *Nuclear Hormone Receptors*, Parker MG, ed. London: Academic Press

Blomhoff R, Green MH, Berg T, Norum KR (1991): Vitamin A metabolism: new perspectives on absorption, transport and storage. *Physiol Rev* 71: 952–990

Blomhoff R, Green MH, Berg T, Norum KR (1990): Transport and storage of vitamin A. *Science* 250: 399–404

Blumberg B, Mangelsdorf DJ, Dyck JA, Bittner DA, Evans RM, De Robertis E (1992): Multiple retinoid-responsive receptors in a single cell: families of RXRs and RARs in the *Xenopus* egg. *Proc Natl Acad Sci USA* 89: 2321–2325

Boylan JF, Lohnes D, Taneja R, Chambon P, Gudas LJ (1993): Loss of retinoic acid receptor gamma function in F9 cells by gene disruption results in aberrant Hoxa-1 expression and differentiation upon retinoic acid treatment. *Proc Natl Acad Sci USA* 90: 9601–9605

Brand N, Petkovich M, Krust A, Chambon P, de Thé H, Marchio A, Tiollais P, Dejean A (1988): Identification of a second human retinoic acid receptor. *Nature* 332: 850–853

Brand NJ, Petkovich M, Chambon P (1990): Characterization of a functional promoter for the human retinoic acid receptor alpha (hRARα). *Nucl Acids Res* 18: 6799–6806

Brockes JP (1989): Retinoids, homeobox genes and limbs morphogenesis. *Neuron* 2: 1285–1294

Brockes JP (1990): Retinoic acid and limb regeneration. *J Cell Sci Suppl* 13: 191–198

Brockes J (1991): We may not have a morphogen. *Nature* 350: 15

Bugge TH, Pohl J, Lonnoy O, Stunnenberg HG (1992): RXR alpha, a promiscuous partner of retinoic acid and thyroid hormone receptors. *EMBO J* 11: 1409–1418

Burnside J, Darling DS, Carr FE, Chin WW (1989): Thyroid hormone regulation of the rat glycoprotein hormone α-subunit gene promoter activity. *J Biol Chem* 264: 6886–6891

Burnstein KL, Cidlowski JA (1993): Multiple mechanisms for regulation of steroid hormone action. *J Cell Biochem* 51: 130–134

Castaigne S, Chomienne C, Daniel MT, Ballerini P, Berger R, Fenaux P, Degos L (1990): All-trans retinoic acid as a differentiation therapy for acute promyelocytic leukemia: Clinical results. *Blood* 76: 1704–1709

Chambon P (1994): The retinoid signalling pathway: molecular and genetic analyses. *Seminars in Cell Biology* 5: 115–125

Chen Z, Brand NJ, Chen A, Chen SJ, Tong JH, Wang ZY, Waxman S, Zelent A (1993): Fusion between a novel Kruppel-like zinc finger gene and the retinoic acid receptor-alpha locus due to a variant k(11:17) translocation associated with acute promyelocytic leukaemia. *EMBO J* 12: 1161–1167

Chomienne C, Balitrand N, Bellerini P, Castaigne S, de Thé H, Degos L (1991): All-*trans* retinoic acid modulates the retinoic acid receptor-α in promyelcytic cells. *J Clin Invest* 88: 2150–2154

Chomienne C, Ballerini P, Balitrand N, Amar M, Bernard JF, Boivin P, Daniel MT, Berger R, Castaigne S, Degos L (1989): Retinoic acid: an alternative therapy of promyelocytic leukaemias. *Lancet* 1: 746–747

Chomienne C, Ballerini P, Balitrand N, Daniel MT, Fenaux P, Castaigne S, Degos, L (1990): All-trans retinoic acid in acute promyelocytic leukemias. II. In vitro studies: structure-function relationship. *Blood* 76: 1710–1717

Chytil F (1984): Retinoic acid: biochemistry, pharmacology, toxicology and therapeutic use. *Pharama Res Suppl* 36: 93–100

Chytil FJ, Ong, DE (1984): Cellular retinoid-binding proteins. In: *The Retinoids*, Sporn MB, Roberts AB, Goodman DS eds. New York: Academic Press

Collins SJ (1987): The HL-60 promyelocytic leukemia cell line: Proliferation, differentiation and cellular oncogene expression. *Blood* 70: 1233–1244

Collins SJ, Robertson KA, Mueller L (1990): Retinoic acid-induced granulocytic differentiation of HL-60 myeloid leukemia cells is mediated directly through the retinoic acid receptor α. *Mol Cell Biol* 10: 2154–2163

Cooney AJ, Tsai SY, O'Malley, BW, Tsai MJ (1992): Chicken ovalbumin upstream promoter transcription factor (COUP-TF) dimers bind to different GGTCA response elements, allowing COUP-TF to repress hormonal induction of the vitamin D_3, thyroid hormone, and retinoic acid receptors. *Mol Cell Biol* 12: 4153–4163

Danielian PS, White R, Lees JA, Parker MG (1992): Identification of a conserved region required for hormone dependent transcriptional activation by steroid hormone receptors. *EMBO J* 11: 1025–1033

Davis KD, Lazar MA (1993): Induction of retinoic acid receptor-beta by retinoic acid is cell specific. *Endocrinol* 132: 1469–1474

Degos L, Chomienne C, Daniel M-T, Berger R, Dombret H, Fenaux P, Castaigne S

(1990): Treatment of first relapse in acute promyelocytic leukemia with all-*trans* retinoic acid. *Lancet* 336: 1440–1441

De Luca LM (1991): Retinoids and their receptors in differentiation, embryogenesis, and neoplasia. *FASEB J* 5: 2924–2933

Demmer LA, Birkenmeier EH, Sweetser DA, Leving MS, Zollman S, Sparkes RS, Mohandas T, Lusis AJ, Gordon JI (1987): The cellular retinol binding protein II gene. *J Biol Chem* 262: 2458–2467

Dencker L, Annerwall E, Busch C, Eriksson U (1990): Localization of specific retinoid-acid-binding protein (CRABP) in the early mouse embryo. *Development* 110: 343–352

Denis M, Gustafsson J, Wikstrom A (1988): Interaction of the MR = 90,000 heat shock protein with the steroid binding domain of the glucocorticoid receptor. *J Biol Chem* 263: 18520–18523

Desbois C, Aubert D, Legrand C, Pain B, Samarut J (1991): A novel mechanism of action of v-*erb* A: Abrogation of the inactivation of transcription factor AP-1 by retinoic acid and thyroid hormone receptors. *Cell* 67: 731–740

De Thé H, Lavau C, Marchio A, Chomienne C, Degos L, Dejean A (1991): The PML-RARα fusion mRNA generated by the t(15; 17) translocation in acute promyelocytic leukemia encodes a functionally altered RAR. *Cell* 66: 675–684

De Thé, H, Marchio A, Tiollas P, Dejean A (1989): Differential expression and ligand regulation of the retinoic acid receptor α and β genes. *EMBO J* 8: 429–433

De Thé H, Vivanco Ruiz MdM, Tiollais, P, Stunnenberg H, Dejean A (1990): Identification of a retinoic acid responsive element in the retinoic acid receptor β gene. *Nature* 343: 177–180

Diamond MI, Miner J, Yoshinaga SK, Yamamoto, KR (1990): Transcription factor interactions: selectors of positive or negative regulation from a single DNA element. *Science* 249: 1266–1272

Dingwell C, Laskey RA (1991): Nuclear target sequences – a consensus? *Trends Biochem Sci* 16: 478–481

Dollé P, Fraulob V, Kastner P, Chambon P (1994): Developmental expression of murine retinoid X receptor (RXR) genes. *Mech of Dev* 45: 91–104

Dollé P, Ruberte E, Kastner P, Petkovich M, Stoner CM, Gudas, LJ, Chambon P (1989): Differential expression of genes encoding α, β and γ retinoic acid receptors and CRABP in the developing limbs of the mouse. *Nature* 342: 702–705

Dollé P, Roberte E, Leroy P, Morriss-Kay G, Chambon P (1990): Retinoic acid receptors and cellular binding proteins. I. A systematic study of their differential pattern of transcription during mouse organogenesis. *Development* 110: 1133–1151

Drouin J, Sun YL, Chamberland M, Gauthier Y, Léan AD, Nemer M, Schmidt TJ (1993): Novel glucocorticoid receptor complex with DNA element of the hormone-repressed POMC gene. *EMBO J* 12: 145–156

Durand B, Saunders M, Leroy P, Leid M, Chambon P (1992): All-trans and 9-cis retinoic acid induction of CRABP II transcription is mediated by RAR-RXR heterodimers bound to DR1 and DR2 repeated motifs. *Cell* 71: 73–85

Dyck JA, Maul GG, Miller WHJ, Chen JD, Kakizuka A, Evans RM (1994): A novel macromolecular structure is a target of the promyelocyte-retinoic acid receptor oncoprotein. *Cell* 76: 333–343

Eichele G (1989a): Retinoic acid induces a pattern of digits in anterior half wing buds that lack the zone of polarizing activity. *Development* 107: 863–867

Eichele G (1989b): Retinoids and vertebrate limb pattern formation. *Trends in Genet.* 5: 246–251

Fawell SE, Lees JA, White R, Parker MG (1990): Characterization and colocalization of steroid binding and dimerization activities in the mouse estrogen receptor. *Cell* 60: 953–962

Folkers GE, van der Leede BJ, van der Saag PT (1993): The retinoic acid receptor-beta 2 contains two separate cell-specific transactivation domains, at the N-terminus and in the ligand-binding domain. *Mol Endocrinol* 7: 616–627

Fondell JD, Roy AL, Roeder RG (1993): Unliganded thyroid hormone receptor inhibits formation of functional preinitiation complex: implications for active repression. *Genes Dev* 7: 1400–1410

Forman BM, Samuels HH (1990a): Dimerization among nuclear hormone receptors. *New Biol* 2: 587–594

Forman BM, Samuels HH (1990b): Interactions among a subfamily of nuclear hormone receptors: The regulatory zipper model. *Mol Endocrinol* 4: 1293–1301

Freemont PS, Hanson IM, Trowsdale J (1991): A novel cysteine-rich motif. *Cell* 64: 483–484

Giguere V, Lyn S, Yip P, Siu CH, Amin S (1990a): Molecular cloning of cDNA encoding a second cellular retinoic acid binding protein. *Proc Natl Acad Sci USA* 87: 6233–6237

Giguere V, Ong ES, Segui P, Evans RM (1987): Identification of a receptor for the morphogen retinoic acid. *Nature* 330: 624–629

Giguere V, Shago M, Zirnibl RTP, Rossant J, Varmuza S (1990b). Identification of a novel isoform of the retinoic acid receptor γ expressed in the mouse embryo. *Mol Cell Biol* 10: 2335–2340

Glass CK, DiRenzo J, Kurokawa R, Han Z (1991): Regulation of Gene Expression by Retinoic Acid Receptors. *DNA Cell Biol* 10: 623–638

Goodman DS (1984): Vitamin A and retinoids in health and disease. *New Engl J Med* 310: 1023–1031

Graupner G, Malle G, Maignan J, Lang G, Pruniéras M, Pfahl M (1991): 6'-substituted naphthalene-2-carboxylic acid analogs, a new class of retinoic acid receptor subtype-specific ligands. *Biochem Biophy Res Comm* 179: 1554–1561

Green S, Chambon P (1988): Nuclear receptors enhance our understanding of transcription regulation. *Trends Genet* 4: 309–314

Grignani F, Fagioli M, Ferrucci PF, Alcalay M, Pelicci PG (1993a): The molecular genetics of acute promyelocytic leukemia. *Blood Rev* 7: 87–93

Grignani F, Ferrucci PF, Testa U, Talamo G, Fagioli M, Alcalay M, Mencarelli A, Grignani F, Peschle C, Nicoletti I, Pelicci PG (1993b): The acute promyelocytic leukemia-specific PML-RAR alpha fusion protein inhibits differentiation and promotes survival of myeloid precursor cells. *Cell* 74: 423–431

Gronemeyer H (1993): Transcription activation by nuclear receptors. *J Recept Res* 13: 667–691

Gronemeyer H, Benhamou B, Berry M, Bocquel MT, Gofflo D, Garcia T, Lerouge T, Metzger D, Meyer ME, Tora, L, Chambon P (1992): Mechanisms of antihormone action. *J Steroid Biochem Mol Biol* 41: 217–221

Guiochon-Mantel A, Loosfelt H, Lescop P, Sar S, Atger M, Perrot-Applanat M, Milgrom E (1989): Mechanisms of nuclear localization of the progesterone receptor: evidence for interaction between monomers. *Cell* 57: 1147–1154

Hanson IM, Poustka A, Trowsdale J (1991): New genes in the class II region of the major histocompatibility complex. *Genomics* 10: 417–424

Hazel TG, Nathans D, Lau LF (1988): A gene inducible by serum growth factors encodes a member of the steroid and thyroid hormone receptor superfamily. *Proc Natl Acad Sci USA* 85: 8444–8448

Hernandez N (1993): TBP, a universal eukaryotic transcription factor? *Genes Dev* 7: 1291–1308

Heyman RA, Mangelsdorf DJ, Dyck JA, Stein R, Eichele G, Evans, RM, Thaller C (1992): *9-Cis* retinoic acid is a high affinity ligand for the retinoid X receptor. *Cell* 68: 397–406

Horikoshi N, Maguire K, Kralli A, Maldonado E, Reinberg D, Weinmann R (1991): Direct interaction between adenovirus E1A protein and the TATA box binding transcription factor IID. *Proc Natl Acad Sci USA* 88: 5124–5128

Houle B, Rochette-Egly C, Bradley WE (1993): Tumor-suppressive effect of the retinoic acid receptor beta in human epidermoid lung cancer cells. *Proc Natl Acad Sci USA* 90: 985–989

Ikeda Y, Lala DS, Luo X, Kim E, Moisan M-P, Parker KL (1993): Characterization of the mouse FTZ-F1 gene, which encodes a key regulator of steroid hydro-oxylase gene expression. *Mol Endocrinol* 7: 852–860

Ing NH, Beekman JM, Tsai SY, Tsai MJ, O'Malley BW (1992): Members of the steroid hormone receptor superfamily interact with TFIIB (S300-II). *J Biol Chem* 267: 17617–17623

Izpisua-Belmonte J-C, Tickle C, Mangelsdorf DF, Dollé P, Wolpert L, Duboule D (1991): Expression of the homeobox Hox-4 genes and the specification of position in chick wing development. *Nature* 350: 585–589

Jonat C, Rahmsdorf HJ, Park KK, Cato ACB, Gebel S, Ponta H, Herrlich P (1990): Antitumour promotion and antiinflammation: down-modulation of AP-1 (Fos/Jun) activity by glucocorticoid. *Cell* 62: 1189–1204

Jones-Villeneuve EM, Rudnicki VMA, Harris JF, McBurney MW (1983: Retinoic acid induced neural differentiation of embryonal carcinoma cells. *Mol Cell Biol* 3: 2271–2279

Kakizuka A, Miller WHJ, Umesono K, Warrell RPJ, Frankel SR, Murty VVVS, Dmitrovsky E, Evans RM (1991): Chromosomal translocation t(15; 17) in human acute promyelocytic leukemia fuses RARα with a novel putative transcription factor, PML. *Cell* 66: 663–674

Kastner P, Krust A, Mendelsohn C, Garnier JM, Zelent A, Leroy P, Staub A, Chambon P (1990): Murine isoforms of retinoic acid receptor γ with specific patterns of expression. *Proc Natl Acad Sci USA* 87: 2700–2704

Kastner P, Perez A, Lutz Y, Rochette-Egly C, Gaub MP, Durand B, Lanotte M, Berger R, Chambon P (1992): Structure, localization and transcriptional proper-ties of two classes of retinoic acid receptor alpha fusion proteins in acute promyelocytic leukemia (APL): structural similarities with a new family of oncoproteins. *EMBO J* 11: 629–642

Keaveney M, Berkenstam A, Feigenbutz M, Vriend G, Stunnenberg HG (1993): Residues in the TATA-binding protein required to mediate a transcriptional response to retinoic acid in EC cells. *Nature* 365: 562–566

Keidel S, LeMotte P, Apfel C (1994): Different agonist- and antagonist-induced conformational changes in retinoic acid receptors analyzed by protease mapping. *Mol Cell Biol* 14: 287–298

Kerppola TK, Curran T (1991): Transcription factor interactions: Basics on zippers. *Curr Opin Struct Biol* 1: 71–79

Kessel M, Gruss P. (1991): Homeotic transformations of murine vertebrae and concomitant alteration of Hox codes induced by retinoic acid. *Cell* 67: 89–104

Kliewer SA, Umersono K, Heyman RA, Mangelsdorf DJ, Dyck JA, Evans RM (1992a): Retinoid X receptor-COUP-TF interactions modulate retinoic acid signalling. *Proc Natl Acad Sci USA* 89: 1448–1452

Kliewer SA, Umesono K, Mangelsdorf DJ, Evans RM (1992b): Retinoid X receptor interacts with nuclear receptors in retinoic acid, thyroid hormone and vitamin D$_3$ signalling. *Nature* 355: 446–449

Koelle MR, Talbot WS, Seagraves WA, Bender MT, Cherbas P, Hogness DS (1991): The Drosophila EcR gene encodes an Ecdysone receptor, a new member of the steroid receptor superfamily. *Cell* 67: 59–77

Krishna V, Chatterjee K, Lee J-K, Rentoumis A, Jameson JL (1989): Negative regulation of the thyroid-stimulating hormone α gene by thyroid hormone: Receptor interaction adjacent to the TATA box. *Proc Natl Acad Sci USA* 86: 9114–9118

Krust A, Kastner PH, Petkovich M, Zelent A, Chambon P (1989): A third human retinoic acid receptor, hRARγ. *Proc Natl Acad Sci USA* 86: 5310–5314

Kumar V, Chambon P (1988): The estrogen receptor binds tightly to its responsive element as a ligand-induced homodimer. *Cell* 55: 145–156

La Thangue NB, Rigby PWJ (1987): An Adenovirus E1A-like transcription factor is regulated during the differentiation of murine Embryonal Carcinoma stem cells. *Cell* 49: 507–513

Ladias JAA, Karathanasis SK (1991): Regulation of the apolipoprotein AI gene by ARP-1, a novel member of the steroid receptor superfamily. *Science* 251: 561–565

Laudet V, Hanni C, Coll J, Catzflis F, Stehelin D (1992): Evolution of the nuclear receptor gene superfamily. *EMBO J* 11: 1003–1013

Lavorgna G, Ueda H, Clos J, Wu C (1991): FTZ-F1, a steroid hormone receptor-like protein implicated in the activation of fushi tarazu. *Science* 252: 848–851

Lazar MM, Hodin RA, Darling DS, Chin WW (1989): A novel member of the thyroid/steroid hormone receptor family is encoded by the opposite strand of the rat c-*erb* A-alpha transcriptional unit. *Mol Cell Biol* 9: 1128–1136

Lee MS, Kliewer SA, Provencal J, Wright PE, Evans RM (1993): Structure of the retinoid X receptor alpha DNA binding domain: a helix required for homo-dimeric DNA binding. *Science* 260: 1117–1121

Lee WS, Kao CC, Bryant GO, Liu X, Berk AJ (1991): Adenovirus E1A activation domain binds the basic repeat in the TATA box transcription factor. *Cell* 67: 365–376

Lehmann JM, Dawson MI, Hobbs PD, Husmann M, Pfahl M (1991a): Identification of retinoids with nuclear receptor subtype-selective activities. *Cancer Res* 51: 4804–4809

Lehmann JM, Hoffmann B, Pfahl M (1991b): Genomic organization of the retinoic acid receptor gamma gene. *Nucl Acids Res* 19: 573–578

Lehmann JM, Jong L, Fanjul A, Cameron JF, Lu XP, Haefner P, Dawson MI, Pfahl M (1992): Retinoids selective for retinoid X receptor response pathways. *Science* 258: 1944–1946

Leid M, Kastner P, Chambon P (1992a): Multiplicity generates diversity in the retinoic acid signalling pathways. *Trends Biochem Sci* 17: 427–433

Leid M, Kastner P, Lyons R, Nakshatri H, Saunders M, Zacharewski T, Chen J-Y, Staub A, Garnier J-M, Mader S, Chambon P (1992b): Purification, cloning, and RXR identity of a HeLa cell factor with which RAR or TR heterodimerizes to bind target sequences efficiently. *Cell* 68: 377–395

Leroy P, Krust A, Zelent A, Mendelsohn C, Garnier JM, Kastner P, Dierich A, Chambon P (1991a): Multiple isoforms of the mouse retinoic acid receptor α are generated by alternative splicing and differential induction by retinoic acid. *EMBO J* 10: 59–69

Leroy P, Nakshatri H, Chambon P (1991b): The mouse retinoic acid receptor α2 isoform is transcribed from a promoter that contains a retinoic acid response element. *Proc Natl Acad Sci USA* 88: 10138–10142

Levin AA, Sturzenbecker LJ, Kazmer S, Bosakowski T, Huselton C, Allenby G, Speck J, Kratzeisen C, Rosenberger M, Lovey A, Grippo JF (1992): 9-cis retinoic acid stereoisomer binds and activates the nuclear receptor RXR alpha. *Nature* 355: 359–361

Li E, Sucov HM, Lee KF, Evans RM, Jaenisch R (1993): Normal development and growth of mice carrying a targeted disruption of the alpha 1 retinoic acid receptor gene. *Proc Natl Acad Sci USA* 90: 1590–1594

Lipkin SM, Nelson CA, Glass CK, Rosenfeld MG (1992): A negative retinoic acid response element in the rat oxytocin promoter restricts transcriptional stimulation by heterologous transactivation domains. *Proc Natl Acad Sci USA* 89: 1209–1213

Lohnes D, Dierich A, Ghyselinck N, Kastner P, Lampron C, LeMeur M, Lufkin T, Mendelsohn C, Nakshatri H, Chambon P (1992): Retinoid receptors and binding proteins. *J Cell Sci Suppl* 16: 69–76

Lohnes D, Kastner P, Dierich A, Mark M, LeMeur M, Chambon P (1993): Function of retinoic acid receptor gamma in the mouse. *Cell* 73: 643–658

Lufkin T, Lohnes D, Mark M, Dierich A, Gorry P, Gaub MP, LeMeur M, Chambon P (1993): High postnatal lethality and testis degeneration in retinoic acid receptor alpha mutant mice. *Proc Natl Acad Sci USA* 90: 7225–7229

Luisi BF, Xu WX, Otwinowski Z, Freedman LP, Yamamoto KR, Sigler PB (1991): Crystallographic analysis of the interaction of the glucocorticoid receptor with DNA. *Nature* 352: 497–505

Maden M, Ong DE, Chytil F (1990): Retinoid-binding protein distribution in the developing mammalian nervous system. *Development* 109: 75–80

Maden M, Ong DE, Summerbell D, Chytil F (1988): Spatial distribution of cellular protein binding to retinoic acid in the chick limb bud. *Nature* 335: 733–735

Maden M, Ong DE, Summerbell D, Chytil F (1989): The role of retinoid-binding proteins in the generation of pattern in the developing limb, the regenerating limb and the nervous system. *Development Suppl* 107: 109–119

Mader S, Leroy P, Chen JY, Chambon P (1993): Multiple parameters control the selectivity of nuclear receptors for their response elements. Selectivity and promiscuity in response element recognition by retinoic acid receptors and retinoid X receptors. *J Biol Chem* 268: 591–600

Mangelsdorf DJ, Evans RM (1992): Retinoic Acid Receptors as Transcription Factors. In: *Transcriptional Regulation*, McKnight S Yamamoto KR, eds. New York: CSHL Press

Mangelsdorf DJ, Borgmeyer U, Heyman RA, Zhou JY, Ong ES, Oro AE, Kakizuka A, Evans RM (1992): Characterization of three RXR genes that mediate the action of 9-cis retinoic acid. *Genes Dev* 6: 329–344

Mangelsdorf DJ, Ong ES, Dyck JA, Evans RM (1990): Nuclear receptor that identifies a novel retinoic acid response pathway. *Nature* 345: 224–229

Marks MS, Hallenbeck PL, Nagata T, Segars JH, Appella E, Nikodem VM, Ozato K (1992): H-2RIIBP (RXRβ) heterodimerization provides a mechanism for combinatorial diversity in the regulation of retinoic acid and thyroid hormone responsive genes. *EMBO J* 11: 1419–1435

Masagué J (1990): The transforming growth factor-β family. *Ann Rev Cell Biol* 6: 597–641

Matsuoka A, Miyamura K, Emi N, Tahara T, Tanimoto M, Naoe T, Ohno R, Kakizuka A, Evans RM, Saito H (1993): Unexpected heterogeneity of PML/RAR alpha fused mRNA detected by nested polymerase chain reaction in acute promyelocytic leukemia. *Leukemia* 7: 1151–1155

Mattei MG, Riviere M, Krust A, Ingvarsson S, Venström B, Islam MW, Levan G, Kastner P, Zelent A, Chambon P, Szpierer J, Szpierer C (1991): Chromosomal assignment of retinoic acid receptor (RAR) genes in the human, mouse and rat genomes. *Genomics* 10: 1061–1069

Mendelsohn C, Larkin S, Mark M, LeMeur M, Clifford J, Zelent A, Chambon P (1994): RARβ isoforms: distinct transcriptional control by retinoic acid and specific spatial patterns of promoter activity during mouse embryonic development. *Mech of Dev* 45: 227–241

Mendelsohn C, Ruberte E, LeMeur M, Morriss-Kay G, Chambon P (1991): Developmental analysis of the retinoic acid-inducible RAR-β2 promoter in transgenic animals. *Development* 113: 723–734

Mendelsohn C, Ruberte, E, Chambon P (1992): Retinoid receptors in vertebrate limb development. *Dev Biol* 152: 50–61

Mietus SM, Sladek FM, Ginsburg GS, Kuo CF, Ladias JA, Darnell JJ, Karathanasis SK (1992): Antagonism between apolipoprotein AI regulatory protein 1, Ear3/COUP-TF, and hepatocyte nuclear factor 4 modulates apolipoprotein CIII gene expression in liver and intestinal cells. *Mol Cell Biol* 12: 1708–1718

Milbrandt J (1988): Nerve growth factor induces a gene homologous to the glucocorticoid receptor gene. *Neuron* 1: 183–188

Miner JN, Yamamoto KR (1991): Regulatory crosstalk at composite response elements. *Trends Biochem Sci* 16: 423–427

Miner JN, Yamamoto KR (1992): The basic region of AP-1 specifies glucocorticoid receptor activity at a composite response element. *Genes Dev* 6: 2491–2501

Miner JN, Diamond MI, Yamamoto KR (1991): Joints in the regulatory lattice: Composite regulation by steroid receptor-AP-1 complexes. *Cell Growth Diff* 2: 525–530

Minucci S, Zand DJ, Dey A, Marks MS, Nagata T, Grippo JF, Ozato K (1994): Dominant negative retinoid X receptor beta inhibits retinoic acid-responsive gene regulation in embryonal carcinoma cells. *Mol Cell Biol* 14: 360–372

Miyajima N, Horiuchi R, Shibuya Y, Fukushige S-I, Matsubara K-I, Toyoshima K, Yamamoto T (1989): Two *erb* A homologs encoding proteins with different T_3 binding capacities are transcribed from opposite DNA strands of the same genetic locus. *Cell* 57: 31–39

Mlodzik M, Hiromi Y, Weber U, Goodman CS, Rubin GM (1990): The Drosophila *seven-up* gene, a member of the steroid receptor gene superfamily, controls photoreceptor cell fates. *Cell* 60: 211–224

Moore DM (1990): Diversity and unity in the nuclear hormone receptors: A terpenoid receptor superfamily. *New Biol* 2: 100–105

Morriss-Kay G (1993): Retinoic acid and craniofacial development: molecules and morphogenesis. *BioEssays* 15: 9–15

Näär AM, Boutin J-M, Lipkin SM, Yu, VC, Holloway JM, Glass, CK, Rosenfeld MG (1991): The orientation and spacing of core DNA-binding motifs dictate selective transcriptional responses to three nuclear receptors. *Cell* 65: 1267–1279

Nagpal S, Friant S, Nakshatri H, Chambon P (1993): RARs and RXRs: evidence for two autonomous transactivation functions (AF-1 and AF-2) and heterodimerization in vivo. *EMBO J* 12: 2349–2360

Nagpal S, Saunders M, Kastner P, Durand B, Nakshatri H, Chambon P (1992a): Promoter context- and response element-dependent specificity of the transcriptional activation and modulating functions of retinoic acid receptors. *Cell* 70: 1007–1019

Nagpal S, Zelent A, Chambon P (1992b): RAR-β, a retinoic acid receptor isoform is generated from RAR-$\beta2$ by alternative splicing and usage of a CUG initiator codon. *Proc Natl Acad Sci USA* 89; 2718–2722

Nakai A, Kartha S, Sakurai A, Toback FG, De Groot LJ (1990): A human early response gene homologous to murine nur77 and rat NGFI-B, and related to the nuclear receptor superfamily. *Mol Endocrinol* 4: 1438–1443

Nakshatri H, Chambon P (1994): The directly repeated RG(G/T)TCA motifs of the rat and mouse cellular retinol-binding protein II genes are promiscuous binding sites for RAR, RXR, HNF-4, and ARP-1 homo- and heterodimers. *J Biol Chem* 269: 890–902

Nicholson RC, Mader S, Nagpal S, Rochette-Egly C, Chambon P (1990): Negative regulation of the rat stromelysin gene promoter by retinoic acid is mediated by an AP1 binding site. *EMBO J* 9: 4443–4454

Noji S, Nohno T, Koyama E, Muto K, Ohyama K, Aoki Y, Ohsugi K, Ide H, Taniguchi S, Saito T (1991): Retinoic acid induces polarizing activity but it is unlikely to be a morphogen in the chick limb bud. *Nature* 350: 83–86

O'Donnell AL, Koenig RJ (1990): Mutational analysis identifies a new functional domain of the thyroid hormone receptor. *Mol Endocrinol* 4: 715–720

O'Malley BW, Conneely OM (1992): Orphan receptors: in search of a unifying hypothesis for activation. *Mol Endocrinol* 6: 1359–1361

Ong DE (1984): A novel retinol-binding protein from rat. *J Biol Chem* 259: 1476–1482

Ong DE (1987): Cellular retinoid binding proteins. *Arch Dermatol* 123: 1693a–1695a

Oro AE, McKeown M, Evans RM (1990): Relationship between the product of the *Drosophila ultraspiracle* locus and vertebrate retinoid X receptor. *Nature* 347: 298–301

Pandolfi PP, Alcalay M, Fagioli M, Zangrilli D, Mencarelli A, Diverio D, Biondi A, Lo Coco F, Rambaldi A, Grignani F, Rochette-Egly C, Gaube M-P, Chambon P, Pelicci PG (1992): Genomic variability and alternative splicing generate multiple PML/RARα transcripts that encode aberrant PML proteins and PML/RARα isoforms in acute promyelocytic leukaemia. *EMBO J* 11: 1397–1407

Parker MG (1993): Steroid and related receptors. *Curr Opin Cell Biol* 5: 499–504

Perez A, Kastner P, Sethi S, Lutz Y, Reibel C, Chambon P (1993): PMLRAR homodimers: distinct DNA binding properties and heteromeric interactions with RXR. *EMBO J* 12: 3171–3182

Perez-Castro AV, Toth-Rogler LE, Wei L-N, Nguyen-Huu MC (1989): Spatial and temporal pattern of expression of the cellular retinoic acid-binding protein and the cellular retinol-binding protein during mouse embryogenesis. *Proc Natl Acad Sci USA* 86: 8813–8817

Perlmann T, Rangarajan PN, Umesono K, Evans RM (1993): Determinants for selective RAR and TR recognition of direct repeat HREs. *Genes Dev* 7: 1411–1422

Petkovich M, Brand NJ, Krust A, Chambon P (1987): A human retinoic acid receptor which belongs to the family of nuclear receptors. *Nature* 330: 444–450

Picard D, Yamamoto KR (1987): Twin signals mediate hormone dependent nuclear localization of the glucocorticoid receptor. *EMBO J* 6: 3333–3340

Pignoni F, Baldarelli RM, Steingrímsson E, Diaz, RJ, Patapoutian A, Merriam JR, Lengyel JA (1990): The Drosophila gene *tailless* is expressed at the embryonic termini and is a member of the steroid receptor superfamily. *Cell* 62: 151–163

Pijnappel WW, Hendriks HF, Folkers GE, van der Brink CE, Dekker EJ, Edelenbosch C, van der Saag PT, Durston AJ (1993): The retinoid ligand 4-oxo-retinoic acid is a highly active modulator of positional specification. *Nature* 366: 340–344

Pratt WB, Jolly DJ, Pratt DV, Hollenberg SM, Giguere, V, Cadepond FM, Schweizer-Groyer G, Catelli M, Evans RM, Baulieu E (1988): A region in the steroid binding domain determines formation of the non-DNA-binding, 9S glucocorticoid receptor complex. *J Biol Chem* 263: 267–273

Ptashne M (1988): How eukaryotic transcriptional activators work. *Nature* 335: 683–689

Ragsdale CW, Brockes JP (1991): Retinoic acid receptors and vertebrate limb morphogenesis. In: *Nuclear Hormone Receptors*, Parker MG, ed. London: Academic Press

Rousselot P, Hardas B, Patel A, Guidez F, Gaken J, Castaigne S, Dejean A, De Thé H, Degos L, Farzaneh F, Chomienne C (1994): The PML-RAR alpha gene product of the t(15; 17) translocation inhibits retinoic acid-induced granulocytic differentiation and mediated transactivation in human myeloid cells. *Oncogene* 9: 545–551

Ruberte E, Dollé P, Krust A, Zelent A, Morriss-Kay G, Chambon P (1990): Specific spatial and temporal distribution of retinoic acid receptor gamma transcripts during mouse embryogenesis. *Development* 108: 213–222

Ruberte E, Dollé P, Chambon P, Morriss-Kay G (1991): Retinoic acid receptors and cellular binding proteins. III. Their differential pattern of transcription during early morphogenesis in mouse embryos. *Development* 111: 45–60

Ruberte E, Friederich V, Morriss-Kay G, Chambon P (1992): Differential distribution patterns of CRABP I and CRABP II transcripts during mouse embryogenesis. *Development* 115: 973–987

Ryseck R-P, MacDonald-Bravo H, Mattéi M-G, Ruppert S, Bravo R (1989): Structure, mapping and expression of a growth factor inducible gene encoding a putative nuclear hormonal binding receptor. *EMBO J* 8: 3327–3335

Saari JC, Bredberg L, Garwin GG (1982): Identification of the endogenous retinoids associated with three cellular retinoid-binding proteins from bovine retina and retinal pigment epithelium. *J Biol Chem* 257: 13329–13333

Sagami I, Tsai SY, Wang H, Tsai M-J, O'Malley BW (1986): Identification of two factors required for the transcription of the ovalbumin gene. *Mol Cell Biol* 6: 4259–4267

Sap J, Muñoz A, Damm K, Goldberg Y, Ghysdael J, Vennström B (1986): The c-*erb* A protein is a high-affinity receptor for thyroid hormone. *Nature* 324: 635–640

Schüle R, Rangarajan P, Kliewer S, Ransone LJ, Bolado J, Yang N, Verma IM, Evans RM (1990): Functional antagonism between oncoprotein c-Jun and the glucocorticoid receptor. *Cell* 62, 1217–1226

Schüle R, Randarajan P, Yang N, Kliewer S, Ransone LJ, Bolado J, Verma IM, Evans RM (1991): Retinoic acid is a negative regulator of AP-1-responsive genes. *Proc Natl Acad Sci USA* 88: 6092–6096

Schwabe JW, Chapman L, Finch JT, Rhodes D (1993): The crystal structure of the estrogen receptor DNA-binding domain bound to DNA: how receptors discriminate between their response elements. *Cell* 75: 567–578

Schwabe JWR, Neuhaus D, Rhodes D (1990): Solution structure of the DNA-binding domain of the estrogen receptor. *Nature* 348: 458–461

Shubeita HE, Sambrook JF, McCormick AM (1987): Molecular cloning and analysis of functional cDNA and genomic clones encoding bovine cellular retinoic-acid binding protein. *Proc Natl Acad Sci USA* 84: 5645–5649

Simeone A, Acampora D, Arcioni L, Andrews PW, Boncinelli E, Mavilio F (1990): Sequential activation of Hox2 homeobox genes by retinoic acid in human embryonal carcinoma cells. *Nature* 346: 763–766

Slack JM (1987): We have a morphogen. *Nature* 317: 553–554

Sladek FM, Zhong W, Lai E, Darnell JE (1990): Liver-enriched transcription factor HNF-4 is a novel member of the steroid receptor superfamily. *Genes Dev* 4: 2353–2365

Smith SM, Eichele G (1991): Temporal and regional difference in the expression pattern of distinct retinoic acid receptor-β transcripts in the chick embryo. *Development* 111: 245–252

Smith SM, Pang, K, Sundin O, Wedden SE, Thaller C, Eichele G (1989): Molecular approaches to vertebrate limb morphogenesis. *Development Suppl* 107: 121–131

Smith WC, Nakshatri H, Leroy P, Rees J, Chambon P (1991): A retinoic acid response element is present in the mouse cellular retinol binding protein I (mCRBPI) promoter. *EMBO J* 10: 2223–2230

Sporn MB, Roberts AB (1990): TGF-β: problems and prospects. *Cell Regulation* 1: 875–882

Sporn MB, Roberts A (1991): Interactions of retinoids and transforming growth factor-β in regulation of cell differentiation and proliferation. *Mol Endocrinol* 5: 3–7

Stoner CM, Gudas LJ (1989): Mouse cellular retinoic acid binding protein: cloning, complementary DNA sequence and messenger RNA expression during the retinoic acid induced differentiation of F9 wild type and RA-3-10 mutant teratocarcinoma cells. *Cancer Res* 49: 1497–1504

Strickland S, Mahdavi V (1978): The induction of differentiation in teratocarcinoma stem cells by retinoic acid. *Cell* 15: 333–343

Stunnenberg HG (1993): Mechanisms of transactivation by retinoic acid receptors. *BioEssays* 15: 309–315

Sucov HM, Murakami KK, Evans RM (1990): Characterization of an autoregulated response element in the mouse retinoic acid receptor type β gene. *Proc Natl Acad Sci USA* 87: 5392–5396

Summerbell D, Maden M (1990): Retinoic acid, a developmental signalling molecule. *Trends-Neurosci* 13: 142–147

Sundelin J, Anundi H, Tragardh L, Eriksson UL, Lind P, Ronne P, Peterson PA, Rask L (1985): The primary structure of rat liver cellular retinol-binding protein. *J Biol Chem* 260: 6488–6493

Tabin CJ (1991): Retinoids, homeoboxes, and growth factors: towards molecular models for limb development. *Cell* 66: 199–217

Thaller C, Eichele G (1987): Identification and spatial distribution of retinoids in the developing chick limb bud. *Nature* 327: 625–628

Thaller C, Eichele G (1990): Isolation of 3,4-didehydroretinoic acid, a novel morphogenetic signal in the chick wing bud. *Nature* 345: 815–819

Thomas HE, Stunnenberg HG, Stewart AF (1993): Heterodimerisation of the Drosophila ecdysone receptor with retinoid X receptor and Ultraspiracle. *Nature* 362: 471–475

Thompson KL, Santon JB, Shephard LB, Walton GM, Gill GN (1992). A nuclear protein is required for thyroid hormone receptor binding to an inhibitory half-site in the epidermal growth factor receptor promoter. *Mol Endocrinol* 6: 627–635

Tickle C, Alberts B, Wolpert L, Lee J (1982): Local application of retinoic acid to the limb bud mimics the action of the polarizing region. *Nature* 296: 564–566

Tomic M, Jiang C-K, Epstein HS, Freedberg IM, Samuels HH, Blumenberg M (1990): Nuclear receptors for retinoic acid and thyroid hormone regulate transcription of keratin genes. *Cell Regulation* 1: 965–973

Tran P, Zhang XK, Salbert G, Hermann T, Lehmann JM, Pfahl M (1992): COUP orphan receptors are negative regulators of retinoic acid response pathways. *Mol Cell Biol* 12: 4666–4676

Tsai SY, Carlstedt-Duke J, Weigel NL, Dahlamn K, Gustaffson JA, Tsai M-J, O'Malley BW (1988): Molecular interactions of steroid hormone receptor with its enhancer element: evidence for receptor dimer formation. *Cell* 55: 361–369

Tsukiyama T, Ueda H, Hirose S, Niwa O (1992): Embryonal long terminal repeat-binding protein is a murine homolog of FTZ-F1, a member of the steroid receptor superfamily. *Mol Cell Biol* 12: 1286–1291

Tugwood JD, Issemann I, Anderson RG, Bundell KR, McPheat WL, Green S (1992): The mouse peroxisome proliferator activated receptor recognizes a response element in the 5' flanking sequence of the rat acyl CoA oxidase gene. *EMBO J* 11: 433–439

Ueda H, Sun G-C, Murata T, Hirose S (1992): A novel DNA-binding motifs abuts the Zinc finger domain of insect nuclear hormone receptor FTZ-F1 and mouse embryonal long terminal repeat-binding protein. *Mol Cell Biol* 12: 5667–5672

Umesono K, Murakami KK, Thompson CC, Evans RM (1991): Direct repeats as selective response elements for the thyroid hormone, retinoic acid and vitamin D receptors. *Cell* 65: 1255–1266

Vaessen M-J, Meijers JHC, Bootsma D, Van Kessel AD (1990): The cellular retinoic-acid binding protein is expressed in tissues associated with retinoic-acid-induced malformations. *Development* 110: 371–378

Valcárcel R, Holz H, García Jiménez C, Barettino D, Stunnenberg HG (1994): Retinoid-dependent in vitro transcription mediated by the RXR/RAR hetero-dimer (submitted)

Vivanco Ruiz MdM, Bugge T, Hirschmann P, Stunnenberg HG (1991): Functional characterization of a natural element for retinoic acid. *EMBO J* 10: 3829–3838

Wanek N, Gardiner DM, Muneoka K, Bryant SV (1991): Conversion by retinoic acid of anterior cells into ZPA cells in the chick wing bud. *Nature* 350: 81–83

Wang C, Kelly J, Bowden-Pope DF, Stiles CD (1990): Retinoic acid promotes transcription of the PDGFα-receptor gene. *Mol Cell Biol* 10: 6781–6784

Wang LH, Tsai SY, Cook RG, Beattie WG, Tsai MJ, O'Malley BW (1989): COUP transcription factor is a member of the steroid receptor superfamily *Nature* 340: 163–166

Warrell RJ, De Thé H, Wang ZY, Degos L (1993): Acute promyelocytic leukemia. *New Engl J Med* 329: 177–189

Wei L-N, Mertz JR, Goodman DS, Nguyen-Huu MC (1987): Cellular retinoic acid- and cellular retinol-binding proteins: complementary deoxyribonucleic acid cloning, chromosomal assignment, and tissue expression. *Mol Endocrinol* 1: 526–534

Weinberger C, Thompson CC, Ong ES, Lebo R, Gruol D, Evans RM (1986): The c-*erb* A gene encodes a thyroid hormone receptor. *Nature* 324: 641–646

Weis K, Rambaud S, Lavau C, Jansen J, Carvalho T, Carmo-Fonseca M, Lamond A, Dejean A (1994): Retinoic acid regulates aberrant nuclear localization of PML-RARα in acute promyelocytic leukemia cells. *Cell* 76: 345–356

Widom RL, Rhee M, Karathanasis SK (1992): Repression by ARP-1 sensitizes apolipoprotein AI gene responsiveness to RXR alpha and retinoic acid. *Mol Cell Biol* 12: 3380–3389

Wilson TE, Paulsen RE, Padgett KA, Milbrandt J (1992): Participation of non-Zinc finger residues in DNA binding by two nuclear orphan receptors. *Science* 256: 107–110

Yang-Yen HF, Chambard JC, Sun YL, Smeal T, Schmidt TJ, Drouin J, Karin M (1990): Transcriptional interference between c-jun and the glucocorticoid receptor: mutual inhibition of DNA binding due to direct protein-protein interaction. *Cell* 62: 1205–1215

Yao T-P, Segraves WA, Oro AE, McKeown M, Evans RM (1992): Drosophila ultraspiracle modulates ecdysone receptor function via heterodimer formation. *Cell* 71: 63–72

Ylikomi T, Bocquel MT, Berry M, Gronemeyer H, Chambon P (1992): Cooperation of proto-signals for nuclear accumulation of estrogen and progesterone receptors. *EMBO J* 11: 3681–3694

Yu VC, Delsert C, Andersen B, Holloway JM, Devary OM, Näär AM, Kim SY, Boutin J-M, Glass CK, Rosenfeld MG (1991): RXRβ: a coregulator that enhances binding of retinoic acid, thyroid hormone and vitamin D receptors to their cognate response elements. *Cell* 67: 1251–1266

Yu VC, Näär AM, Rosenfeld MG (1992): Transcriptional regulation by the nuclear receptor superfamily. *Curr Opin Biotechnol* 3: 597–602

Zechel C, Shen X-Q, Chambon P, Gronemeyer H (1944a): Dimerization interfaces formed between the DNA binding domains determine the cooperative binding of RXR/RAR and RXR/TR heterodimers to DR5 and DR4 elements. *EMBO J* 13: 1414–1424

Zechel C, Shen X-Q, Chen J-Y, Chen Z-P, Chambon P, Gronemeyer H (1994b): The dimerization interfaces formed between the DNA binding domains of RXR, RAR and TR determine the binding specificity and polarity of the full-length receptors to direct repeats. *EMBO J* 13: 1425–1433

Zelent A, Krust A, Petkovich M, Kastner P, Chambon P (1989): Cloning of murine α and β retinoic acid receptors and a novel receptor γ predominantly expressed in skin. *Nature* 339: 714–717

Zelent A, Mendelsohn C, Kastner P, Krust A, Garnier JM, Ruffenach F, Leroy P, Chambon P (1991): Differentially expressed isoforms of the mouse retinoic acid receptor β are generated by usage of two promoters and alternative splicing. *EMBO J* 10: 71–81

Zenke M, Muñoz A, Sap J, Vennström B, Beug H (1990): v-*erb* A oncogene activation entails the loss of hormone-dependent regulator activity of c-*erb* A. *Cell* 61: 1035–1049

Zhang X, Lehmann J, Hoffmann B, Dawson MI, Cameron J, Graupner G, Hermann T, Tran P, Pfahl M (1992): Homodimer formation of retinoid X receptor induced by 9-*cis* retinoic acid. *Nature* 358–591

8

Regulation of Nuclear Transport and Activity of the *Drosophila* Morphogen Dorsal

Jacqueline L. Norris and James L. Manley

The *Drosophila* Dorsal (dl) protein is the maternal morphogen responsible for establishing dorsal-ventral (D/V) polarity in the early embryo. The dl protein is localized in a nuclear concentration gradient with the highest levels of dl present in the ventral most region of the embryo and the lowest levels in the dorsal most region (Steward et al, 1988). The dl protein is a member of the rel/NF-κB family of transcription factors and, like other members of this family, is regulated by its subcellular localization (Rushlow and Warrior, 1992). The dl morphogen is uniformly cytoplasmic in the early embryo, where it is held in an inactive state, until shortly after fertilization when it is transported into ventral but not dorsal nuclei (Rushlow et al, 1989; Roth et al, 1989; Steward et al, 1989). Once localized in the nucleus dl acts as both a transcriptional activator and repressor (Ip et al, 1991, 1992a, b; Jiang et al, 1991; Thisse et al, 1991; Pan et al, 1991; Kirov et al, 1994). This review will focus on the signaling pathway that regulates dl subcellular localization. Particular attention will be paid to the events that occur after the signal for dl nuclear uptake is received. Similarities between the regulation of dl and the regulation of NF-κB will also be emphasized.

Generation of the Signal for dl Nuclear Uptake

The activity of at least eleven maternal genes, listed in the Table, is required for the proper localization of dl (Anderson and Nüsslein-Volhard, 1984; Anderson et al, 1985a, b). Ten of these genes, and *dl*, comprise the dorsal group. Recessive mutations in any of these give rise to dorsalized embryos where all cells follow the developmental fate of dorsal cells (Anderson and Nüsslein-Volhard, 1986; Schüpbach and Wiechaus, 1989). The twelfth gene in the Table, *cactus*, is not consid-

INDUCIBLE GENE EXPRESSION, VOLUME 2
P.A. Baeuerle, Editor
© 1995 Birkhäuser Boston

Table. Dorsal-Ventral Patterning Genes*

Gene	Cloned	Protein Homology
pipe	no	
nudel	no	
gastrulation defective	yes	serine protease
snake	yes	serine protease
easter	yes	serine protease
spatzle	yes	secreted processed protein
Toll	yes	transmembrane protein Interleukin 1 receptor
tube	yes	
pelle	yes	protein kinase *raf/mos* family
cactus	yes	IkB, ankryin repeats
dorsal	yes	*rel*/NF-kB family

*Maternal genes required for establishing dorsal-ventral polarity as identified by screens for defects in dorsal-ventral polarity. References for cloned genes are given in the text.

ered a member of the dorsal group, because recessive mutations in *cactus* give rise to ventralized embryos (Schüpbach and Wieschaus, 1989; Roth et al, 1991). However, as discussed in a later section, cactus does play a critical role in dl regulation. Genetic epistasis studies have provided an approximate order for the activity of the dorsal group genes in the signaling pathway that we now know ultimately results in the graded nuclear distribution of the dorsal protein (Anderson et al, 1985a, b). The products of five of these genes, *gastrulation defective (gd), windbeutel (wind), nudel (ndl), pipe (pip),* and *spätzle (spz)*, are required late in oogenesis, while the activity of the other five genes, *easter (ea), snake (snk), Toll (T1), pelle (pll),* and *tube (tub)*, are not required until after fertilization (Anderson, 1987).

The dl protein is uniformally cytoplasmic until cleavage cycle 10, which represents the syncytial blastoderm stage just prior to cellularization, while by cleavage cycle 11, dl protein is found in ventral but not dorsal nuclei (Rushlow et al, 1989; Roth et al, 1989; Steward, 1989). The activities of seven dorsal group genes, *gd, wind, ndl, pip, spz, ea,* and *snk*, are required prior to cleavage cycle 10 to generate the signal that results in dl nuclear uptake (Chasan et al, 1992; Morisato and Anderson, 1994). As discussed further below, the Toll protein is also required at this time to transmit the signal to dl. Two observations suggest that this signal is localized in the perivitelline fluid, the space between the vitelline and plasma membranes, and is recognized by Toll on the ventral side of the embryo (Stein et al, 1991). First, the dorsalized phenotype observed for *nud, pip,* or *wind* mutant embryos can be rescued by injecting perivitelline fluid from Toll⁻ but not Toll⁺ embryos. It may be that the products of the *nud, pip,* and *wind* genes are involved in establishing polarity, as expression of these genes is required in the adult female, probably in the follicle cells surrounding the oocyte. Also, the site of injection of this fluid defines the orientation of the D/V axis. For example, injection of fluid on the dorsal surface results in the development of ventral structures on the dorsal side and dorsal structures on the ventral side of the embryo (Stein et al, 1991). Therefore, the Toll protein recognizes a signal generated on the ventral side of the embryo early in embryogenesis. If Toll is not present, this signal is free in the perivitelline space and can be recognized if isolated and injected into other embryos. However, the presence of Toll results in the signal being bound and, therefore, unavailable for rescue experiments. In addition, production of a limited amount of signal would ensure that it would not diffuse throughout the perivitelline fluid to the dorsal side of the embryo (Stein et al, 1991).

The nature of several of the dorsal group proteins, deduced from gene cloning experiments (see Table), suggests a model for the generation of the signal bound by Toll, which is also largely consistent with additional embryo injection experiments (Stein and Nüsslein-Volhard, 1992). Three genes, *gd* (Govind and Steward, 1991), *snk* (DeLotto and Spierer, 1986), and *ea* (Chasan and Anderson, 1989), encode serine proteases that are likely synthesized as zymogens and secreted into the perivitelline space (Stein and Nüsslein-Volhard, 1992). A proteolytic cascade is then activated, resulting ultimately in processing and activation of a protein that forms the Toll ligand (Stein and Nüsslein-Volhard, 1992). Recent cloning experiments strongly suggest that this is the product of *spz* (Morisato and Anderson, 1994). Isolation of the *spz* gene suggests that it encodes a secreted polypeptide, and experiments with anti-spz antisera suggest that the protein is processed in a manner dependent on *ea* activity and that this processed polypeptide can associate with the plasma membrane in the presence of Toll. An important remaining issue deals with the basis for polarity, an attractive model being that the ea zymogen, and, therefore, spz, is activated only on the ventral side of the embryo (Chasan and Anderson, 1992).

The Receptor: Toll

The importance of the Toll protein in defining the D/V axis was shown first in studies where cytoplasm from wild-type embryos was injected into Toll⁻ embryos, which showed that the site of injection ultimately becomes the area where ventral structures arise (Anderson et al, 1985a). This result was explained when the *Toll* gene was cloned and found to encode a protein with a possible membrane spanning domain (Hashimoto et al, 1988). These findings suggest that Toll is a receptor protein localized in the plasma membrane and so is unable to diffuse throughout the embryo once translated (Hashimoto et al, 1988). The localization of Toll in the plasma membrane was confirmed by both immunofluorescence studies that showed Toll staining uniformly around the plasma membrane of the embryo and by the presence of Toll protein in embryo membrane fractions (Hashimoto et al, 1991). These findings, taken together with the notion, discussed above, of a ventrally localized signal, support the idea that Toll does, in fact, serve as the receptor for this signal and is therefore activated on the ventral side of the embryo. Toll is then responsible for transmitting an extracellular signal into the embryo so that dl is transported into ventral nuclei. Indeed, expression of Toll enhances the nuclear localization of dl in cotransfected *Drosophila* Schneider cells (Norris and Manley, 1992). The Toll protein has a large extracellular domain consisting of 803 amino acids and a small intracellular domain consisting of 269 amino acids (Hashimoto et al, 1988) (Figure 8.1). The extracellular domain is exposed to the perivitelline space and so is accessible to interact with the extracellular signal. Several observations suggest how this domain interacts with its ligand. The extracellular domain contains 15 repeats of a 22–26 amino acid leucine-rich sequence localized in two blocks (Hashimoto et al, 1988). Adjacent to each leucine-rich block is a 20 amino acid sequence containing cysteine residues which can form intramolecular disulfide bonds (Keith and Gay, 1990; Schneider et al, 1991). Leucine-rich repeats (LRRs) are found in other proteins with Toll being most similar to the α chain of human platelet glycoprotein 1b (GP1bα) (Hashimoto et al, 1988). Both proteins contain a large extracellular domain with LRRs, two blocks in Toll and one in GP1bα, and cysteine-containing sequences, two in Toll and one in GP1bα (Lopez et al, 1987; Hashimoto et al, 1988). The LRRs have been proposed to be important for protein-protein interactions. For example, GP1b serves as a receptor for von Willebrand factor and for thrombin, with the binding sites for each mapping to the LRRs and cysteine-containing sequence (Harmon and Jamieson, 1986; Vicente et al, 1990). In addition, the *Drosophila* protein chaoptin, required for photoreceptor cell morphogenesis, contains 41 LRRs (Reinke et al, 1988) and can promote cell adhesion in transfected cells (Krantz and Zipursky, 1990). Like chaoptin, Toll can promote cell aggregation in cultured cells (Keith and Gay, 1990). The LRRs in Toll's extracellular

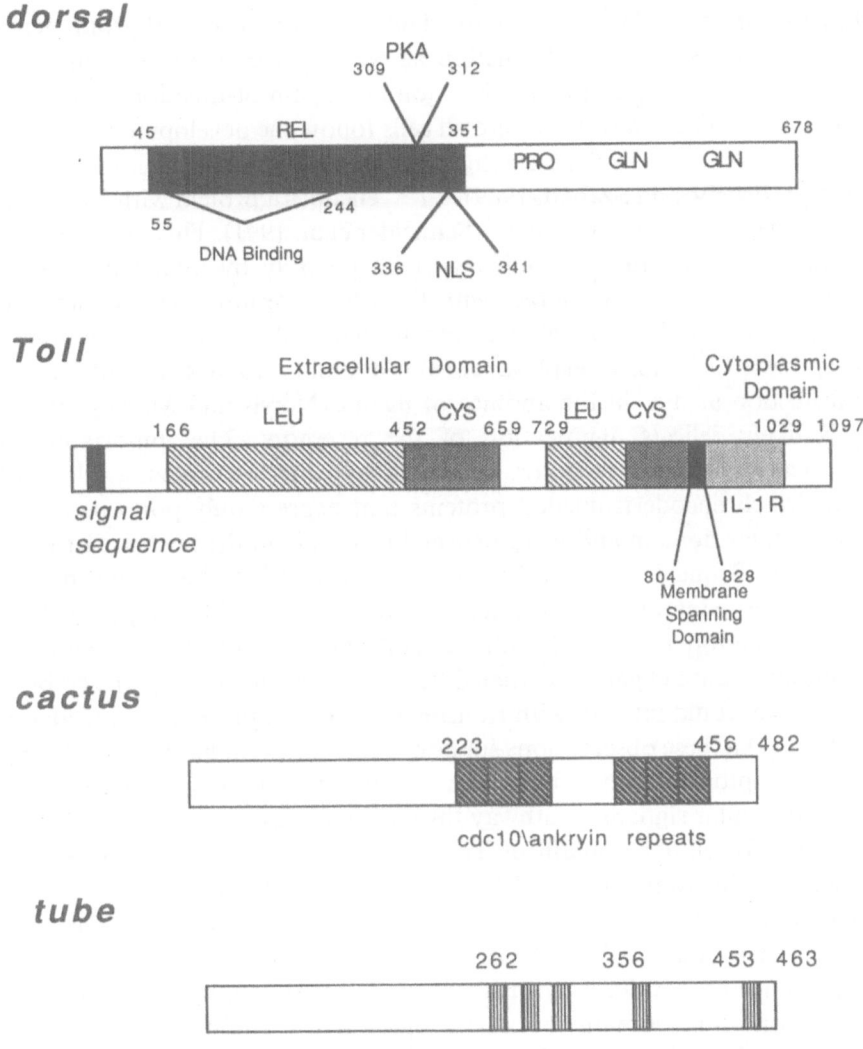

Figure 8.1 Protein structure of four of the D/V patterning genes. The dl protein shares homology with Rel and NF-κB over the amino terminal 300 amino acids (solid area). This region includes DNA binding and dimerization domains as well as a NLS and a conserved PKA site. The C-terminus contains proline (PRO)- and glutamine (GLN)-rich regions. The Toll protein has an extracellular domain, with two blocks of leucine rich repeats (LEU) and cysteine motifs (CYS), and an intracellular domain with a region that is similar to the interleukin-1 receptor cytoplasmic domain (IL-1R). The cactus protein contains six ankryin repeats and the tube protein contains five tube repeats (shaded areas). All numbers refer to amino acid residues.

domain may mediate Toll's interaction with the ventral signal as well as adjacent Toll proteins.

Genetic studies have also shown the importance of the extracellular domain for regulating Toll activity. Toll is one of the dorsal group genes that has dominant gain-of-function as well as recessive loss-of-function alleles (Anderson et al, 1985b). The dominant gain-of-function alleles give rise to ventralized embryos where all cells follow the developmental fate of the ventral-most cells. The strongest of these dominant alleles is $Toll^{10b}$ (Tl^{10b}) (Erdélyi and Szabad, 1989). Tl^{10b} encodes a protein with a tyrosine at residue 781 instead of cysteine (Schneider et al, 1991). This single change results in a constitutively active protein, possibly by mimicking ligand binding. Ligand could interact with the LRRs, resulting in a conformational change in the extracellular domain that could induce aggregation of receptors. Tl^{10b} also displays enhanced ability to activate dl in cell transfection assays, in the absence of ligand (Norris and Manley, 1992). This likely reflects aggregation of the receptors. The importance of interactions between receptors is shown by other ventralizing alleles of *Toll,* which encode truncated proteins that express only portions of the extracellular domain and ventralize embryos only in the presence of wild-type Toll (Schneider et al, 1991). These truncated proteins do not have a membrane spanning domain and so should be localized in the perivitelline space. Therefore they could ventralize embryos by first interacting with the ventrally localized ligand and then diffusing to the dorsal side of the embryo where they could interact with transmembrane Toll proteins (Schneider et al, 1991). All these observations support Toll's extracellular domain serving as the receptor for a ventrally localized ligand that, upon binding, activates an intracellular signaling pathway through Toll's intracellular domain.

The extracellular domain of Toll contains regions of homology with other proteins that provide clues to its function. However, the intracellular domain has no known functional domains, so the mechanism by which it transmits a signal to activate dl is unknown. Nevertheless, the importance of this domain for Toll function is shown by several results. The cytoplasmic domain contains a region that is similar to the intracellular domain of the interleukin-1 receptor (IL-1R) (Schneider et al, 1991). The two share homology over the entire IL-1R intracellular domain and 201 amino acids of the Toll intracellular domain, with Toll containing an additional 68 C-terminal amino acids (Schneider et al, 1991). Three recessive loss-of-function *Toll* alleles have point mutations within this region of similarity (Schneider et al, 1991), and mutations in IL-1R residues that are conserved between the two receptors result in a loss of IL-1R function (Heguy et al, 1992). In addition, deletion into this region of similarity also results in a loss of Toll's ability to enhance dl nuclear localization in cotransfected Schneider cells (Norris and Manley, 1992, 1994). Interestingly, expression of only the intracellular domain of Toll can still enhance the nuclear localization of dl in cotransfected cells

(Norris and Manley, 1994). Therefore, while the extracellular domain is required to regulate activity, the intracellular domain is required for activity and is, in fact, active by itself.

IL1 can activate NF-κB in cells expressing IL-1R (Shirakawa et al, 1989). The similarities between dl and NF-κB and Toll and IL-1R suggest that the two receptors use similar pathways for transmitting their signals. IL1 stimulates intracellular cAMP levels, suggesting that IL-1R may stimulate cAMP-dependent protein kinase (Shirakawa et al, 1988), although there is data arguing that the levels of cAMP generated are not sufficient to induce NF-κB (Bomsztyk et al, 1990). The next section discusses the possibility that Toll may utilize a second messenger system involving protein kinase A.

The Signaling Pathway: Protein Kinase A

Members of the rel/NF-κB family share homology over approximately 300 N-terminal amino acids. This region, the rel homology domain (RHD), contains DNA binding and dimerization domains as well as a nuclear localization sequence (NLS) and a protein kinase A (PKA) phosphorylation site (Steward, 1987; Ghosh et al, 1990; Kieran et al, 1990; Gilmore, 1991; Ip et al, 1991; Isoda et al, 1992; Norris and Manley, 1992), (see Figure 8.1). Therefore, the RHD contains the elements necessary for regulating the activity and subcellular localization of these proteins. Phosphorylation appears to be important for the nuclear uptake of several proteins. For example, both SV40 large T antigen and p53 have phosphorylation sites adjacent to their NLSs and, for SV40 large T antigen, phosphorylation of a serine residue increases the rate of nuclear localization of the protein (Rihs and Peters, 1989; Addison and Sturzbecher, 1990). The localization of dl's PKA site, adjacent to its NLS, suggests that phosphorylation by PKA could be involved in regulating the subcellular localization of the dl protein.

Several observations suggest that PKA activity is important for dl activity. First, the ability of a constitutively nuclear dl mutant, lacking approximately 100 C-terminal residues, to activate transcription is enhanced by expression of Toll in cotransfected cells (Norris and Manley, 1992). This and other data suggest that Toll can enhance the activity of dl, not only by inducing nuclear localization, but also by bringing about a posttranslational modification of dl. Second, expression of the catalytic subunit of PKA enhances the nuclear localization and transcriptional activity of dl just as does expression of Toll (Norris and Manley, 1992). These results suggest that Toll utilizes activation of PKA to transmit the signal for nuclear uptake to dl. Also supporting this are experiments in which the dl expression vector is cotransfected with plasmids encoding Toll and a peptide inhibitor of PKA, PKi, which inhibits activation of dl by Toll (Norris and Manley, 1992).

Both Toll and PKA appear to exert their effects through the conserved PKA consensus site in dl. When this site is mutated, dl's response to Toll and PKA is affected. First, disruption of the site by changing the serine (S) residue at position 312 to glutamine both reduces dl's ability to activate transcription and results in a primarily cytoplasmic protein, in the presence or absence of either Toll or PKA (Norris and Manley, 1992). In addition, the placement of a negative charge at this site, by changing the S residue to aspartic acid, also affects dl subcellular localization. The mutant protein is localized both in the nucleus and cytoplasm when expressed by itself, showing that the need for the signaling pathway is at least partially overcome (Norris and Manley, 1992). Therefore, dl's PKA site is required for the protein to respond to Toll in cell transfection assays. The role of this site in dl's function in embryos has not yet been determined.

Experiments with v-Rel and c-Rel have shown that the conserved PKA site is important for regulating subcellular localization and activity of these proteins. v-*rel* is an oncogene that encodes a truncated and mutated form of the avain proto-oncogene c-*rel* (Capobianco et al, 1990; Hannink and Temin, 1989; Wilhelmsen et al, 1984). c-Rel is localized primarily in the cytoplasm of cells while v-Rel is primarily nuclear (Capobianco et al, 1990; Simek and Rice, 1988). v-Rel can repress transcription from promoters containing NF-κB binding sites, probably due to the loss of most of the activation domain from c-Rel (Richardson and Gilmore, 1991; Inoue et al, 1991). When the structure of the PKA site is disrupted by insertion of 2 amino acids, v-Rel no longer represses transcription and c-Rel is localized in the nucleus (Mosialos et al, 1991). When a negative charge is placed at this site by changing the S residue to aspartic or glutamic acid, the repression of v-Rel is greatly reduced and the nuclear concentration of c-Rel is enhanced (Mosialos et al, 1991). These alterations in the PKA site have been shown to affect both the DNA binding and dimerization properties of c-Rel and v-Rel. Disruption of the structure of the PKA site or placement of a negative charge at this site abolishes c-Rel binding to DNA containing κB sites and reduces or eliminates the ability of the protein to activate transcription (Mosialos and Gilmore, 1993). These changes also have been shown to inhibit or abolish the ability of c-rel and v-rel to form homodimers (Mosialos and Gilmore, 1993). Therefore, the PKA site is important for DNA binding, homodimer formation, transcriptional activity, and nuclear localization of the Rel proteins.

Studies of phosphorylation of dl in embryos have shown that dl is a phosphoprotein and that its phosphorylation state is developmentally regulated (Whalen and Steward, 1993; Gillespie and Wasserman, 1994). When dl from 0–3 hr embryos is separated on SDS-PAGE gels, several bands can be resolved with the slowest migrating forms appearing at the time when dl nuclear uptake begins (Whalen and Steward, 1993; Gillespie and Wasserman, 1994). The appearance of these modified forms of dl, which have been shown to reflect phosphorylation, is dependent on the signaling pathway. For example, dl protein isolated from *gd* mutant embryos lacks the slowest migrating forms, while dl isolated from Tl^{10b} embryos consists of predominantly these forms (Gillespie and Wasserman, 1994). Although the sites of phosphorylation have not been determined, these studies suggest that phosphorylation of dl is important for regulating dl in the embryo.

One of the dorsal group genes has been found to encode a protein kinase. The product of the *pelle* gene encodes a protein with homology to the raf/mos family of kinases (Shelton and Wasserman, 1993). The slowest migrating forms of dl are absent in *pelle* mutant embryos, suggesting that pelle is responsible, directly or indirectly, for some dl phosphorylation (Gillespie and Wasserman, 1994). However, how the pelle-induced modifications function in the dl signaling pathway is unknown. No genetic evidence exists implicating PKA in dl activation. However, since PKA would be required in many pathways, and thus could have eluded detection in genetic screens for D/V patterning genes, this is not surprising.

The Inhibitor: Cactus

The cactus protein acts downstream of all the dorsal group proteins except dl, with its activity required at the syncytial blastoderm stage at the time the dl nuclear gradient is established (Roth et al, 1991). Two observations show that, while the other D/V patterning genes positively regulate dl nuclear transport, the *cactus* gene negatively regulates dl. First, dl is found in dorsal as well as ventral nuclei in embryos containing loss-of-function *cactus* alleles (Roth et al, 1989; Steward, 1989). Second, gain-of-function *cactus* alleles result in dorsalized embryos due to an exclusion of dl from ventral nuclei (Roth et al, 1991). Therefore, *cactus* appears to inhibit dl nuclear transport, perhaps in a manner similar to the NF-κB inhibitor IκB.

The IκB protein was first identified as an activity that could inhibit NF-κB DNA binding (Baeuerle and Baltimore, 1988a). IκB was not found associated with nuclear NF-κB suggesting it might be responsible for retaining NF-κB in the cytoplasm of unstimulated cells (Baeuerle and Baltimore, 1988b). Several IκBs have been identified and cloned, with each containing five to seven ankryin repeats (Beg and Baldwin, 1993, Gilmore and Morin, 1993). These repeats, first identified in the yeast cell-cycle protein cdc10 and SW16 (Breeden and Nasmyth, 1987), have been shown to be important for interactions between IκB and NF-κB subunits. Mutations within the ankryin repeats of one form of IκB, IκBα, abolished IκBα's ability to interact with NF-κB (Inoue et al, 1992). The interaction between IκB and NF-κB also requires sequences within the RHD of NF-κB, specifically the region around the NLS (Beg et al, 1992; Ganchi et al, 1992; Zable et al, 1993; Nolan et al 1993; Wulczyn et al, 1992; Hatada et al, 1993). This suggests that IκB could retain NF-κB in the cytoplasm by masking NF-κB's NLS, with activation of NF-κB requiring a dissociation of the IκB/NF-κB complex. Phosphorylation of IκB has been proposed to be important for this dissociation (Shirakawa and Mizel, 1989; Ghosh and Baltimore, 1990; Kerr et al, 1991; Link et al, 1992; Beg et al, 1993; Cordle et al, 1993). However, the exact role that phosphorylation plays in activating NF-κB is still not understood.

The cloning of the *cactus* gene has shown that it encodes an IκB-like protein with six ankryin repeats (Geisler et al, 1992; Kidd, 1992), (see Figure 8.1). In fact, cactus appears to interact with dl much like IκB interacts with NF-κB. Like IκB, the cactus protein inhibits dl DNA binding in vitro (Geisler et al, 1992; Kidd, 1992). Deletion of the cactus ankryin repeats abolishes cactus's ability to inhibit dl DNA binding (Geisler et al, 1992; Kidd, 1992) while the cactus ankryin repeats alone can inhibit DNA binding (Kidd, 1992). Deletion analysis also showed that a region of dl's RHD adjacent to the NLS is required for the dl/cactus interaction, suggesting that cactus might retain dl in the cytoplasm by masking the NLS (Kidd, 1992). However, unlike a IκB, phosphorylation of cactus does not appear to be responsible for disrupting the dl/cactus complex, since dl can interact with the phosphorylated form of cactus (Kidd, 1992).

Expression of cactus is required both maternally and zygotically (Kidd, 1992). The zygotically expressed form is expressed in Schneider cells (Kidd, 1994) and dl appears able to interact with this form. When low amounts of dl expression vector are used, the dl protein is localized primarily in the cytoplasm, but when higher amounts of dl are transfected, an increasing amount of the protein is found in the nucleus (Rushlow et al, 1989; Norris and Manley, 1992). These results can be explained as a dl/cactus interaction. There is sufficient endogenous cactus to interact with low levels of dl protein, retaining it in the cytoplasm, but when the level of dl protein exceeds the level of endogenous cactus, the unbound dl is free to move into the nucleus. When Toll or PKA are cotransfected with dl, the dl/cactus complex is disrupted so that even low levels of dl can be localized in the nucleus.

A Model for dl Nuclear Transport

The model presented in Figure 8.2 summarizes what is currently known about the regulation of dl nuclear transport. Prior to cleavage cycle 10, when dl begins to move into ventral nuclei, dl is complexed in the cytoplasm with cactus, and the ventrally localized signal that will result in dl nuclear transport is being generated. The nud, pip, wind, gd, and snk proteins interact in a proteolytic processing cascade leading to the activation of the easter zymogen on the ventral side of the embryo (Chasen and Anderson, 1992). The activation of easter results in a proteolytic processing of spätzle to generate an active Toll ligand (Morisato and Anderson, 1994). The processed spätzle protein interacts immediately with the Toll extracellular domain, preventing diffusion of the ligand to the dorsal side of the embryo. Once Toll binds spätzle the signal is transmitted into the embryo through Toll's intracellular domain. How this transmission occurs is unknown, although the 201 amino acids of Toll's intracellular domain that are similar to the IL-1R receptor intracellular domain are required for Toll activity. There is evidence that the 68 amino acids found in Toll but not IL-1R can function to negatively regulate Toll activity, since deletion of these amino acids results in a protein that is more active than wild-type in cotransfection assays (Norris and Manley, 1994). Thus it could be that the Toll intracytoplasmic domain consists of an active sub-domain and a regulatory sub-domain. The regulatory sub-domain might physically mask the active sub-domain, and binding of spätzle could lead to a conformational change that would unmask the active sub-domain.

The intracellular signaling pathway activated by Toll results in the activation of PKA, at least in cultured cells, and the phosphorylation of dl, apparently at multiple sites. Phosphorylation of dl could disrupt dl's ability to interact with cactus, thereby unmasking dl's NLS. Phosphorylation of cactus might also be involved in disrupting this interaction since this seems to be the predominant way of disrupting the IκB/NF-κB complex. However, since dl can interact with phosphorylated cactus, modification of dl may be more important for disrupting the complex.

Genetic studies have identified two genes, *tube* and *pelle*, that, in addition to cactus, act downstream of Toll and upstream of or in parallel

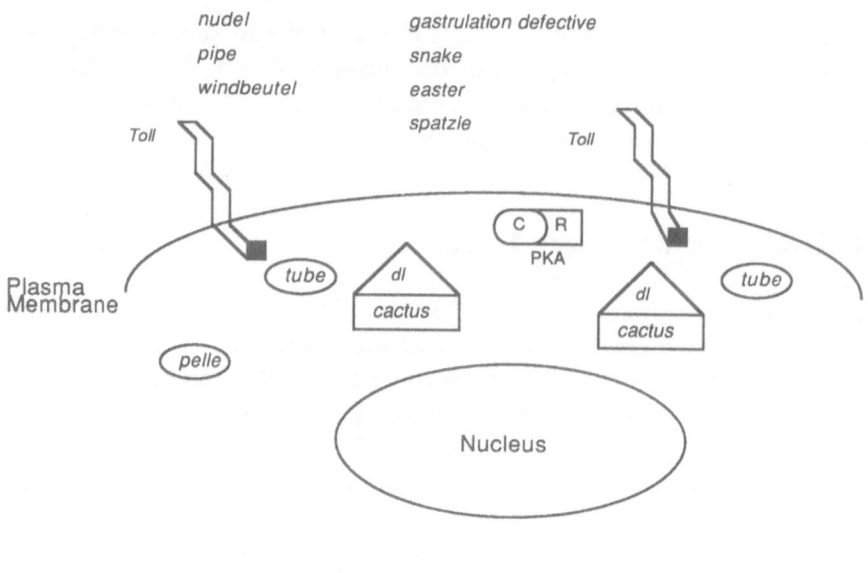

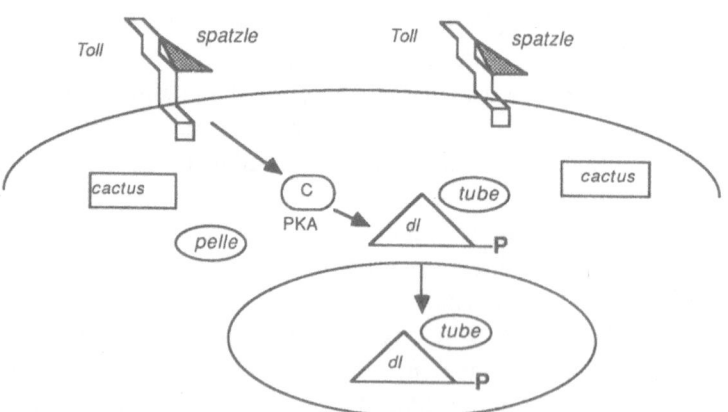

Figure 8.2 A model for dl nuclear transport. The activity of six genes is required to generate the Toll ligand spätzle. Prior to ligand binding, Toll is inactive and dl is held in the cytoplasm through association with cactus. After ligand binding, Toll is activated and results in the phosphorylation of dl and its release form cactus. The tube protein accompanies dl into the nucleus where the expression of zygotic genes is activated or repressed.

to dl (See Table). The functions of the protein products of these genes in the signaling pathway are not yet clearly understood. As mentioned above, phosphorylation of dl induced by the pelle protein kinase appears to occur after the disruption of the dl/cactus complex (Gillespie and Wasserman, 1994). This phosphorylation could nonetheless be important for dl nuclear transport or for dl activity once in the nucleus, or for both.

The tube protein shares no homology with any known proteins. The tube C-terminus contains five copies of an eight amino acid repeat sequence, the tube repeats, (Letsou et al, 1991) (see Figure 8.1), however the C-terminus is not required to rescue tube null embryos in RNA injection experiments (Letsou et al, 1993). Cotransfections in Schneider cells have recently show that tube can activate dl (Norris and Manley, 1994). This activation does not result from an increase in dl nuclear localization, but from colocalization of tube with dl. Specifically, when dl is cytoplasmic, or not present, tube is cytoplasmic, but when dl is nuclear so is tube. Therefore, tube may function after disruption of the dl/cactus complex by directly interacting with dl, chaperoning it to the nucleus, and enhancing dl's transcriptional activity once in the nucleus.

Transcriptional Activity of dl

Establishment of the dl concentration gradient is crucial for proper zygotic expression of a number of D/V patterning genes. The dl nuclear gradient is responsible for initiating the differentiation of mesoderm, neuroectoderm, and dorsal ectoderm. The highest levels of dl are present in ventral nuclei where dl initiates expression of *twist (twi)* and *snail (sna)*, which are themselves transcription factors responsible for meso- derm differentiation (Simpson, 1983; Boulay et al, 1987; Thisse et al, 1987, 1988, 1991; Jiang et al, 1991; Pan et al, 1991) (see Figure 8.3). The levels of nuclear dl decrease progressively in dorsal regions allowing expression of *zerknüllt (zen)*, *decapentaplegic (dpp)*, and *tolloid (tld)*, which are repressed by dl in ventral regions (Rushlow et al, 1987; St. Johnston and Gelbart, 1987; Ip et al, 1991; Shimell et al, 1991; Huang et al, 1993; Kirov et al, 1994) (see Figure 8.3). Intermediate levels of nuclear dl are found in ventrolateral regions where *rhomboid (rho)* expression is activated by dl to give rise to the neuroectoderm (Ip et al, 1992) (see Figure 8.3). Therefore, expression of all these genes is sensitive to the exact concentration of dl, with higher levels activating *twi* and *sna*, lower levels activating *rho*, and both high and low levels repressing *zen, dpp,* and *tld*. Understanding how the dl concentration gradient results in both repression and activation of transcription requires an analysis both of the promoters that respond to dl and other proteins with which dl may interact.

The dl protein is a sequence specific DNA-binding protein that recognizes sequences similar to those bound by NF-κB (Ip et al, 1991, 1992a, b; Thisse et al, 1991; Jiang et al, 1991; Pan et al, 1991; Huang et al, 1993; Kirov et al, 1994). Like NF-κB, the RHD is required for DNA

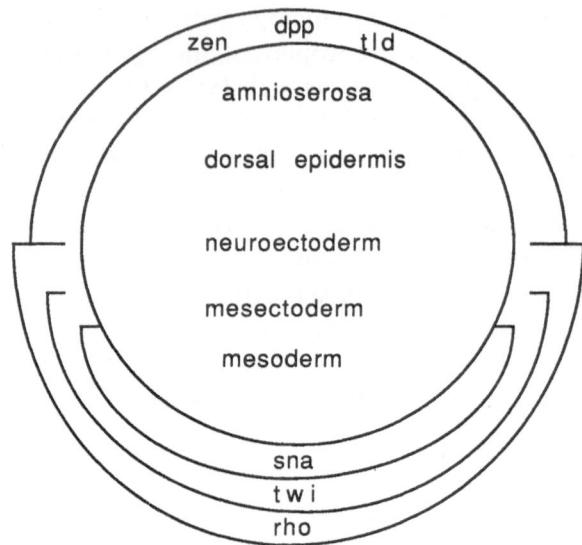

Figure 3 Cross section of the embryo indicating the regions of expression of known D/V patterning zygotic genes regulated by dl and the tissues that are derived from these regions. The bottom is the ventral surface where the concentration of dl is the highest while the top is the dorsal surface where nuclear dl is absent.

binding since this region alone can bind DNA and deletions in it abolish binding (Ip et al, 1991). However, the DNA-binding activities of NF-κB and dl are different since oligos containing high affinity NF-κB sites can not compete for dl binding to high affinity dl binding sites (Ip et al, 1991). There is considerable sequence variation among identified dl binding sites, suggesting that the context of the promoter is important for regulation of expression by dl.

Differentiation of dorsal tissues requires the expression of *zen*, *dpp*, and *tld* in the dorsal-most regions of the embryo. Expression of all three of these genes is dependent on the absence of dl, since their expression is repressed in ventral and lateral regions of the embryo. Therefore, all three must be responsive to high as well as low levels of dl. Analysis of each of these promoters by P-element transformation experiments has shown that each contains a region required for repression, the ventral repression element (VRE). Deletion of the VRE leads to expression in ventral regions while placement of the VRE upstream of a heterologous promoter results in repression (Doyle et al, 1989; Ip et al, 1991; Huang et al, 1993; Kirov et al, 1994). The *zen* VRE contains four dl binding sites (Ip et al, 1991); the *dpp* VRE has multiple dl binding sites (Huang et al, 1993); and the *tld* VRE contains two dl binding sites (Kirov et al, 1994). The presence of dl binding sites is the only obvious similarity between the

VRE's, although there is evidence that there are binding sites, adjacent to the dl sites, that are also required for repression.

When dl binding sites from the *zen* promoter are placed upstream of a minimal heat shock promoter, they can mediate activation (Jiang et al, 1992; Pan and Courey, 1992). This suggests that dl is by default an activator of transcription and that repression can occur only when the dl binding sites are in the proper promoter context. This is consistent with the results of transfection experiments (Rushlow et al, 1989; Norris and Manley, 1992), in which dl has only been observed to function as an activator. Therefore, dl appears to require the presence of other factors to repress transcription. Evidence for the existence of such a factor, called a corepressor, comes from analysis of the *zen* and *dpp* promoters. Identification of a minimal element required for repression of *zen* has shown that sequences adjacent to dl binding sites are important for repression (Kirov et al, 1993; Jiang et al, 1993). A minimal element, that can repress transcription from a heterologous promoter when triplicated, contains one dl binding site and an adjacent short A + T − rich sequence (Kirov et al, 1993). Mutations in the A + T − rich sequence result in expression from this promoter in ventral regions, again suggesting that dl is by default an activator (Kirov et al, 1993; Jiang et al, 1993). In addition, analysis of the *dpp* VRE reveals that a fragment containing multiple dl binding sites can activate expression of a heterologous promoter in ventral regions but addition of approximately 200 base pairs restores ventral repression, suggesting that a binding site(s) for another protein(s) is required for dl-mediated repression (Huang et al, 1993). The existence of a corepressor could also explain the results from the transfection studies, in which dl was found to activate, not repress, expression from the *zen* promoter (Rushlow et al, 1989; Norris and Manley, 1992). Since dl is by default an activator, the absence of the putative corepressor in Schneider cells could result in the observed activation.

Activation by dl is also dependent on the concentration of dl, as well as the presence of other proteins. Analysis of the *twi* promoter has shown that it contains multiple potential dl binding sites (Thisse et al, 1991; Jiang et al, 1991; Pan et al, 1991). Two regions, each containing two dl binding sites, appear to be most important for proper *twi* expression (Jiang et al, 1991). These sites differ from those found in the *zen* and *tld* VREs in that their affinity for dl is much lower (Jiang et al, 1991; Thisse et al, 1991; Pan et al, 1991). The presence of low affinity sites in *twi* restricts its expression to the ventral-most region of the embryo, where the dl concentration is highest, while the high affinity sites in *zen* and *tld* allow repression in lateral as well as ventral regions. However, these differences in activity are not due simply to sequence variation because *twi* dl binding sites can promote repression when placed in the *zen* VRE (Jiang et al, 1992). This suggests that the difference between activation and repression by dl is due to dl's interaction with other proteins on a particular promoter.

One promoter for which dl interactions have been analyzed is *sna* (Ip et al, 1992a). The *sna* promoter contains ten dl binding sites. Deletion of, or mutations in, several of these sites leads to reduction or loss of *sna* expression suggesting that dl directly controls expression from this promoter. Interestingly, the *sna* promoter also contains twi binding sites and these sites, like the dl sites, are required for optimal *sna* expression. The *sna* expression pattern is responsible for defining the boundary between the mesoderm and the neuroectoderm (see below). The dl-twi interaction is important for defining this sharp boundary by allowing high and intermediate levels of dl to promote the same level of *sna* expression.

Protein-protein interactions are also important for dl-mediated activation of *rho* expression in the presumptive neuroectoderm (Ip et al, 1992b). Low levels of nuclear dl extend to the boundary between the ventral neuroectoderm and the dorsal ectoderm (Rushlow et al, 1989; Roth et al, 1989; Steward, 1989). The *rho* promoter contains four dl binding sites, four sna binding sites, and two twi binding sites. Mutations in the dl sites abolish *rho* expression and mutations in the twi sites reduce it. In contrast, mutations within the sna sites result in expression of *rho* in ventral regions, where the gene is normally inactive, showing that sna functions as a repressor of *rho* expression in ventral regions. Therefore, interactions between dl and twi are responsible for activation of *rho*, in the absence of sna, in the presumptive neuroectoderm, just as they are for the expression of *sna*. *Rho* expression is thus limited to regions where dl levels are sufficient to bind its high affinity sites but not the lower affinity *sna* sites.

The twi protein contains a basic-helix-loop-helix (bHLH) DNA binding domain. Interactions between dl and proteins belonging to the bHLH family provide a mechanism for the control of expression of genes in areas where levels of nuclear dl are low (Jiang and Levine, 1993). Studies of the *rho* and *sna* promoters, which contain both dl and bHLH protein binding sites (E boxes), have established a general model for activation of target gene expression by dl. In this model, expression in the presumptive mesoderm only requires low affinity dl binding sites, since the level of nuclear dl protein is high. In the mesectoderm, where there are intermediate levels of dl, either high affinity dl binding sites alone are required or low affinity plus E boxes must be present. Finally, expression in the presumptive neuroectoderm, where nuclear dl levels are low, requires both high affinity dl binding sites and E boxes. It is likely that a key feature of dl bHLH interactions involves cooperative DNA binding (Jiang and Levine, 1993).

In summary, three factors are important for activation of transcription by dl. First, the concentration of nuclear dl establishes different regions of target gene expression. Second, binding site affinities define the limits of gene expression. Finally, protein-protein interactions refine the boundaries of expression. Transcriptional repression by dl, which is currently less well understood, appears to require yet unidentified corepressors.

Concluding Remarks

The cloning of many of the maternal genes required for establishing D/V polarity has allowed the study of how dl subcellular localization is regulated. Four of these gene products, Toll, tube, pelle, and cactus, participate in an intracellular signaling pathway that activates dl nuclear transport. Analysis of these proteins in embryos and cultured cells has begun to define the nature of the signaling pathway that they participate in. Further analysis is required to define the precise protein-protein interactions and protein modifications resulting in dl nuclear localization. Once localized in the nucleus, dl is responsible for both activating and repressing transcription of zygotic D/V patterning genes. Analysis of promoters regulated by dl has shown that multiple protein-protein interactions are required for both activation and repression of transcription. Further analysis is required to identify the exact protein-protein interactions required for the regulation of zygotic gene expression. The study of the *Drosophila* dl protein will help define both regulation of subcellular localization and transcriptional activity of proteins in other systems, most notably other members of the rel/NF-κB family.

References

Addison C, Jenkins JR, Sturzbecher H-W (1990): The p53 nuclear localization signal is structurally linked to a p34 ^{cdc2} motif. *Oncogene* 5: 423–426.

Anderson KV (1987): Dorsal-ventral embryonic pattern genes of *Drosophila*. *Trends Genet* 3: 91–97

Anderson KV, Nüsslein-Volhard C (1984): Genetic analysis of dorsal-ventral embryonic pattern in *Drosophila*. In: *Pattern Formation: A Primer in Developmental Biology*, Malacinski GM, Bryant SV, eds. New York: Macmillan

Anderson KV, Nüsslein-Volhard C (1986): Dorsal-group genes of *Drosophila*. In: *Gametogenesis and the Early Embryo*, Gall J, ed. New York: Alan R. Liss

Anderson KV, Bokla L, Nüsslein-Volhard C (1985a): Establishment of dorsal-ventral polarity in the *Drosophila* embryo: the induction of polarity by the *Toll* gene product. *Cell* 42: 791–798

Anderson KV, Jürgens G, Nüsslein-Volhard C (1985b): Establishment of dorsal-ventral polarity in the *Drosophila* embryo: genetic studies on the role of the *Toll* gene product. *Cell* 42: 779–789

Baeuerle PA, Baltimore D (1988a): Activation of DNA-binding activity in an apparently cytoplasmic precursor of the NF-κB transcription factor. *Cell* 53: 211-217

Baeuerle PA, Baltimore D (1986): IκB: a specific inhibitor of the NF-κB transcription factor. *Science* 242: 540–546

Beg AA, Baldwin AS (1993): The NF-κB proteins: multifunctional regulators of Rel/ NF-κB transcription factors. *Genes Dev* 7: 2064–2070

Beg AA, Finco TS, Nantermet PV, Baldwin AS, (1993): Tumor necrosis factor and interleukin-1 lead to phosphorylation and loss of IκBα: A mechanism for NF-κB activation. *Mol Cell Biol* 13: 3301–3310

Beg AA, Ruben SM, Scheinman RI, Haskill S, Rosen CA, Baldwin AS (1992): IκB interacts with the nuclear localization sequences of the subunits of NF-κB: a mechanism for cytoplasmic retention. *Genes Dev* 6: 1899–1913

Bomsztyk K, Toivola B, Emery DW, Rooney JW, Dower SK, Rachie NA, Sibley CH (1990): Role of cAMP in interleukin-1 induced κ light chain gene expression in murine B cell line. *J Biol Chem* 265: 9413–9417

Boulay JL, Dennefeld C, Alberga A (1987): The *Drosophila* developmental gene *snail* encodes a protein with nucleic acid binding fingers. *Nature* 330: 395–398.

Breeden L, Nasmyth K (1987): Similarity between cell-cycle genes of budding yeast and the *Notch* gene of *Drosophila*. *Nature* 329: 651–654

Capobianco AJ, Simmons DL, Gilmore TD (1990): Cloning and expression of a chicken c-*vel* cDNA: unlike p59[v-rel] p68 [c-rel] is a cytoplasmic protein in chicken embryo fibroblasts. *Oncogene* 5: 257–265

Chasan R, Anderson KV (1989): The role of easter, an apparent serine protease, in organizing the dorsal-ventral pattern of the *Drosophila* embryo. *Cell* 56: 391–400

Chasan R, Jin Y, Anderson KV (1992): Activation of the easter zymogen is regulated by five other genes to define dorsal-ventral polarity in the *Drosophila* embryo. *Development* 115: 607–616

Cordle SR, Donald R, Read MA, Hawiger J (1993): Lipopolysaccharide induces phosphorylation of MAD3 and activation of c-Rel and related NF-κB proteins in human monocytic THP-1 cells. *J Biol Chem* 268: 11803–11810

DeLotto R, Spierer P (1986): A gene required for the specification of dorsal-ventral pattern in *Drosophila* appears to encode a serine protease. *Nature* 323: 688–692

Doyle HJ, Kraut R, Levine M (1989): Spatial regulation of *zerknüllt*: a dorsal-ventral patterning gene in *Drosophila*. *Genes Dev* 3: 1518–1533

Erdélyi M, Szabad J (1989): Isolation and characterization of dominant female sterile mutations of *Drosophila melanogaster*. I. Mutations on the third chromosome. *Genetics* 122: 111–127

Ganchi PA, Sun S-C, Green WC, Ballard DW (1992): IκB/MAD-3 masks the nuclear localization signal of NF-κB p65 and requires the transactivation domain to inhibit NF-κB p65 DNA binding. *Mol Biol Cell* 3: 1339–1352

Geisler R, Bergmann A, Hiromi Y, Nüsslein-Volhard C (1992): *cactus*, a gene involved in dorsoventral pattern formation of *Drosophila*, is related to the IκB gene family of vertebrates. *Cell* 71: 613–621

Ghosh S, Baltimore D (1990): Activation in vitro of NF-κB by phosphorylation of its inhibitor Iκb. *Nature* 344: 678–682

Ghosh S, Gifford AM, Riviere LR, Tempst P, Nolan GP, Baltimore D (1990):

Cloning of the p50 DNA binding subunit of NF-κB: homology to *rel* and *dorsal*. *Cell* 62: 1019–1029

Gillespie SKH, Wasserman SA (1994): *Dorsal*, a *Drosophila* rel-like protein, is phosphorylated upon activation of the transmembrane protein Toll. *Mol Cell Biol* 14: 3559–3568

Gilmore TD (1991): Malignant transformation by mutant Rel proteins. *Trends Genet* 7: 318–322

Gilmore TD, Morin PJ (1993): The IκB proteins: members of a multifunctional family. *Trends Genet* 9: 427–433.

Govind S, Steward R (1991): Dorsoventral pattern formation in *Drosophila*: signal transduction and nuclear targeting. *Trends Genet* 7: 119–125

Hannink M, Temlin HM (1989): Transactivation of gene expression by nuclear and cytoplasmic Rel proteins. *Mol Cell Biol* 9: 4323–4336

Harmon JT, Jamieson GA (1986): The glycocalicin portion of platelet glycoprotein 1b expresses both high and moderate affinity receptor sites for thrombin. *J Biol Chem* 261: 13224–13229

Hashimoto C, Gerttula S, Anderson KV (1991): Plasma membrane localization of the Toll protein in the syncytial *Drosophila* embryo: importance of transmembrane signaling for dorsal-ventral pattern formation. *Development* 111: 1021–1028

Hashimoto C, Hudson KL, Anderson KV (1988): The *Toll* gene of *Drosophila*, required for dorsal-ventral embryonic polarity, appears to encode a transmembrane protein. *Cell* 52: 269–279

Hatada EN, Nieters A, Wulczyn FG, Naumann M, Meyer R, Nucifora G, McKeithan TW, Scheidereit C (1992): The ankryin repeat domains of the NF-κB precursor p105 and the proto-oncogene bcl-3 act as specific inhibitors of NF-κB DNA binding. *Proc Natl Acad Sci USA* 89: 2489–2493

Heguy A, Baldari CT, Macchia G, Telford JL, Melli M (1992): Amino acids conserved in interleukin-1 receptors (IL-1Rs) and the *Drosophila* Toll protein are essential for IL-1Rs) and the Drosophila Toll protein are essential for IL-1R signal transduction. *J Biol Chem* 267: 2605–2609

Huang JD, Schwyter DH, Shirokawa JM, Courey AJ (1993): The interplay between multiple enhance and silencer elements defines the pattern of *decapentaplegic* expression. *Genes Dev* 7: 694–704

Ip YT, Kraut R, Levine M, Rushlow C (1991): The dorsal morphogen is a sequence-specific DNA-binding protein that interacts with a long-range repression element in *Drosophila*. *Cell* 64: 429–446

Ip YT, Park RE, Kosman D, Yazdanbakhsh K, Levine M (1992a): Dorsal-twist interactions establish *snail* expression in the presumptive mesoderm of the *Drosophila* embryo. *Genes Dev* 6: 1518–1530

Ip YT, Park RE, Kosman D, Bier E, Levine M (1992b): The dorsal gradient morphogen regulates stripes of *rhomboid* expression in the presumptive neuroectoderm of the *Drosophila* embryo. *Genes Dev* 6: 1728–1739

Inoue J-I, Kerr LD, Ransone LJ, Bengal E, Hunter T, Verma IM (1991): c-Rel activates but v-Rel suppresses transcription from κB sites. *Proc Natl Acad Sci USA* 88: 3715–3719

Inoue J-I, Kerr LD, Rashid D, Davis N, Bose HR Jr, Verma IM (1992): Direct association of pp40/IκBβ with rel/NF-κB transcription factors: role of ankryin repeats in the inhibition of DNA binding activity. *Proc Natl Acad Sci USA* 89: 4333–4337

Isoda K, Roth S, Nüsslein-Volhard C (1992): The functional domains of the *Drosophila* morphogen dorsal: evidence from the analysis of mutants. *Genes Dev* 6: 619–630

Jiang J, Levine M (1993): Binding affinities and cooperative interactions with bHLH activators delimit threshold responses to the dorsal gradient morphogen. *Cell* 72: 741–752

Jiang J, Cai H, Zhou Q, Levine M (1993): Conversion of a dorsal-dependent silencer into an enhancer: evidence for dorsal co-repressors. *EMBO J* 12: 3201–3210

Jiang J, Ip YT, Kosman D, Levine M (1991): The dorsal morphogen gradient regulates the mesoderm determinant *twist* in early *Drosophilia* embryos. *Genes Dev* 5: 1881–1891

Jiang J, Rushlow CA, Zhou Q, Small S, Levine M (1992): Individual dorsal morphogen binding sites mediate activation and repression in the *Drosophila* embryo. *EMBO J* 11: 3147–3154

Keith FJ, Gay NJ (1990): The *Drosophila* membrane receptor Toll can function to promote cellular adhesion. *EMBO J* 9: 4299–4306

Kerr LD, Inoue JI, Davis N, Link E, Baeuerle PA, Bose HAJ, Verma IM (1991): The Rel-associated pp40 protein prevents DNA binding of Rel and NF-κB: relationship with IκBβ and regulation by phosphorylation. *Genes Dev* 5: 1464–1476

Kidd S (1992): Characterization of the *Drosophila cactus* locus and analysis of interactions between cactus and dorsal proteins. *Cell* 71: 623–635

Kidd S (1994): Personal Communication

Kieran M, Blank V, Logeat F, Vandekerckhove J, Lottspeich F, Le Bail O, Urban MB, Kourilsky P, Baeuerle PA, Israël A (1990): The DNA binding subunit of NF-κB is identical to factor KBF1 and homologous to the *rel* oncogene product *Cell* 62: 1007–1018

Kirov N, Childs S, O'Connor M, Rushlow C (1994): The *Drosophila* dorsal morphogen represses the *tolloid* gene by interacting with a silencer element. *Mol Cell Biol* 14: 713–722

Kirov N, Zhelnin L, Shah J, Rushlow C (1993): Conversion of a silencer into an enhancer: evidence for a co-repressor in dorsal-mediated repression in *Drosophila*. *EMBO J* 12: 3193–3200

Krantz DE, Zipursky SC (1990): *Drosophila* chaoptin, a member of the leucine-rich repeat family, is a photoreceptor cell-specific adhesion molecule. *EMBO J* 9: 1969–1977

Letsou A, Alexander S, Orth K, Wasserman SA (1991): Genetic and molecular characterization of *tube*, a *Drosophila* gene maternally required for embryonic dorsoventral polarity. *Proc Natl Acad Sci USA* 88: 810–814

Letsou A, Alexander S, Wasserman SA (1993): Domain mapping of tube, a protein essential for dorsoventral patterning of the *Drosophila* embryo. *EMBO J* 12: 3449–3458

Link E, Kerr LD, Schreck R, Zabel U, Vermer I, Baeuerle PA (1992): Purified IκB-β is inactivated upon dephosphorylation. *J Biol Chem* 267: 239–246

Lopez JA, Chung DW, Fujikawa K, Hagen FS, Papayannopoulou T, Roth GJ (1987): Cloning of the α chain of human platelet glycoprotein 1b: a transmembrane protein with homology to leucine-rich α2-glycoprotein. *Proc Natl Acad Sci USA* 84: 5615–5619

Morisato D, Anderson KV (1994): The *spätzle* gene encodes a component of the extracellular signaling pathway establishing the dorsal-ventral pattern of the *Drosophila* embryo. *Cell* 76: 677–688

Mosialos G, Gilmore TD (1993): vRel and cRel are differentially affected by mutations at a consensus protein kinase recognition sequence. *Oncogene* 8: 721–730

Mosialos G, Hamer P, Capobianco AJ, Laursen RA, Gilmore TD (1991): A protein kinase-A recognition sequence is structurally linked to transformation by p59$^{v\text{-rel}}$ and cytoplasmic retention of p68$^{c\text{-rel}}$ *Mol Cell Biol* 11: 5867–5877

Nolan GP, Fujita T, Bhatia K, Huppi C, Liou HC, Scott M, Baltimore D (1993): The *bcl-3* proto-oncogene encodes a nuclear IκB-like molecule that preferentially interacts with NF-κB p50 and p52 in a phosphorylation dependent manner. *Mol Cell Biol* 13: 3557–3566

Norris JL, Manley JL (1992): Selective nuclear transport of the *Drosophila* morphogen dorsal can be established by a signaling pathway involving the transmembrane protein Toll and protein kinase A. *Genes Dev* 6: 1654–1667

Norris JL, Manley JL (1994): Unpublished results

Pan D, Courey AJ (1992): The same dorsal binding site mediates both activation and repression in a context-dependent manner. *EMBO J* 11: 1837–1842

Pan D, Huang DJ, Courey AJ (1991): Functional analysis of the *Drosophila twist* promoter reveals a dorsal-binding ventral activator region. *Genes Dev* 5: 1892–1901

Reinke R, Krantz DE, Yen D, Zipurksy SL (1988): Chaoptin, a cell surface glycoprotein required for *Drosophila* photoreceptor cell morphogenesis, contains a repeat motif found in yeast and human. *Cell* 52: 291–301

Richardson PM, Gilmore TD (1991): vRel is an inactive member of the Rel family of transcriptional activating proteins. *J Virol* 65: 3122–3130

Rihs HP, Peters R (1989): Nuclear transport kinetics depend on phosphorylation-site-containing sequences flanking the karyophilic signal of the simian virus 40 T-antigen. *EMBO J* 8: 1479–1484

Roth S, Hiromi Y, Godt D, Nüsslein-Volhard C (1991): *cactus*, a maternal gene required for proper formation of the dorsoventral morphogen gradient in *Drosophila* embryos. *Development* 112: 371–388

Roth S, Stein D, Nüsslein-Volhard C (1989): A gradient of nuclear localization of the dorsal protein determines dorsoventral pattern in the *Drosophila* embryo. *Cell* 59: 1189–1202

Rushlow C, Warrior R (1992): The Rel family of proteins. *Bioessays* 14: 89–95

Rushlow C, Frasch M, Doyle H, Levine M (1987): Maternal regulation of *zerknüllt:* a homeobox gene controlling differentiation of dorsal tissues in *Drosophila*. *Nature* 330: 583–586

Rushlow C, Han K, Manley JL, Levine M (1989): The graded distribution of the dorsal morphogen is initiated by selective nuclear transport in *Drosophila*. *Cell* 59: 1165–1177

Schneider DS, Hudson KL, Lin TY, Anderson KV (1991): Dominant and recessive mutation define functional domains of Toll, a transmembrane protein required for dorsal-ventral polarity in the *Drosophila* embryo. *Genes Dev* 5: 797–807

Schüpbach T, Wieschaus E (1989): Female sterile mutations on the second chromosome of *Drosophila melanogaster*. I. Maternal effect mutations. *Genetics* 121: 101–117

Shelton CA, Wasserman SA (1993): *pelle* encodes a protein kinase required to establish dorsoventral polarity in the *Drosophila* embryo. *Cell* 72: 515–525

Shimell MJ, Ferguson EL, Childs SR, O'Connor MB (1991): The *Drosophila* dorsal-ventral pattern gene *tolloid* is related to human bone morphogenic protein 1. *Cell* 67: 469–481

Shirakawa F, Mizel SB (1989): In vitro activation and nuclear translocation of NF-κB catalyzed by cyclic AMP-dependent protein kinase and protein kinase C. *Mol Cell Bio* 9: 2424-2430

Shirakawa F, Chedid M, Suttles J, Pollok BA, Mizel SB (1989): Interleukin 1 and cyclic AMP induce κ immunoglobulin light-chain expression via activation of an NF-κB-like DNA-binding protein. *Mol Cell Biol* 9: 959–964

Shirakawa F, Yamashita U, Chedid M, Mizel SB (1988): Cyclic AMP_an intracellular second messenger for interleukin 1. *Proc Natl Acad Sci USA* 85: 8201–8205

Simek SL, Rice NR (1988): p59^{v-rel}, the transforming protein of reticuloendotheliosis virus, is complexed with at least four other proteins in transformed chicken lymphoid cells. *J Virol* 62: 4730–4736

Simpson P (1983): Maternal-zygotic gene interactions during formation of the dorsoventral pattern in *Drosophila* embryos. *Genetics* 105: 615–632

Stein D, Nüsslein-Volhard C (1992): Multiple extracellular activities in *Drosophila* egg perivitelline fluid are required for establishment of embryonic dorsal-ventral polarity. *Cell* 68: 429–440

Stein D, Roth S, Vogelsang E, Nüsslein-Volhard C (1991): The polarity of the dorsoventral axis in the *Drosophila* embryo is defined by an extracellular signal. *Cell* 65: 725–735

St Johnston RD, Gelbart WM (1987): *decapentaplegic* transcripts are localized along the dorsal-ventral axis of the *Drosophila* embryo. *EMBO J* 6: 2785–2791

Steward R (1987): Dorsal, an embryonic polarity gene in *Drosophila*, is homologous to the vertebrate proto-oncogene, c-*rel*. *Science* 238: 692-694

Steward R (1989): Relocalization of the dorsal protein from the cytoplasm to the nucleus correlates with its function. *Cell* 59: 1179–1188

Steward R, Zusman SB, Huang LH, Schedl P (1988): The dorsal protein is distributed in a gradient in early *Drosophila* embryos. *Cell* 55: 487–495

Thisse C, Perrin-Schmitt F, Stoetzel C, Thisse B (1991): Sequence-specific transactivation of the *Drosophila twist* gene by the *dorsal* gene product. *Cell* 65: 1191–1201

Thisse B, Stoetzel C, El Messal M, Perrin-Schmitt F (1987): Genes of the *Drosophila* maternal dorsal group control the specific expression of the zygotic gene *twist* in presumptive mesodermal cells. *Genes Dev* 1: 709–715

Thisse B, Stoetzel C, El Messal M, Perrin-Schmitt F (1988): Sequence of the *twist* gene and nuclear localization of its protein in endomesodermal cells of early *Drosophila* embryos. *EMBO J* 7: 2175–2183

Vicente V, Houghten R, Ruggeri ZM (1990): Identification of a site in the α chain of platelet glycoprotein 1b that participates in von Willebrand factor binding. *J Biol Chem* 265: 274–280

Walen AH, Steward R (1993): Dissociation of the dorsal-cactus complex and phosphorylation of the dorsal protein correlate with the nuclear localization of dorsal. *J Cell Biol* 123: 523–534

Wilhelmsen KC, Eggleton K, Temin HM (1984): Nucleic acid sequences of the oncogene v-*rel* in reticuloendotheliosis virus strain T and its cellular homolog, the proto-oncogene c-*rel*. *J Virol* 52: 172–182

Wulczyn FG , Naumann M, Scheidereit C (1992): Candidate proto-oncogene *bcl-3* encodes a subunit-specific inhibitor of transcription factor NF-κB. *Nature* 358: 597–599

Zabel U, Henkel T, dos Santos Silva M, Baeuerle PA (1993): Nuclear uptake control of NF-κB by MAD-3, and IκB protein present in the nucleus. *EMBO J* 12: 201–211

Index

Combined Index for Volumes 1 and 2

278   Index